Spring Boot+Spring Cloud+Spring Cloud Alibaba
微服务训练营

黄文毅 编著

清华大学出版社
北京

内 容 简 介

本书以分布式微服务项目需求为主线，系统地介绍了 Spring Boot、Spring Cloud、Spring Cloud Alibaba 的生产级特性、重要组件及核心技术，包括服务治理、服务注册与发现、负载均衡等分布式框架所需要的各种功能。本书共分为 14 章。第 1 章主要讲解 Spring Boot 的重要功能特性以及阅读本书之前需要准备的环境。第 2 章主要介绍 Spring Cloud 和 Spring Cloud Alibaba 模块、版本以及之间的关系。第 3 章主要介绍注册中心和配置中心 Nacos 以及其他开源的组件。第 4 章主要讲解微服务网关 Spring Cloud Gateway 和 Zuul。第 5、6 章主要讲解负载均衡组件 Ribbon 和微服务调用组件 OpenFeign。第 7 章主要讲解服务限流、降级、容错以及熔断等技术，包括 Hystrix 组件和 Sentinel 组件。第 8、9 章主要介绍 Spring Cloud Bus 消息总线、分布式事务解决方案 Seata。第 10、11 章主要讲解链路追踪组件 Spring Cloud Sleuth 和 Zipkin，以及 Spring Cloud Commons 基础包。第 12 章主要讲解如何通过 OAuth 2.0 进行授权。第 13 章主要讲解微服务和组件容器化。第 14 章主要介绍分布式微服务架构的具体案例。

本书技术先进，贴近实践，讲练结合，适合具有 Java 基础或 1~2 年开发经验的读者使用，也可作为网课、培训机构和大专院校的教学用书。

本书封面贴有清华大学出版社防伪标签，无标签者不得销售。
版权所有，侵权必究。举报：010-62782989，beiqinquan@tup.tsinghua.edu.cn。

图书在版编目（CIP）数据

Spring Boot+Spring Cloud+Spring Cloud Alibaba 微服务训练营 / 黄文毅编著.—北京：清华大学出版社，2021.7
 ISBN 978-7-302-58223-6

Ⅰ.①S… Ⅱ.①黄… Ⅲ.①JAVA 语言－程序设计 Ⅳ.①TP312.8

中国版本图书馆 CIP 数据核字（2021）第 096243 号

责任编辑：王金柱
封面设计：王 翔
责任校对：闫秀华
责任印制：宋 林

出版发行：清华大学出版社
网　　址：http://www.tup.com.cn, http://www.wqbook.com
地　　址：北京清华大学学研大厦 A 座　　邮　编：100084
社 总 机：010-62770175　　邮　购：010-62786544
投稿与读者服务：010-62776969，c-service@tup.tsinghua.edu.cn
质 量 反 馈：010-62772015，zhiliang@tup.tsinghua.edu.cn

印 装 者：三河市铭诚印务有限公司
经　　销：全国新华书店
开　　本：190mm×260mm　　印　张：23.75　　字　数：608 千字
版　　次：2021 年 7 月第 1 版　　印　次：2021 年 7 月第 1 次印刷
定　　价：99.00 元

产品编号：088602-01

前　　言

Spring Cloud/Spring Cloud Alibaba 是一系列框架的有序集合。它利用 Spring Boot 的开发便利性巧妙地简化了分布式系统基础设施的开发，如服务发现注册、配置中心、消息总线、负载均衡、断路器、数据监控等，都可以用 Spring Boot 的开发风格做到一键启动和部署。Spring Cloud 并没有重复制造轮子，它只是将各家公司开发的比较成熟、经得起实际考验的服务框架组合起来，通过 Spring Boot 风格进行再封装，屏蔽掉了复杂的配置和实现原理，最终给开发者留出了一套简单易懂、易部署和易维护的分布式系统开发工具包。

本书理论与实践并重，采用一步一步的教学方法，较为系统地介绍 Spring Boot、Spring Cloud、Spring Cloud Alibaba 的生产级特性、重要组件及核心技术，并通过大量生动形象的原理图以及实践案例加深读者对分布式微服务架构的理解，希望对于拥有 Java 基础或 1~2 年开发经验、想突破开发瓶颈、进阶架构师的读者有所帮助。

让我们开始 Spring Boot、Spring Cloud、Spring Cloud Alibaba 的探险之旅吧！

本书结构

本书共 14 章，以下是各章节的内容概要。

第 1 章首先介绍环境准备，包括安装 JDK、安装 Intellij IDEA、安装 Maven、Docker 概述等；紧接着讲述 Spring Boot 核心功能和生产级特性、快速搭建第一个 Spring Boot 项目、Spring Boot 原理解析、SpringApplication 执行流程以及如何自定义 starter 组件。

第 2 章主要介绍 Spring Cloud 功能特性、模块以及版本信息，Spring Cloud 和 Spring Boot 之间的关系，Spring Cloud Alibaba 简介、主要功能以及相关组件，最后介绍 Netflix、Spring Cloud 以及 Spring Cloud Alibaba 之间的关系。

第 3 章主要介绍 Spring Cloud Alibaba 的服务注册和配置中心组件 Nacos、Nacos 单机模式/集群模式以及 Nacos+Nginx 集群模式搭建，Spring Boot 如何注册到 Nacos 以及如何将配置文件抽到 Nacos 配置中心，Nacos 配置中心和服务发现原理分享，Eureka 简介以及如何通过 Eureka 搭建注册中心集群，Spring Cloud Consul 简介、安装与启动，Spring Cloud Config 简介和原理等内容。

第 4 章首先介绍 Zuul 网关、快速搭建 Zuul 网关、Zuul 网关路由配置/过滤器/管理端点等内容；接着介绍 Spring Cloud Gateway 相关内容，包括如何快速入门 Spring Cloud Gateway、Gateway 路由断言工厂、全局过滤器、跨域、HTTP 超时配置、TLS/SSL 配置、Gateway 底层原理等；最后对比 Gateway 和 Zuul 网关的区别。

第 5 章主要介绍 Ribbon 负载均衡器、常用负载均衡算法、如何自定义负载均衡算法、如何自定义 Ribbon 客户端，最后介绍 Eureka/Nacos 如何整合 Ribbon 客户端。

第 6 章主要介绍 Spring Cloud OpenFeign 声明式调用，包括 OpenFeign 简介、快速创建第一个

Feign 程序、@FeignClient 注解详解、Feign @QueryMap 支持、Feign 请求响应压缩、Feign 日志配置（Java 方式、配置文件方式以及全局日志配置）、Feign 自定义错误、Feign 拦截器以及如何自定义 Feign 客户端等内容。

第 7 章主要介绍熔断、限流以及降级相关组件，包括 Hystrix 简介、Hystrix 请求缓存和请求合并、Spring Boot 应用配置 Hystrix 仪表盘、Turbine 集群监控、阿里 Sentinel 组件简介、常用的限流算法、Sentinel 与 Hystrix 的区别、Sentinel 如何进行限流和熔断降级等内容。

第 8 章主要介绍 Spring Cloud Bus 消息总线、Spring 事件机制、Spring Cloud Bus 原理、如何使用 Kafka 实现消息总线、Kafka 介绍与安装、Spring Cloud Stream 简介和核心概念讲解、Stream 应用编程模型/Binder 抽象、Stream 快速入门、Stream 原理等内容。

第 9 章主要介绍 Spring Cloud Alibaba Seata 分布式事务组件，包括 Seata 简介、Seata 部署、Seata 原理与设计以及如何通过 Seata 解决分布式事务问题等。

第 10 章主要介绍 Spring Cloud Sleuth 服务链路追踪，包括 Sleuth 和 Zipkin 简介、Zipkin 安装与快速启动、Spring Cloud Sleuth 整合 Zipkin、Spring Cloud Sleuth 整合 ELK、Sleuth 原理浅析等内容。

第 11 章主要介绍 Spring Cloud Commons 公共包、Spring Cloud Context 功能、Spring Cloud Commons 功能、Spring Cloud LoadBalance 负载均衡、Spring Cloud Circuit Breaker 断路器介绍和核心概念等内容。

第 12 章主要介绍 OAuth 2.0 核心概念、OAuth 2.0 协议流程、OAuth 2.0 四种授权方式、快速搭建 OAuth 2.0 服务、授权码模式实现、JWT 简介、JWT 结构和应用，最后结合 Spring Security + OAuth 2.0 + JWT 开发的具体案例。

第13章主要介绍 Spring Boot 项目容器化、Spring Cloud Alibaba 组件容器化，包括 Nacos、Sentinel 以及 Seata 等组件。

第 14 章主要介绍使用 Spring Cloud、Spring Cloud Alibaba 以及开源技术框架，一步一步搭建分布式微服务架构和服务治理平台，并提供具体的架构图和原理图，帮助读者理解分布式架构的具体细节。

学习本书的预备知识

Java 基础

读者需要掌握 J2SE 基础知识，这是最基本的，也是最重要的。

Java Web 开发技术

在项目实战中需要用到 Java Web 的相关技术，比如 Spring、Spring MVC、Tomcat 等技术。

Spring Boot 技术

本书的很多内容都是建立在读者了解 Spring Boot 的基础上展开的，如果读者对微服务脚手架 Spring Boot 的知识和功能特性有更多的了解，会更顺利地阅读本书。

其他技术

读者需要了解目前主流的技术，比如数据库 MySQL、缓存 Redis、消息中间件 Kafka、容器技术 Docker 等。

本书使用的软件版本

本书项目实战开发环境为：

- 操作系统 Mac Pro
- 开发工具 Intellij IDEA 2019.3
- JDK 1.8 版本以上
- Spring Boot 2.2x 以上
- Spring Cloud Hoxton 版本
- Spring Cloud Alibaba 2.2.0 RELEASE
- 其他主流技术基本使用最新版本

读者对象

- 具有 Java 基础的大学生。
- 拥有 1~2 年开发经验的从业人员和运维人员。
- 网课、培训机构、大专院校教学用书。

源代码下载

可以用微信扫描下面的二维码获取本书配套的源文件。

如果阅读过程中遇到问题，请联系 booksaga@126.com，邮件主题为"Spring Boot+Spring Cloud+Spring Cloud Alibaba 微服务训练营"。

致谢

本书能够顺利出版，首先感谢清华大学出版社的王金柱老师及背后的团队为本书的辛勤付出，这是我第六次和王金柱老师合作，每次合作都能让我感到轻松和快乐，我很享受写作的过程。

感谢厦门海西医药交易有限公司，书中很多的知识点和项目实战经验都来源于贵公司，如果没有贵公司提供的实战案例，这本书就不可能问世。感谢技术总监赵定益认可和栽培，以及同事涂勇的鼎力支持解答。

  感谢我的妻子郭雅苹，感谢她一路不离不弃的陪伴和督促，感谢她对我工作的理解和支持。感谢家人对我生活无微不至的照顾，使我没有后顾之忧，全身心投入本书的写作中。

  由于水平所限，书中所存不足敬请广大读者不吝指正。

黄文毅

2021 年 4 月 1 日

目　　录

第 1 章　从 Spring Boot 开始 1

1.1　环境准备 1
1.1.1　安装 JDK 1
1.1.2　安装 Intellij IDEA 4
1.1.3　安装 Maven 4
1.1.4　Docker 概述 5
1.2　Spring Boot 简介 14
1.3　第一个 Spring Boot 项目 16
1.3.1　使用 Spring Initializr 新建项目 16
1.3.2　测试 18
1.4　Spring Boot 目录介绍 19
1.4.1　Spring Boot 工程目录 19
1.4.2　Spring Boot 入口类 20
1.4.3　Spring Boot 测试类 20
1.4.4　pom.xml 文件 21
1.5　Spring Boot 生产级特性 23
1.5.1　应用监控 23
1.5.2　健康检查 26
1.5.3　跨域访问 27
1.5.4　外部配置 28
1.6　Spring Boot 原理解析 29
1.6.1　DemoApplication 入口类 29
1.6.2　@SpringBootApplication 的原理 29
1.6.3　SpringApplication 的 run 方法 31
1.6.4　SpringApplicationRunListener 监听器 32
1.6.5　ApplicationContextInitializer 接口 32
1.6.6　ApplicationRunner 与 CommandLineRunner 34
1.7　SpringApplication 的执行流程 35
1.7.1　spring-boot-starter 原理 36
1.7.2　Bean 参数获取 39
1.7.3　Bean 的发现与加载 40

1.7.4 自定义 starter 46

第 2 章 Spring Cloud/Spring Cloud Alibaba 52

2.1 Spring Cloud 介绍 52
- 2.1.1 Spring Cloud 的特性 52
- 2.1.2 Spring Cloud 的模块 53
- 2.1.3 Spring Cloud 版本介绍 54
- 2.1.4 Spring Cloud 与 Spring Boot 的关系 55

2.2 Spring Cloud Alibaba 简介 55
- 2.2.1 Spring Cloud Alibaba 的主要功能 55
- 2.2.2 Spring Cloud Alibaba 组件 56
- 2.2.3 Spring Cloud Alibaba 版本简介 57

2.4 Netflix/Spring Cloud/Spring Cloud Alibaba 的关系 58

第 3 章 注册中心/配置管理 59

3.1 Nacos 简介 59
3.2 Nacos 快速开始 60
- 3.2.1 Nacos Server 单机模式 60
- 3.2.2 Nacos Server 集群模式 63
- 3.2.3 Nacos+Nginx 集群模式 66

3.3 Spring Boot 注册到 Nacos 67
- 3.3.1 Nacos 配置管理 67
- 3.3.2 Nacos 服务注册 69

3.4 Nacos Spring Cloud 70
- 3.4.1 Nacos 配置管理 70
- 3.4.2 Nacos 服务注册 72

3.5 Nacos 原理解析 75
- 3.5.1 Nacos 配置中心原理分析 75
- 3.5.2 Nacos 服务发现原理分析 84

3.6 Eureka 服务发现 86
- 3.6.1 Eureka 简介 86
- 3.6.2 如何看待 Eureka 停产 88
- 3.6.3 搭建 Eureka 注册中心 88
- 3.6.4 搭建 Eureka 注册中心集群 92

3.7 Spring Cloud Consul 95
- 3.7.1 Consul 简介 95
- 3.7.2 Consul 安装与启动 95
- 3.7.3 Consul 服务注册与发现 96
- 3.7.4 Consul 配置中心 100

　　　　3.7.5　Consul 简单架构 103
　3.8　Spring Cloud Config 104
　　　　3.8.1　Spring Cloud Config 简介 104
　　　　3.8.2　Spring Cloud Config 快速入门 105
　　　　3.8.3　Spring Cloud Config 配置中心原理 108

第 4 章　微服务网关 109

　4.1　Zuul 网关 109
　　　　4.1.1　Zuul 概述 109
　　　　4.1.2　Zuul 快速入门 110
　　　　4.1.3　Zuul 路由配置 111
　　　　4.1.4　Zuul 过滤器 112
　　　　4.1.5　管理端点 114
　　　　4.1.6　禁用 Zuul 过滤器 115
　　　　4.1.7　启用 Zuul 跨域请求 115
　　　　4.1.8　Eureka 整合 Zuul 116
　4.2　Spring Cloud Gateway 120
　　　　4.2.1　Gateway 简介 120
　　　　4.2.2　Gateway 快速入门 121
　　　　4.2.3　Gateway 路由断言工厂 123
　　　　4.2.4　Gateway 过滤器工厂 127
　　　　4.2.5　Gateway 全局过滤器 128
　　　　4.2.6　Gateway 跨域 131
　　　　4.2.7　Gateway Actuator API 132
　　　　4.2.8　HTTP 超时配置 134
　　　　4.2.9　TLS / SSL 设置 135
　　　　4.2.10　Gateway 底层原理 136
　4.3　Gateway 与 Zuul 的区别 137

第 5 章　Ribbon 负载均衡 138

　5.1　Ribbon 基础知识 138
　　　　5.1.1　Ribbon 简介 138
　　　　5.1.2　负载均衡算法 140
　　　　5.1.3　第一个 Ribbon 程序 144
　5.2　Ribbon 实战 147
　　　　5.2.1　Ribbon 自定义负载均衡策略 147
　　　　5.2.2　Ribbon 饥饿加载 151
　　　　5.2.3　Ribbon 默认配置 151
　　　　5.2.4　配置文件定义 Ribbon 客户端 152

第6章 Spring Cloud OpenFeign 声明式调用 155

6.1 Spring Cloud Feign 155
- 6.1.1 Feign 简介 155
- 6.1.2 第一个 Feign 程序 156

6.2 FeignClient 详解与配置 161
- 6.2.1 @FeignClient 详解 161
- 6.2.2 Feign Hystrix 错误回退 166
- 6.2.3 Feign @QueryMap 支持 167
- 6.2.4 HATEOAS 支持 167
- 6.2.5 Spring @MatrixVariable 支持 168
- 6.2.6 Feign 继承支持 168
- 6.2.7 Feign CollectionFormat 支持 169
- 6.2.8 Feign 请求响应压缩 169

6.3 Feign 日志配置 170
- 6.3.1 Java 代码方式 170
- 6.3.2 配置文件方式 171
- 6.3.3 全局日志配置 171

6.4 自定义处理 172
- 6.4.1 Feign 自定义错误 172
- 6.4.2 Feign 拦截器 176
- 6.4.3 自定义 Feign 客户端 177

第7章 熔断、限流、降级 179

7.1 Spring Cloud Hystrix 179
- 7.1.1 Hystrix 简介 179
- 7.1.2 Hystrix 初体验 182
- 7.1.3 Hystrix 请求缓存 184
- 7.1.4 Hystrix 请求合并 187
- 7.1.5 Hystrix 默认配置 190
- 7.1.6 Hystrix 配置详解 191

7.2 Hystrix 工作流程 194

7.3 Hystrix 监控 196
- 7.3.1 Spring Boot 应用配置 Hystrix 仪表板 197
- 7.3.2 Turbine 集群监控 200

7.4 Sentinel 204
- 7.4.1 Sentinel 简介 204

（接上页）
- 5.2.5 直接使用 Ribbon API 153
- 5.2.6 Eureka/Nacos 整合 Ribbon 153

7.4.2　限流算法 204
　　7.4.3　Sentinel 项目结构 206
　　7.4.4　Sentinel 与 Hystrix 的区别 207
　　7.4.5　Sentinel 控制台 207
　　7.4.6　客户端接入控制台 209
　　7.4.7　Sentinel 微服务限流 210

第 8 章　Spring Cloud Bus 消息总线 214

8.1　Kafka 实现消息总线 214
　　8.1.1　Kafka 概述 214
　　8.1.2　Kafka 安装 217
　　8.1.3　Docker 安装 ZooKeeper 和 Kafka 219
8.2　Stream 简介 219
　　8.2.1　核心概念 219
　　8.2.2　Stream 应用编程模型 220
　　8.2.3　Binder 抽象 220
　　8.2.4　发布—订阅 221
　　8.2.5　消费组 221
　　8.2.6　分区支持 221
　　8.2.7　健康指标 221
8.3　Spring Cloud Stream 实战 222
　　8.3.1　Stream 快速入门 222
　　8.3.2　生产者的另一种实现 227
　　8.3.3　生产和消费消息 229
8.4　Bus 简介 232
　　8.4.1　Bus 消息总线 232
　　8.4.2　Spring 事件机制 232
　　8.4.3　Spring Cloud Bus 实战 235
　　8.4.4　Spring Cloud Bus 原理 239
　　8.4.5　Spring Cloud Bus 端点 240
　　8.4.6　Bus 事件追踪 240

第 9 章　Spring Cloud Alibaba Seata 分布式事务 243

9.1　Seata 基础知识 243
　　9.1.1　Seata 简介 243
　　9.1.2　Seata 部署 244
　　9.1.3　Seata 原理与设计 246
9.2　Seata 使用 247
　　9.2.1　数据库准备 247

9.2.2 创建微服务 248

第 10 章 Spring Cloud Sleuth 服务链路追踪 257

10.1 Spring Cloud Sleuth 简介 257
10.2 Zipkin 简介 259
10.3 Spring Cloud Sleuth 整合 Zipkin 261
10.3.1 整合 Zipkin 261
10.3.2 MySQL 存储链路数据 265
10.3.3 Sleuth 抽样采集 267
10.3.4 Trace 和 Span 268
10.4 Spring Cloud Sleuth 整合 ELK 271
10.5 Sleuth 原理浅析 275
10.5.1 TraceId 传递 275
10.5.2 spring.factories 配置文件 276
10.5.3 TraceEnvironmentPostProcessor 处理日志 278
10.5.4 TraceAutoConfiguration 279
10.5.5 TracingFilter 过滤器 280
10.5.6 TraceWebClientAutoConfiguration 283

第 11 章 Spring Cloud Commons 286

11.1 Spring Cloud Commons 简介 286
11.2 Spring Cloud Context 功能 286
11.2.1 bootstrap 应用程序上下文 286
11.2.2 修改 bootstrap.properties 位置 287
11.2.3 覆盖远程属性的值 287
11.2.4 自定义 bootstrap 配置 287
11.2.5 刷新范围 288
11.2.6 加密与解密 288
11.2.7 Endpoints 端点 288
11.3 Spring Cloud Commons 功能 289
11.3.1 @EnableDiscoveryClient 注解 289
11.3.2 服务注册 ServiceRegistry 290
11.3.3 多个 RestTemplate 实例 290
11.3.4 多个 WebClient 实例 291
11.3.5 忽略网卡 293
11.3.6 HTTP 客户端工厂 293
11.3.7 启用功能特性 294
11.3.8 Spring Cloud 兼容性验证 295
11.4 Spring Cloud LoadBalancer 295

- 11.4.1 LoadBalancer 简介 ... 295
- 11.4.2 Spring Cloud LoadBalancer 缓存 ... 296
- 11.4.3 Spring Cloud LoadBalancer Starter ... 296
- 11.4.4 自定义 Spring Cloud LoadBalancer 配置 ... 297
- 11.5 Spring Cloud Circuit Breaker ... 297
  - 11.5.1 Circuit Breaker 介绍 ... 297
  - 11.5.2 核心概念 ... 298
  - 11.5.3 配置断路器 ... 299
- 11.6 具备缓存功能随机数 ... 300

第 12 章 Spring Cloud OAuth 2.0 保护 API 安全 ... 301

- 12.1 使用 OAuth 2.0 进行授权 ... 301
  - 12.1.1 OAuth 2.0 简介 ... 301
  - 12.1.2 OAuth 2.0 协议流程 ... 302
  - 12.1.3 认证与授权 ... 302
  - 12.1.4 OAuth 2.0 的授权方式 ... 303
  - 12.1.5 Spring Cloud Security OAuth 2.0 认证流程 ... 305
- 12.2 搭建 OAuth 2.0 服务 ... 306
  - 12.2.1 快速搭建 OAuth 2.0 服务 ... 306
  - 12.2.2 授权码模式实现 ... 308
- 12.3 JWT 简介 ... 313
  - 12.3.1 JWT 的结构 ... 313
  - 12.3.2 JWT 的应用 ... 315
  - 12.3.3 Spring Security+OAuth 2.0+JWT 应用 ... 315

第 13 章 Spring Cloud 组件容器化 ... 336

- 13.1 Spring Boot 项目容器化 ... 336
  - 13.1.1 制作镜像 ... 336
  - 13.1.2 使用 Dockerfile 构建镜像 ... 338
  - 13.1.3 Spring Boot 集成 Docker ... 341
- 13.2 Spring Cloud Alibaba 组件容器化 ... 345
  - 13.2.1 Nacos Docker ... 345
  - 13.2.2 Sentinel Docker ... 346
  - 13.2.3 Seata Docker ... 346

第 14 章 使用 Spring Cloud 构建微服务综合案例 ... 348

- 14.1 案例介绍 ... 348
- 14.2 技术选型 ... 348
  - 14.2.1 Spring Boot 构建微服务 ... 348

14.2.2 Nacos 注册/配置中心 350
14.2.3 Spring Cloud Gateway 网关 352
14.2.4 OpenFeign 服务调用 355
14.2.5 Ribbon 负载均衡 355
14.2.6 Sentinel 熔断/降级/限流 356
14.2.7 ELK+FileBeat 日志系统 357
14.2.8 Promethous+Grafana+InfluxDB 监控系统 359
14.2.9 SkyWalking 链路追踪系统 363
14.3 总结 365
参考文献 366

第 1 章

从 Spring Boot 开始

本章首先介绍环境准备，包括安装 JDK、安装 Intellij IDEA、安装 Maven、Docker 概述等；紧接着讲述 Spring Boot 核心功能和生产级特性、快速搭建第一个 Spring Boot 项目、Spring Boot 原理解析、SpringApplication 执行流程以及如何自定义 starter 组件。

1.1 环境准备

1.1.1 安装 JDK

JDK（Java SE Development Kit）建议使用 1.8 及以上的版本，可从官网下载，下载界面如图 1-1 所示。可以根据计算机的操作系统配置选择合适的 JDK 安装包，笔者的计算机操作系统是 MacBook Pro，因此下载安装包 jdk-11.0.1_osx-x64_bin.dmg。

安装包下载完成之后，双击下载软件，按照提示安装即可，如图 1-2 所示。

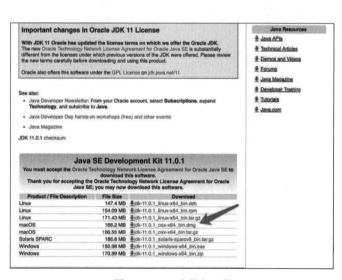

图 1-1　JDK 安装包下载

图 1-2 JDK 安装

打开 Finder,找到安装好的 JDK 路径,具体如图 1-3 所示。

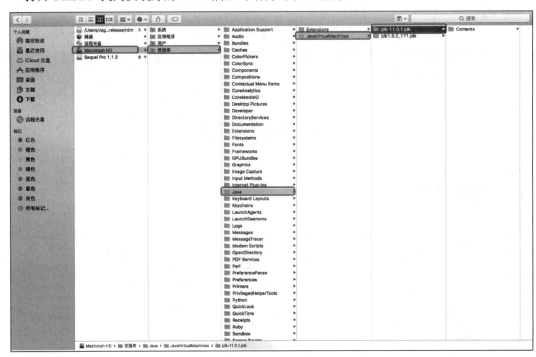

图 1-3 JDK 安装路径图

其中,Contents 下的 Home 文件夹是该 JDK 的根目录,具体如图 1-4 所示。

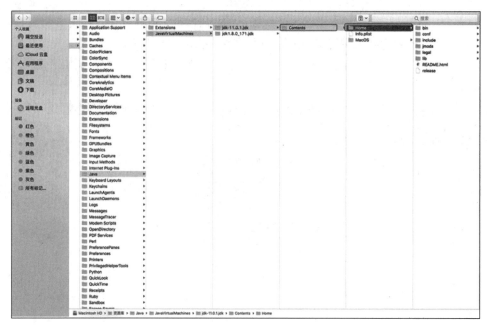

图 1-4　JDK Home 文件夹

在英文输入法的状态下，按键盘上的"Command + 空格"组合键，调出 Spotlight 搜索，输入"ter"，选择"终端"选项，然后按回车键，便可以快速启动终端，具体如图 1-5 所示。

图 1-5　从 Spotlight 搜索启动终端

在"终端"中输入"java –version"，如果能看到 JDK 版本 11.0.1，就说明 JDK 安装成功，具体如图 1-6 所示。

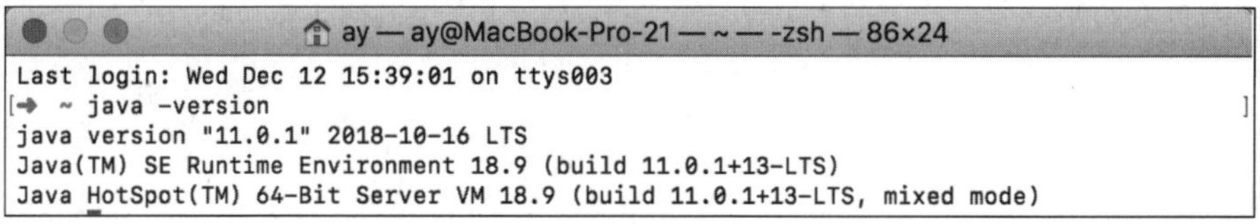

图 1-6　JDK 安装成功

如果是第一次配置环境变量，那么可以使用 touch .bash_profile 创建一个 .bash_profile 的隐藏配置文件。如果是编辑已存在的配置文件，那么可以使用 open -e .bash_profile 命令。假设配置文件已存在，这里我们使用 open -e .bash_profile 命令打开配置文件。在配置文件中添加如下代码，具体如图 1-7 所示。

```
//JAVA_HOME 是 Java 的安装路径（注意该行注释不可加到 .bash_profile 配置文件中）
JAVA_HOME=/Library/Java/JavaVirtualMachines/jdk-11.0.1.jdk/Contents/Home
PATH=$JAVA_HOME/bin:$PATH:.
CLASSPATH=$JAVA_HOME/lib/tools.jar:$JAVA_HOME/lib/dt.jar:.
export JAVA_HOME
export PATH
export CLASSPATH
```

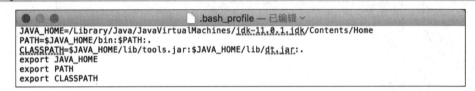

图 1-7　bash_profile 添加 JDK 配置

保存并关闭 .bash_profile 文件，在命令行"终端"输入命令"source .bash_profile"使配置文件生效。同时在"终端"输入"echo $JAVA_HOME"显示刚才配置的路径，具体如图 1-8 所示。

```
Last login: Tue Dec 18 00:21:53 on ttys005
~ source .bash_profile
~ echo $JAVA_HOME
/Library/Java/JavaVirtualMachines/jdk-11.0.1.jdk/Contents/Home
```

图 1-8　验证 JDK 配置是否添加成功

1.1.2　安装 Intellij IDEA

在 Intellij IDEA 的官方网站可以免费下载 IDEA。下载 IDEA 后，运行安装程序，按提示安装即可。本书使用的是 Intellij IDEA 2019.3 版本，当然也可以使用其他版本的 IDEA，版本不要过低即可。

1.1.3　安装 Maven

Apache Maven 是目前流行的项目管理和构建自动化工具。虽然 IDEA 已经包含 Maven 插件，但是还是希望大家在工作中能够安装自己的 Maven 插件，以便满足以后的项目配置需要。可以通过 Maven 的官方网站下载最新版的 Maven。本书的 Maven 版本为 apache-maven-3.6.0，如图 1-9 所示。

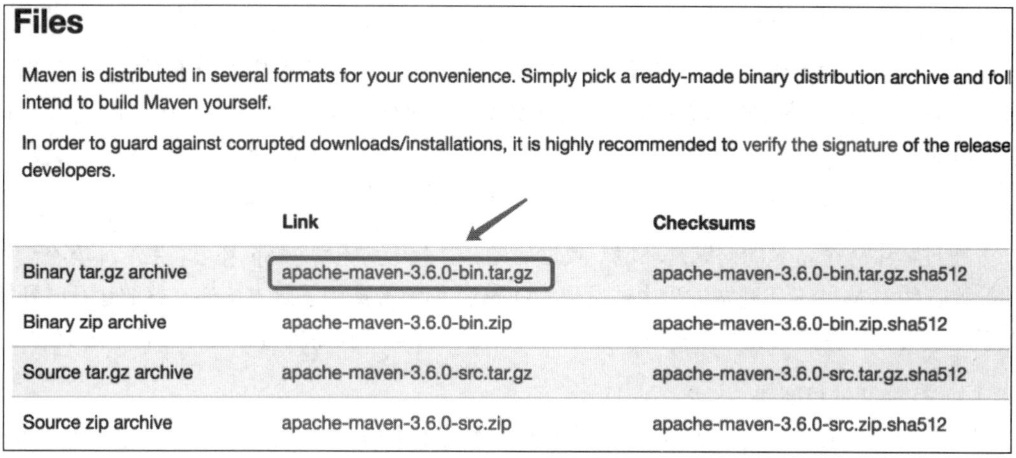

图 1-9 apache-maven-3.6.0 下载页面

打开命令行"终端",输入"open -e .bash_profile"命令打开配置文件,然后输入 Maven 的环境变量,具体代码如下:

```
###MAVEN_HOME 是 Maven 的安装路径（注意该行注释不可加到.bash_profile 配置文件中）
MAVEN_HOME=/Users/ay/Downloads/soft/apache-maven-3.6.0

###JAVA_HOME 是 Java 的安装路径（注意该行注释不可加到.bash_profile 配置文件中）
JAVA_HOME=/Library/Java/JavaVirtualMachines/jdk-11.0.1.jdk/Contents/Home
###在原有的基础上添加:$M2_HOME/bin（注意该行注释不可加到.bash_profile 配置文件中）
PATH=$JAVA_HOME/bin:$PATH:.:$M2_HOME/bin
CLASSPATH=$JAVA_HOME/lib/tools.jar:$JAVA_HOME/lib/dt.jar:.
export JAVA_HOME
export PATH
export CLASSPATH
```

Maven 环境变量添加完成之后,保存并退出.bash_profile 文件。在命令行"终端"输入"source ~/.bash_profile"命令使环境变量生效,然后输入"mvn –v"查看 Maven 是否安装成功。

1.1.4 Docker 概述

Docker 是一个开源的应用容器引擎,基于 Go 语言并遵从 Apache 2.0 协议开源。Docker 可以让开发者打包他们的应用以及依赖包到一个轻量级、可移植的容器中,然后发布到任何流行的 Linux 机器上,也可以实现虚拟化。容器完全使用沙箱机制,相互之间不会有任何接口,更重要的是容器性能开销极低。

作为一种新兴的虚拟化方式,Docker 跟传统的虚拟化方式相比具有众多优势。

(1) 高效利用系统资源

由于容器不需要进行硬件虚拟以及运行完整操作系统等额外开销,因此 Docker 对系统资源的利用率更高。无论是应用执行速度、内存损耗或者文件存储速度,都要比传统虚拟机技术更高效。因此,相比虚拟机技术,一个相同配置的主机往往可以运行更多数量的应用。

（2）快速的启动时间

传统的虚拟机技术启动应用服务往往需要数分钟，而 Docker 容器应用由于直接运行于宿主内核，无须启动完整的操作系统，因此可以做到秒级甚至毫秒级的启动时间，大大节约了开发、测试、部署的时间。

（3）一致的运行环境

开发过程中一个常见的问题是环境一致性问题。由于开发环境、测试环境、生产环境不一致，因此导致有些 Bug 并未在开发过程中被发现。Docker 的镜像提供了除内核外完整的运行时环境，确保了应用运行环境的一致性，从而不会再出现"这段代码在我的机器上没有问题"之类的问题。

（4）持续交付和部署

对开发和运维人员来说，最希望的就是一次创建或配置，以在任意地方正常运行。使用 Docker 可以通过定制应用镜像来实现持续集成、持续交付和部署。开发人员可以通过 Dockerfile 来进行镜像构建，并结合持续集成（Continuous Integration）系统进行集成测试，而运维人员则可以直接在生产环境中快速部署该镜像，甚至结合持续部署（Continuous Delivery/Deployment）系统进行自动部署。使用 Dockerfile 使镜像构建透明化不仅可以让开发团队理解应用运行环境，还可以方便运维团队理解应用运行所需条件，帮助用户更好地在生产环境中部署该镜像。

（5）迁移简单

由于 Docker 确保了执行环境的一致性，因此使得应用的迁移更加容易。Docker 可以在很多平台上运行，无论是物理机、虚拟机、公有云、私有云还是笔记本，其运行结果都是一致的。因此，用户可以很轻易地将在一个平台上运行的应用迁移到另一个平台上，而不用担心运行环境的变化导致应用无法正常运行的情况。

（6）容易维护和扩展

Docker 使用的分层存储及镜像技术使得应用重复部分的复用更为容易，也使得应用的维护更新更加简单，基于基础镜像进一步扩展镜像也变得非常简单。此外，Docker 团队同各个开源项目团队一起维护了一大批高质量的官方镜像，既可以直接在生产环境中使用，又可以作为基础进一步定制，大大降低了应用服务的镜像制作成本。

当需要在宿主机器上运行一个虚拟操作系统时，往往需要安装虚拟软件，如 Oracle VirtualBox 或者 VMware。然后，在虚拟软件上安装操作系统的时候虚拟机软件需要模拟 CPU、内存、I/O 设备和网络资源等，为了能运行应用程序，除了需要部署应用程序本身及其依赖外，还需要安装整个操作系统和驱动，会占用大量的系统开销。下面简单对比一下传统虚拟机和 Docker，具体见表 1-1。

表 1-1 Docker 容器与虚拟机对比

特性	Docker 容器	虚拟机
性能	接近原生	弱于原生
启动	秒级	分钟级
占用的硬盘存储空间	一般为 MB	一般为 GB
系统支持量	单机支持上千个容器	一般支持几十个

从表 1-1 中的数据来看，虚拟机和 Docker 容器虽然可以提供相同的功能，但是优缺点显而易

见。

下面我们来理解Docker中的4个基本概念：镜像（Image）、容器（Container）、仓库（Repository）、镜像注册中心（Docker Registry）。

（1）镜像

Docker镜像可以理解为一个Linux文件系统，并且是一个特殊的文件系统，除了提供容器运行时所需的程序、库、资源、配置等文件外，还包含一些为运行时准备的配置参数（如匿名卷、环境变量、用户等）。

（2）容器

镜像和容器的关系就像是面向对象程序设计中的类和实例一样，镜像是静态的定义，容器是镜像运行时的实体。容器可以被创建、启动、停止、删除、暂停等。

容器可以拥有自己的root文件系统、网络配置、进程空间以及用户ID空间。容器内的进程运行在一个隔离的环境里，使用起来就好像是在一个独立于宿主的系统下操作一样。这种特性使得容器封装的应用比直接在宿主中运行更加安全。

（3）仓库与镜像注册中心

镜像构建完成后，如果需要在其他服务器上使用，就需要一个集中存储、分发镜像的服务。镜像注册中心就是这样的服务。

一个镜像注册中心可以包含多个仓库，每个仓库可以包含多个标签（Tag），每个标签对应一个镜像。通常，一个仓库会包含同一个软件不同版本的镜像，而标签就常用于对应该软件的各个版本。我们可以通过<仓库名>:<标签>的格式来指定具体是这个软件哪个版本的镜像。如果不给出标签，就以latest作为默认标签。

Docker官方提供了一个名为Docker Hub的镜像注册中心，用于存放公有和私有的Docker镜像仓库。我们可以通过Docker Hub下载Docker镜像，也可以将自己创建的Docker镜像上传到Docker Hub上。

Docker引擎可以理解为一个运行在服务器上的后台进程，主要包括三大组件，如图1-10所示。

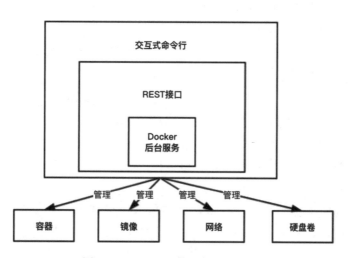

图1-10　Docker引擎三大组件架构图

- Docker 后台服务（Docker Daemon）：长时间运行在后台的守护进程，是 Docker 的核心服务，可以通过命令 dockerd 与它交互通信。
- REST 接口（REST API）：程序可以通过 REST 的接口来访问后台服务或向它发送操作指令。
- 交互式命令行界面（Docker CLI）：我们使用命令行界面与 Docker 进行交互，例如以 docker 开头的所有命令的操作，而命令行界面是通过调用 REST 的接口来控制和操作 Docker 后台服务的。

Docker 官方文档中有一张架构图，如图 1-11 所示。

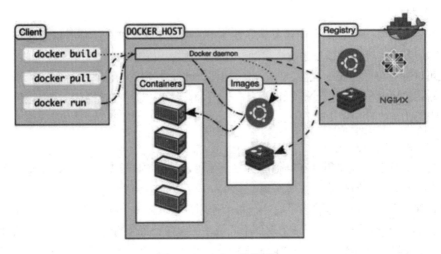

图 1-11 Docker 系统架构

- Docker 客户端（Docker Client）：与 Docker 后台服务交互的主要工具，在使用 docker run 命令时，客户端会把命令发送到 Docker 后台服务，再由后台服务执行该命令。我们可以使用 docker build 命令创建 Docker 镜像，使用 docker pull 命令拉起 Docker 镜像，使用 docker run 命令运行 Docker 镜像，从而启动 Docker 容器。
- Docker Host：表示运行 Docker 引擎的宿主机，包括 Docker Daemon 后台进程，可通过该进程创建 Docker 镜像，并在 Docker 镜像上启动 Docker 容器。
- Registry：表示 Docker 官方镜像注册中心，包含大量的 Docker 镜像仓库，可通过 Docker 引擎拉取所需的 Docker 镜像到宿主机上。

Docker 分为两个版本：社区版（Community Edition，CE）和企业版（Enterprise Edition，EE）。Docker 社区版主要提供给开发者学习和练习，企业版主要提供给企业级开发和运维团队对线上产品进行编译、打包和运行，有很高的安全性和扩展性。Docker 的社区版和企业版都支持 Linux、Cloud、Windows 和 Mac OS 平台等。这里我们以 Mac OS 操作系统为例演示 Docker 的安装，具体步骤如下所示。

步骤01 下载 Mac 版本的 Docker 安装器，这里使用 17.03 版本。

步骤02 双击 Docker.dmg 文件进行安装，拖曳蓝鲸图标到应用程序目录，如图 1-12 所示。

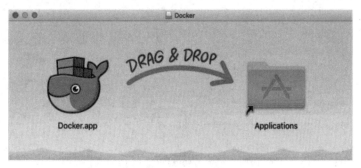

图 1-12　Docker 安装界面

步骤 03　在应用程序中双击 Docker 程序启动 Docker，程序会提示获取访问权限，然后输入系统密码即可。此时在状态栏会显示一个小蓝鲸图标，如图 1-13 所示。

步骤 04　单击状态栏上的小蓝鲸图标会弹出一个菜单项，如图 1-14 所示。选择 Restart 可以重启 Docker 服务，选择 About Docker 可以查看当前 Docker 的版本信息，选择 Preferences… 可以对 Docker 进行一些特定的设置。

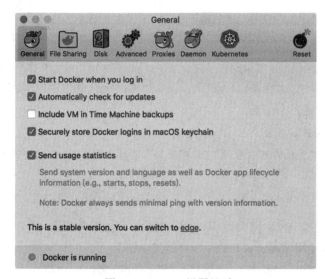

图 1-13　Docker 设置界面　　　　　　　图 1-14　Docker Preferences 设置界面

步骤 05　至此，我们已成功安装和运行 Docker。

Docker 安装成功后，接下来学习 Docker 的常用命令。

（1）查看版本信息

```
### 查看 Docker 版本
→ docker --version
Docker version 18.06.1-ce, build e68fc7a
### 查看 Docker 的更多信息
→ docker version
Client:
 Version:        18.06.1-ce
 API version:    1.38
```

```
 Go version:      go1.10.3
 Git commit:      e68fc7a
 Built:           Tue Aug 21 17:21:31 2018
 OS/Arch:         darwin/amd64
 Experimental:    false

Server:
 Engine:
  Version:        18.06.1-ce
  API version:    1.38 (minimum version 1.12)
  Go version:     go1.10.3
  Git commit:     e68fc7a
  Built:          Tue Aug 21 17:29:02 2018
  OS/Arch:        linux/amd64
  Experimental:   true
```

（2）镜像相关命令

使用 docker images 查看本地镜像：

```
### 查看本地镜像
➜ docker images
REPOSITORY              TAG         IMAGE ID            CREATED             SIZE
centos                  latest      1e1148e4cc2c        8 weeks ago         202MB
jenkinsci/jenkins       latest      b589aefe29ff        2 months ago        703MB
```

- REPOSITORY：表示镜像仓库名称。
- TAG：表示标签名称，latest 表示最新版本。
- IMAGE ID：镜像 ID，唯一的，这里我们可以看到 12 位字符串，实际上它是 64 位完整镜像 ID 的缩略表达式。
- CREATE：表示镜像的创建时间。
- SIZE：表示镜像的字节大小。

要拉取 Docker Hub 中的镜像，可以使用 docker pull 命令，具体代码如下：

```
### 从 Docker Hub 拉取 tomcat 镜像
➜ docker pull tomcat
Using default tag: latest
latest: Pulling from library/tomcat
ab1fc7e4bf91: Pull complete
35fba333ff52: Pull complete
f0cb1fa13079: Pull complete
3d79c18d1bc0: Pull complete
ff1d0ae4641b: Pull complete
8883e662573f: Pull complete
adab760d76bd: Pull complete
86323b680e93: Pull complete
14a2c1cdce1c: Pull complete
ee59bf8c5470: Pull complete
067f988306af: Pull complete
Digest:
```

```
sha256:ca621745cd6f7b2511970b558b4bc6501b0997ea4b6328b3902c0a9d67ee8ddf
    Status: Downloaded newer image for tomcat:latest
```

要在 Docker Hub 中搜索镜像,可以使用 docker search 命令,具体代码如下:

```
### 搜索 tomcat 镜像
→ docker search tomcat
NAME             DESCRIPTION                STARS         OFFICIAL       AUTOMATED
tomcat  Apache Tomcat is an open source implementati…    2271          [OK]
tomee   Apache TomEE is an all-Apache Java EE certif…    60            [OK]
dordoka/tomcat   Ubuntu 14.04, Oracle JDK 8 and Tomcat 8 base…  52     [OK]
davidcaste/alpine-tomcat   Apache Tomcat 7/8 using Oracle Java 7/8 with…  34  [OK]
# 省略内容
```

- NAME:表示镜像名称。
- DESCRIPTION:表示镜像仓库的描述。
- STARS:表示镜像仓库的收藏数,用户可以在 Docker Hub 上对镜像仓库进行收藏,一般可以通过收藏数判断该镜像的受欢迎程度。
- OFFICIAL:表示是否为官方仓库,官方仓库具有更高的安全性。
- AUTOMATED:表示是否自动构建镜像仓库。用户可以将自己的 Docker Hub 绑定到 GitHub 账号上,当代码提交后,可自动构建镜像仓库。

要导入/导出镜像,可以使用 docker save 或者 docker load 命令,具体代码如下:

```
### 导出 centos 镜像为一个 tar 文件,若不指定导出的 tar 路径,则默认为当前目录
→ docker save centos > centos.jar
```

导出的 CentOS 镜像包文件可随时在另一台 Docker 机器导入,命令如下:

```
→ docker load < centos.jar
```

要删除镜像,可以使用 docker rm 命令,具体代码如下:

```
### 删除镜像,镜像 ID 为 3b6d83b7b6e8
→ docker rmi 3b6d83b7b6e8
Untagged: harbor.meitu-int.com/library/meitu-tomcat-centos7:8.5.5
Untagged: harbor.meitu-int.com/library/meitu-tomcat-centos7@sha256:
11c06a560f90dde859e72a84826faadffbf3eff6a2a097d3fbd21b80ad7c9e5b
    Deleted: sha256:
3b6d83b7b6e896041b12bc7c057539c0c5a7ae6606d7da32d9d2491427fb02da
    Deleted: sha256:
9c197a80cdc1c6394e486c1e6e9c0659d4818d36d91dbcca48f97290652884fc
    Deleted: sha256:
6e90a04b31888bc7be6a8b11fcd65b1dcaa187ccecaa86ee77a82d82084092c0
    Deleted: sha256:
563db4d832259cce189af4a8f8d5055b11e7463ea804a4d345c8a3eda48536ad
    Deleted: sha256:
29c44132179f678ff7638afa185c1e9ac2da3fd09ae5d13c0053a54055603d03
    Deleted: sha256:
```

```
8739ec81aa64f7b0928890424b8adbeaed1ccda01685878efb92455657d981ec
  Deleted: sha256:
88d9fdb1acd458424c06679e4b3e201f9f44f2c2baa35c0fdc3d7668cce026df
  Deleted: sha256:
d1be66a59bc56bb90e92c3d4742ce73dcb5f62acc6e92de55039e21957ed5d23
```

（3）容器相关命令

要查看运行的容器，可以使用如下命令：

```
### 查看运行的容器，由于我们还没有启动任何容器，因此没有相关的记录
➜ docker ps
CONTAINER ID    IMAGE       COMMAND         CREATED         STATUS          PORTS       NAMES
### 列出最近创建的容器，包括所有的状态
➜ docker ps -l
CONTAINER ID    IMAGE       COMMAND         CREATED         STATUS          PORTS       NAMES
7735516adfbc    centos      "/bin/bash"     10 minutes ago  Exited (127) 6 seconds ago              cocky_mahavira
### 列出所有的容器，包括所有的状态
➜ docker ps -a
CONTAINER ID    IMAGE       COMMAND         CREATED         STATUS          PORTS       NAMES
7735516adfbc    centos      "/bin/bash"     12 minutes ago  Exited (127) 2 minutes ago              cocky_mahavira
1db7702d3fb7    centos      "/bin/bash"     17 minutes ago  Exited (0) 16 minutes ago               adoring_knuth
3eb65674f664    hello-world "/bin/bash"     17 minutes ago  Created                                 pensive_saha
06f90bf8759f    centos      "/bin/bash"     6 weeks ago     Exited (0) 6 weeks ago                  epic_snyder
6afa1f4ddbd7    centos      "-i -t /bin/bash" 6 weeks ago   Created
### -q 表示仅列出 CONTAINER ID
➜ docker ps -a -q
```

- CONTAINER ID：表示容器 ID。
- IMAGE：表示镜像名称。
- COMMAND：表示启动容器运行的命令，Docker 容器要求我们在启动容器时运行一个命令。
- CREATE：表示容器创建的时间。
- STATUS：表示容器运行的状态，例如 UP 表示运行中、Exited 表示已退出。
- PORTS：表示容器需要对外暴露的端口。
- NAMES:表示容器的名称,由 Docker 引擎自动生成,也可以在 docker run 命令中通过 --name 选项来指定。

要创建并启动容器，可以使用如下命令：

```
### 启动 centos 容器，并进入到容器里
➜ docker run -i -t centos /bin/bash
[root@d666c2e2d235 /]#
```

- -i 选项：表示启动容器后打开标准收入设备（STDIN），可使用键盘输入。
- -t 选项：表示启动容器后分配一个伪终端，将与容器建立会话。
- centos 参数：表示要运行的镜像名称，标准格式为 centos:latest。若为 latest 版本，则可省略

latest。
- /bin/bash 参数：表示运行容器中的 bash 应用程序。

> **注　意**
>
> 上述命令首先从本地获取 CentOS 镜像，若本地没有此镜像，则从 Docker Hub 拉取 CentOS 镜像并放入本地，随后根据 CentOS 镜像创建并启动 CentOS 容器。

除了使用该命令创建和进入容器外，还可以使用如下命令进入运行中的容器：

```
### 进入启动中的容器，但是不能进入已停止的容器
docker attach d666c2e2d235
### root 表示以超级管理员身份进入容器，d666c2e2d235 表示容器的 ID，/表示当前路径
[root@d666c2e2d235 /]#
```

还可以使用如下命令让运行中的容器执行具体命令：

```
### d666c2e2d235 容器 ID，ls -l 表示列出容器中当前的目录结构
docker exec -i -t d666c2e2d235 ls -l
total 56
-rw-r--r--   1 root root 12076 Dec  5 01:37 anaconda-post.log
lrwxrwxrwx   1 root root     7 Dec  5 01:36 bin -> usr/bin
drwxr-xr-x   5 root root   360 Feb  1 04:09 dev
drwxr-xr-x   1 root root  4096 Feb  1 04:09 etc
drwxr-xr-x   2 root root  4096 Apr 11  2018 home
lrwxrwxrwx   1 root root     7 Dec  5 01:36 lib -> usr/lib
lrwxrwxrwx   1 root root     9 Dec  5 01:36 lib64 -> usr/lib64
drwxr-xr-x   2 root root  4096 Apr 11  2018 media
drwxr-xr-x   2 root root  4096 Apr 11  2018 mnt
drwxr-xr-x   2 root root  4096 Apr 11  2018 opt
dr-xr-xr-x 166 root root     0 Feb  1 04:09 proc
dr-xr-x---   2 root root  4096 Dec  5 01:37 root
drwxr-xr-x  11 root root  4096 Dec  5 01:37 run
lrwxrwxrwx   1 root root     8 Dec  5 01:36 sbin -> usr/sbin
drwxr-xr-x   2 root root  4096 Apr 11  2018 srv
dr-xr-xr-x  13 root root     0 Feb  1 04:09 sys
drwxrwxrwt   7 root root  4096 Dec  5 01:37 tmp
drwxr-xr-x  13 root root  4096 Dec  5 01:36 usr
drwxr-xr-x  18 root root  4096 Dec  5 01:36 var
```

可以使用 docker stop 和 docker kill 命令停止或者终止容器，具体代码如下：

```
### 停止运行中的容器，d666c2e2d235 为容器 ID
docker stop d666c2e2d235
d666c2e2d235
### 再次创建和启动容器
docker run -i -t centos /bin/bash
[root@fd35f5e95fe9 /]# %
### 终止运行中的容器，fd35f5e95fe9 为容器 ID
docker kill fd35f5e95fe9
fd35f5e95fe9
```

可以使用 docker start 和 docker restart 命令启动或者重启容器，具体代码如下：

```
### 启动已停止的容器，fd35f5e95fe9 为容器 ID
➜  docker start fd35f5e95fe9
fd35f5e95fe9
### 重启运行中的容器，fd35f5e95fe9 为容器 ID
➜  docker restart fd35f5e95fe9
fd35f5e95fe9
```

可以使用 docker rm 命令来删除已经停止的容器，具体代码如下：

```
### 停止容器 ID 为 fd35f5e95fe9 的容器
➜  docker stop fd35f5e95fe9
fd35f5e95fe9
### 删除已停止的容器，fd35f5e95fe9 为容器 ID
➜  docker rm  fd35f5e95fe9
fd35f5e95fe9
###强制删除所有运行中的容器，docker ps -a -q 命令将返回所有的容器 ID
➜  docker rm -f $(docker ps -a -q)
```

可以使用 docker import 或者 docker export 命令，导入/导出容器，具体代码如下：

```
### 导出容器为 TAR 文件，若不指定导出的 tar 路径，则默认为当前目录
➜  docker export 913111e2d596 > centos.tar
```

导出的 CentOS 容器包可随时在另一台 Docker 机器上导入为镜像，具体命令如下：

```
➜  docker import centos.jar  centos:latest
```

1.2 Spring Boot 简介

Spring Boot 是目前流行的微服务框架，倡导"约定优先于配置"，其设计目的是用来简化新 Spring 应用的初始化搭建以及开发过程。Spring Boot 是一个典型的"核心 + 插件"的系统架构，提供了很多核心的功能，比如自动化配置、提供 starter 简化 Maven 配置、内嵌 Servlet 容器、应用监控等功能，让我们可以快速构建企业级应用程序。Spring Boot 使编码、配置、部署和监控变得简单。

Spring Boot 提供的特性如下：

（1）遵循习惯优于配置的原则

Spring Boot 的配置都在 application.properties 中，但是并不意味着在 Spring Boot 应用中就必须包含该文件。application.properties 配置文件包含大量的配置项，而大多数配置项都有其默认值，很多配置项不用我们去修改，使用默认值即可。这类行为叫作"自动化配置"。

（2）提供了"开箱即用"的 Spring 插件

Spring Boot 提供了大量的 starter，当我们需要整合其他技术（比如 Redis、MQ 等）时，只需要添加一段 Maven 依赖配置即可开启使用。每个 starter 都有自己的配置项，而这些配置都可以在

application.properties 配置文件中进行统一配置，例如常用的 spring-boot-starter-web、spring-boot-starter-tomcat、spring-boot-starter-actuator 等。

（3）内嵌 Servlet 容器

传统的项目都需要将项目打包成 War 包部署到 Web 服务器，比如 Tomcat、Jetty、Undertow。Spring Boot 应用程序启动后会在默认端口 8080 下启动嵌入式 Tomcat，执行 Spring Boot 项目的主程序 main()函数便可以快速运行项目。

（4）倡导 Java Config

Spring Boot 可以完全不使用 XML 配置，并倡导我们使用 Java 注解方式开发项目。

（5）多环境配置

在项目开发过程中，项目不同的角色会使用不同的环境，比如开发人员会使用开发环境、测试人员会使用测试环境、性能测试会使用性能测试环境、项目开发完成之后会把项目部署到线上环境，等等。不同的环境往往会连接不同的 MySQL 数据库、Redis 缓存、MQ 消息中间件等。环境之间相互独立与隔离才不会相互影响。隔离的环境便于部署，提高工作效率。假设项目 my-spring-boot 需要 3 个环境，即开发环境、测试环境、性能测试环境，我们复制 my-spring-boot 项目配置文件 application.properties，分别取名为 application-dev.properties、application-test.properties、application-perform.properties，以作为开发环境、测试环境、性能测试环境。多环境的配置文件开发完成之后，我们在 my-spring-boot 的配置文件 application.properties 中添加配置激活选项，具体代码如下所示：

```
### 激活开发环境配置
spring.profiles.active=dev
```

如果我们想激活测试环境的配置，可修改为：

```
### 激活测试环境配置
spring.profiles.active=test
```

如果我们想激活性能测试环境的配置，可修改为：

```
### 激活性能测试环境配置
spring.profiles.active=test
```

（6）提供大量生产级特性

Spring Boot 提供大量的生产级特性，例如应用监控、健康检查、外部配置和核心指标等。我们可以给 Spring Boot 应用发送 /metrics 请求获取 Json 数据，该数据包含内存、Java 堆、类加载器、处理器、线程池等信息。我们还能在 Java 命令（备注：java -jar xxx.jar）上直接运行 Spring Boot 应用，并带上外部配置参数，这些参数将覆盖已有的默认配置参数。我们甚至可以通过发送一个 URL 请求去关闭 Spring Boot 应用。Spring Boot 提供了基于 HTTP、SSH、Telnet 等方式对运行时的项目进行监控。

Spring Boot 为我们带来诸多便利的同时，也存在如下缺点：

- 高度集成，开发人员不知道底层实现。
- 如果开发人员不了解 Spring Boot 底层，那么项目出现问题就会很难排查。

- 相对来说，将现有或传统的 Spring Framework 项目转换为 Spring Boot 应用程序比较困难和耗时。Spring Boot 适用于全新的 Spring 项目。

Spring Boot 整合公司自研的框架和组件相对比较麻烦，例如 Spring Boot 整合公司自研的 RPC 框架等。

1.3　第一个 Spring Boot 项目

1.3.1　使用 Spring Initializr 新建项目

使用 Intellij IDEA 创建 Spring Boot 项目有多种方式，比如使用 Maven 和 Spring Initializr。这里只介绍 Spring Initializr 方式，因为这种方式不但为我们生成完整的目录结构，还为我们生成一个默认的主程序，节省时间。我们的目的是掌握 Spring Boot 知识，而不是学一堆花样。具体步骤如下：

步骤 01 在 Intellij IDEA 界面中，单击 File→New→Product，在弹出的窗口中选择 Spring Initializr 选项，在 Product SDK 中选择 JDK 的安装路径，如果没有，就新建一个，然后单击 Next 按钮，具体如图 1-15 所示。

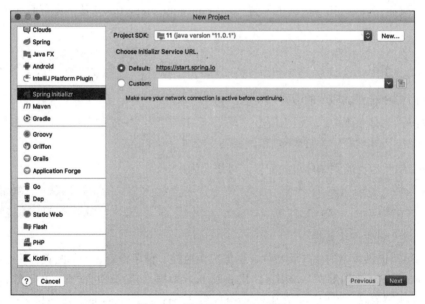

图 1-15　新建 Spring Boot 项目

步骤 02 选择 Spring Boot 版本，这里采用默认版本（本书使用的 Spring Boot 版本为 2.2.6）即可，勾选 Spring Web 选项，然后单击 Next 按钮，具体如图 1-16 所示。

第 1 章　从 Spring Boot 开始　｜　17

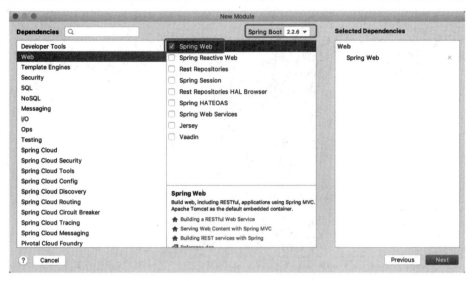

图 1-16　选择 Spring Boot 版本

步骤 03 填写项目名称"my-spring-boot"，其他保持默认即可，然后单击 Finish 按钮。至此，一个完整的 Spring Boot 创建完成，具体如图 1-17 所示。

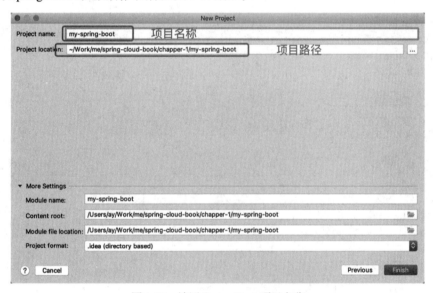

图 1-17　填写 Spring Boot 项目名称

步骤 04 在 IDEA 开发工具上，找到刷新依赖的按钮（Reimport All Maven Projects），下载相关的依赖包，这时开发工具开始下载 Spring Boot 项目所需的依赖包，如图 1-18 所示。

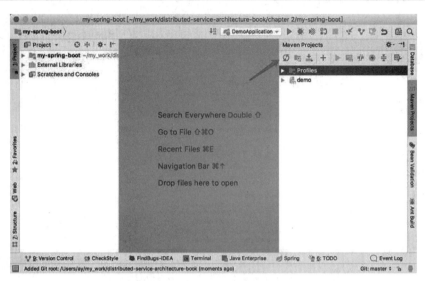

图 1-18　刷新 Spring Boot 项目依赖

步骤 05 在项目的目录 com.example.demo.controller 下创建 HelloController 控制层类，具体代码如下：

```java
package com.example.demo.controller;
import org.springframework.web.bind.annotation.RequestMapping;
import org.springframework.web.bind.annotation.RestController;

/**
 * 描述：控制层类
 * @author ay
 * @date 2020-04-01
 */
@RestController
public class HelloController {

    @RequestMapping("/hello")
    public void say(){
        //打印信息
        System.out.println("hello ay");
    }
}
```

1.3.2　测试

Spring Boot 项目创建完成之后，找到入口类 DemoBootApplication 中的 main 方法并运行。当看到如图 1-19 所示的界面时，表示项目启动成功。同时还可以看出项目启动的端口（8080）以及启动时间。在浏览器中输入访问地址"http://localhost:8080/hello"，便可以在控制台打印信息"hello ay"。

第 1 章 从 Spring Boot 开始

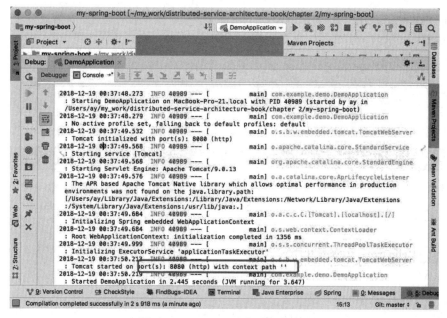

图 1-19　Spring Boot 启动成功界面

1.4　Spring Boot 目录介绍

1.4.1　Spring Boot 工程目录

Spring Boot 的工程目录如图 1-20 所示。

- /src/main/java：目录下放置所有的 Java 文件（源代码文件）。
- /src/main/resources：用于存放所有的资源文件，包括静态资源文件、配置文件、页面文件等。
- /src/main/resources/static：用于存放各类静态资源。
- /src/main/resources/application.properties：配置文件，这个文件非常重要。Spring Boot 默认支持两种配置文件类型（.properties 和.yml）。
- /src/main/resources/templates：用于存放模板文件，如 Thymeleaf 模板文件。
- /src/test/java：放置单元测试类 Java 代码。
- /target：放置编译后的.class 文件、配置文件等。

Spring Boot 将很多配置文件进行了统一管理，且配置了默认值。Spring Boot 会自动在/src/main/resources 目

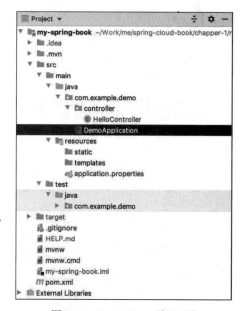

图 1-20　Spring Boot 项目目录

录下找 application.properties 或者 application.yml 配置文件。找到后将运用此配置文件中的配置，否则使用默认配置。这两种类型的配置文件有其一即可，也可以两者并存。两者的区别如下：

```
application.properties:
server.port = 8080
application.yml:
server:
port:8080
```

> **注 意**
>
> .properties 配置文件的优先级高于 .yml。例如，在 .properties 文件中配置了 server.port = 8080，同时在 .yml 中配置了 server.port = 8090，Spring Boot 将使用 .properties 中的 8080 端口。

1.4.2　Spring Boot 入口类

入口类的代码很简单，代码如下：

```java
import org.springframework.boot.SpringApplication;
import org.springframework.boot.autoconfigure.SpringBootApplication;

@SpringBootApplication
public class DemoApplication {

  public static void main(String[] args) {
    SpringApplication.run(DemoApplication.class, args);
  }
}
```

- @SpringBootApplication：一个组合注解，包含@EnableAutoConfiguration、@ComponentScan 和@SpringBootConfiguration 三个注解，是项目启动注解。如果使用三个注解，那么项目依旧可以启动起来，只是过于烦琐，所以用@SpringBootApplication 来简化。
- @SpringApplication.run：应用程序开始运行的方法。

> **注 意**
>
> DemoApplication 入口类需要放置在包的最外层，以便能够扫描到所有子包中的类。

1.4.3　Spring Boot 测试类

Spring Boot 的测试类主要放置在/src/test/java 目录下。项目创建完成后，Spring Boot 会自动为我们生成测试类 DemoApplicationTests.java。测试类的代码如下：

```java
import org.junit.Test;
import org.junit.runner.RunWith;
import org.springframework.boot.test.context.SpringBootTest;
import org.springframework.test.context.junit4.SpringRunner;
```

```
@RunWith(SpringRunner.class)
@SpringBootTest
public class DemoApplicationTests {

    @Test
    public void contextLoads() {

    }
}
```

- @RunWith(SpringRunner.class)：@RunWith(Parameterized.class) 参数化运行器，配合 @Parameters 使用 Junit 的参数化功能。查源码可知，SpringRunner 类继承自 SpringJUnit4ClassRunner 类，此处表明使用 SpringJUnit4ClassRunner 执行器。此执行器集成了 Spring 的一些功能。如果只是简单的 Junit 单元测试，该注解可以去掉。
- @SpringBootTest：此注解能够测试我们的 SpringApplication，因为 Spring Boot 程序的入口是 SpringApplication，所以基本上所有配置都会通过入口类去加载，而该注解可以引用入口类的配置。
- @Test：Junit 单元测试的注解，注解在方法上，表示一个测试方法。

当我们右击执行 DemoApplicationTests.java 中的 contextLoads 方法时，大家可以看到控制台打印的信息和执行入口类中的 SpringApplication.run() 方法打印的信息是一致的。由此便知 @SpringBootTest 是引入了入口类的配置。

1.4.4 pom.xml 文件

Spring Boot 项目下的 pom.xml 文件主要用来存放依赖信息，具体代码如下：

```xml
<?xml version="1.0" encoding="UTF-8"?>
<project xmlns="http://maven.apache.org/POM/4.0.0"
xmlns:xsi="http://www.w3.org/2001/XMLSchema-instance"
         xsi:schemaLocation="http://maven.apache.org/POM/4.0.0
https://maven.apache.org/xsd/maven-4.0.0.xsd">
    <modelVersion>4.0.0</modelVersion>
    <parent>
        <groupId>org.springframework.boot</groupId>
        <artifactId>spring-boot-starter-parent</artifactId>
        <version>2.2.6.RELEASE</version>
        <relativePath/> <!-- lookup parent from repository -->
    </parent>
    <groupId>com.example</groupId>
    <artifactId>demo</artifactId>
    <version>0.0.1-SNAPSHOT</version>
    <name>demo</name>
    <description>Demo project for Spring Boot</description>

    <properties>
```

```xml
        <java.version>1.8</java.version>
    </properties>

    <dependencies>
        <dependency>
            <groupId>org.springframework.boot</groupId>
            <artifactId>spring-boot-starter-web</artifactId>
        </dependency>

        <dependency>
            <groupId>org.springframework.boot</groupId>
            <artifactId>spring-boot-starter-test</artifactId>
            <scope>test</scope>
            <exclusions>
                <exclusion>
                    <groupId>org.junit.vintage</groupId>
                    <artifactId>junit-vintage-engine</artifactId>
                </exclusion>
            </exclusions>
        </dependency>
    </dependencies>

    <build>
        <plugins>
            <plugin>
                <groupId>org.springframework.boot</groupId>
                <artifactId>spring-boot-maven-plugin</artifactId>
            </plugin>
        </plugins>
    </build>
</project>
```

- spring-boot-starter-parent：一个特殊的 starter，用来提供相关的 Maven 默认依赖，使用它之后，常用的包依赖可以省去 version 标签。
- spring-boot-starter-web：只要将其加入项目的 Maven 依赖中，就得到了一个可执行的 Web 应用。该依赖中包含许多常用的依赖包，比如 spring-web、spring-webmvc 等。我们不需要做任何 Web 配置，便能获得相关 Web 服务。
- spring-boot-starter-test：这个依赖和测试相关，只要引入它，就会把所有与测试相关的包全部引入。
- spring-boot-maven-plugin：一个 Maven 插件，能够以 Maven 的方式为应用提供 Spring Boot 的支持，即为 Spring Boot 应用提供了执行 Maven 操作的可能，能够将 Spring Boot 应用打包为可执行的 JAR 或 WAR 文件。

1.5 Spring Boot 生产级特性

1.5.1 应用监控

Spring Boot 大部分模块都是用于开发业务功能或连接外部资源的。除此之外，Spring Boot 还为我们提供了 spring-boot-starter-actuator 模块，该模块主要用于管理和监控应用。这是一个用于暴露自身信息的模块。spring-boot-starter-actuator 模块可以有效地减少监控系统在采集应用指标时的开发量。spring-boot-starter-actuator 模块提供了监控和管理端点以及一些常用的扩展和配置方式，具体如表 1-2 所示。

表 1-2 监控和管理端点

路径（端点名）	描述	鉴权
/actuator/health	显示应用监控指标	false
/actuator/beans	查看 bean 及其关系列表	true
/actuator/info	查看应用信息	false
/actuator/trace	查看基本追踪信息	true
/actuator/env	查看所有环境变量	true
/actuator/env/{name}	查看具体变量值	true
/actuator/mappings	查看所有 url 映射	true
/actuator/autoconfig	查看当前应用的所有自动配置	true
/actuator/configprops	查看应用的所有配置属性	true
/actuator/shutdown	关闭应用（默认关闭）	true
/actuator/metrics	查看应用基本指标	true
/actuator/metrics/{name}	查看应用具体指标	true
/actuator/dump	打印线程栈	true

在 Spring Boot 中使用监控，首先需要在 pom.xml 文件中引入所需的依赖 spring-boot-starter-actuator，具体代码如下：

```xml
<dependency>
    <groupId>org.springframework.boot</groupId>
    <artifactId>spring-boot-starter-actuator</artifactId>
</dependency>
```

在 pom.xml 文件引入 spring-boot-starter-actuator 依赖包之后，需要在 application.properties 文件

中添加如下的配置信息：

```
### 应用监控配置
#指定访问这些监控方法的端口
management.server.port=8099
```

management.port 用于指定访问这些监控方法的端口。spring-boot-starter-actuator 依赖和配置都添加成功之后，重新启动 my-spring-boot 项目。项目启动成功之后，在浏览器测试各个端点。比如在浏览器中输入"http://localhost:8099/actuator/health"，可以看到如图 1-21 所示的应用健康信息；在浏览器中输入"http://localhost:8099/actuator/env"，可以查看所有环境变量，如图 1-22 所示。

图 1-21 应用健康信息

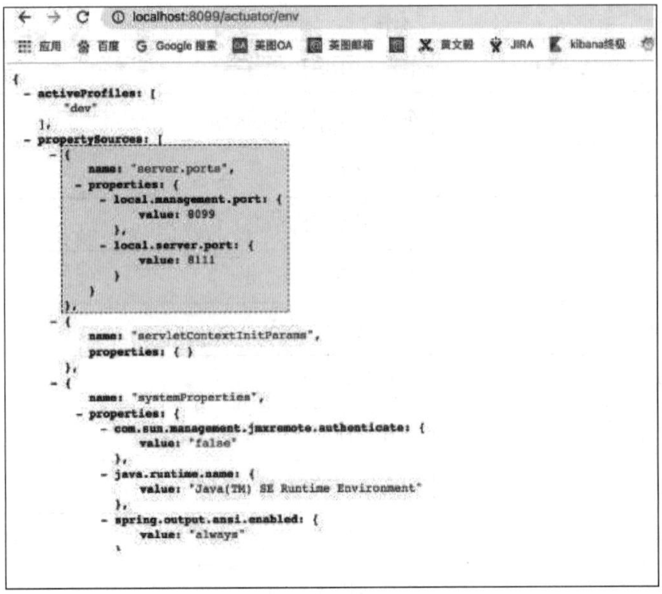

图 1-22 应用环境变量

其他端点测试可以按照表 1-3 所示的访问路径依次访问测试。

表 1-3 监控和管理端点

路径（端点名）	描述
http://localhost:8099/actuator/health	显示应用监控指标
http://localhost:8099/actuator/beans	查看 bean 及其关系列表
http://localhost:8099/actuator/info	查看应用信息
http://localhost:8099/actuator/trace	查看基本追踪信息

路径（端点名）	描述
http://localhost:8099/actuator/env	查看所有环境变量
http://localhost:8099/actuator/env/{name}	查看具体变量值
http://localhost:8099/actuator/mappings	查看所有 url 映射
http://localhost:8099/actuator/autoconfig	查看当前应用的所有自动配置
http://localhost:8099/actuator/configprops	查看应用的所有配置属性
http://localhost:8099/actuator/shutdown	关闭应用（默认关闭）
http://localhost:8099/actuator/metrics	查看应用基本指标
http://localhost:8099/actuator/metrics/{name}	查看应用具体指标
http://localhost:8099/actuator/dump	打印线程栈

在浏览器中可以把返回的数据格式化成 JSON 格式，因为笔者的 Google Chrome 浏览器安装了 JsonView 插件，具体安装步骤如下：

步骤01 在浏览器中输入链接 "https://github.com/search?utf8=%E2%9C%93&q=jsonview"，在弹出的页面中单击 gildas-lormeau/JSONView-for-Chrome，如图 1-23 所示。

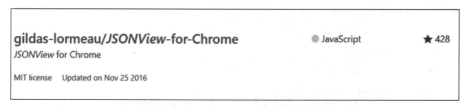

图 1-23 JsonView-for-Chrome 界面

步骤02 单击 Download Zip 按钮，插件下载完成，解压缩到相应目录。

步骤03 在浏览器右上角单击 "更多工具" → "扩展程序" → "加载已解压的扩展程序"，选择插件目录。

步骤04 安装完成，重新启动浏览器（快捷键：Ctrl+R）。

除了 spring-boot-starter-actuator 提供的默认端点外，我们还可以定制端点（一般通过 "management.endpoint + 端点名 + 属性名" 来设置）。比如，我们可以在配置文件 application.properties 中把端点名 health 的更多详细信息打印出来，具体代码如下：

```
management.endpoint.health.show-details=always
```

配置添加完成之后，重新启动 my-spring-boot 项目，在浏览器中输入访问地址 "http://localhost:8099/actuator/health"，可以获得应用健康的更多详细信息。如果想关闭端点 beans，那么可以在配置文件 application.properties 中添加如下代码：

```
management.endpoint.beans.enabled=false
```

配置添加完成之后，重新启动 my-spring-boot 项目，在浏览器中输入访问地址

"http://localhost:8099/actuator/beans"，返回 404 错误信息，具体代码如下：

```
{
    timestamp: "2018-12-23T09:04:34.759+0000",
    status: 404,
    error: "Not Found",
    message: "No message available",
    path: "/actuator/beans"
}
```

如果想知道 Spring Boot 提供了哪些端点，那么可以引入 hateoas 依赖，具体代码如下：

```xml
<dependency>
    <groupId>org.springframework.boot</groupId>
    <artifactId>spring-boot-starter-hateoas</artifactId>
</dependency>
```

hateoas 是一个超媒体技术，通过它可以汇总端点信息，包含各个端点的名称与链接。hateoas 依赖添加完成之后，在浏览器中输入访问地址"http://localhost:8099/actuator"，将看到所有的端点及其访问链接，如图 1-24 所示。

图 1-24 端点信息汇总

1.5.2 健康检查

在浏览器中输入访问地址"http://localhost:8099/actuator/health"，将可以看到如图 1-25 所示的图片。

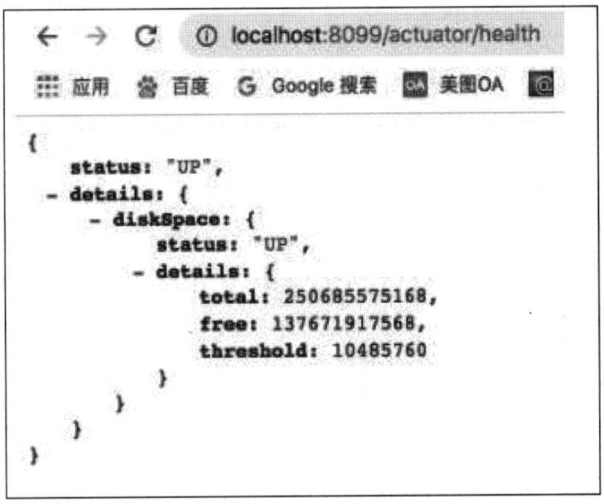

图 1-25 端点信息汇总

health 端点用于查看当前应用的运行状况，即应用的健康状况，简称"健康检查"。status: "UP" 代表应用正处于运行状态。diskSpace 表示磁盘空间的使用情况。默认端点 health 的信息是从 HealthIndicator 的 bean 中收集的。Spring 中内置了一些 HealthIndicator，如表 1-4 所示。

表 1-4 监控和管理端点

路径（端点名）	描述
CassandraHealthIndicator	检测 Cassandra 数据库是否在运行
DiskSpaceHealthIndicator	检测磁盘空间
DataSourceHealthIndicator	检测 DataSource 连接是否能获得
ElasticsearchHealthIndicator	检测 Elasticsearch 集群是否在运行
JmsHealthIndicator	检测 JMS 消息代理是否在运行
MailHealthIndicator	检测邮箱服务器是否在运行
MongoHealthIndicator	检测 Mongo 是否在运行
RabbitHealthIndicator	检测 RabbitMQ 是否在运行
RedisHealthIndicator	检测 Redis 是否在运行
SolrHealthIndicator	检测 Solr 是否在运行

我们可以利用 Spring Boot 的健康检查特性开发一个微服务系统监控平台，用于获取每个微服务的运行状态和性能指标。也可以使用现有的解决方案，比如 spring-boot-admin，这是一款基于 Spring Boot 的开源监控平台。

1.5.3 跨域访问

对于前后端分离的项目来说，如果前端项目与后端项目部署在两个不同的域下，那么势必会引起跨域问题的出现。针对跨域问题，我们可能第一个想到的解决方式就是 JSONP，但是 JSONP

方式有一些不足，JSONP 方式只能通过 Get 请求方式来传递参数，当然还有其他的不足之处。在 Spring Boot 中通过 CORS（Cross-Origin Resource Sharing，跨域资源共享）协议解决跨域问题。CORS 是一个 W3C 标准，它允许浏览器向不同源的服务器发出 XmlHttpRequest 请求，我们可以继续使用 Ajax 进行请求访问。Spring MVC 4.2 版本增加了对 CORS 的支持，具体做法如下：

```
@Configuration
public class MyWebAppConfigurer extends WebMvcConfigurerAdapter{

    @Override
    public void addCorsMappings(CorsRegistry registry) {
        registry.addMapping("/**");
    }
}
```

我们可以在 addMapping 方法中配置路径，/** 代表所有路径。当然也可以修改其他属性，例如：

```
@Configuration
public class MyWebAppConfigurer extends WebMvcConfigurerAdapter{

    @Override
    public void addCorsMappings(CorsRegistry registry) {
        registry.addMapping("/api/**")
            .allowedOrigins("http://192.168.1.97")
            .allowedMethods("GET", "POST")
            .allowCredentials(false).maxAge(3600);
    }
}
```

以上两种方式都是针对全局配置的。如果想做到更细致的控制，那么可以使用@CrossOrigin 注解在 Controller 类中使用。

```
@CrossOrigin(origins = "http://192.168.1.97:8080", maxAge = 3600)
@RequestMapping("rest_index")
@RestController
public class AyController{}
```

这样就可以指定 AyController 中所有方法都能处理来自 http://19.168.1.97:8080 的请求。

1.5.4　外部配置

Spring Boot 支持通过外部配置覆盖默认配置项，具体优先级如下：

（1）Java 命令行参数。
（2）JNDI 属性。
（3）Java 系统属性（System.getProperties()）。
（4）操作系统环境变量。
（5）RandomValuePropertySource 配置的 random.*属性值。

（6）JAR 包外部的 application-{profile}.properties 或 application.yml（带 spring.profile）配置文件。

（7）JAR 包内部的 application-{profile}.properties 或 application.yml（带 spring.profile）配置文件。

（8）JAR 包外部的 application.properties 或 application.yml（不带 spring.profile）配置文件。

（9）JAR 包内部的 application.properties 或 application.yml（不带 spring.profile）配置文件。

（10）@Configuration 注解类上的@PropertySource。

（11）通过 SpringApplication.setDefaultProperties 指定的默认属性。

以 Java 命令行参数为例，运行 Spring Boot jar 包时，指定如下参数：

```
### 参数用--xxx=xxx 形式传递
java -jar app.jar --name=spring-boot --server.port=9090
```

应用启动的时候，就会覆盖默认的 Web Server 8080 端口，改为 9090。

1.6　Spring Boot 原理解析

1.6.1　DemoApplication 入口类

我们先来回顾一下项目 my-spring-boot 的入口类 DemoApplication，具体代码如下：

```
@SpringBootApplication
public class DemoApplication {
    public static void main(String[] args) {
        SpringApplication.run(MySpringBootApplication.class, args);
    }
}
```

在入口类 DemoApplication 中，@SpringBootApplication 和 main 方法是 Spring Boot 为我们自动生成的，其他注解都是我们在学习 Spring Boot 整合其他技术添加上去的。接下来就和大家一起看看@SpringBootApplication 和 SpringApplication.run 方法到底为我们做了些什么。

1.6.2　@SpringBootApplication 的原理

@SpringBootApplication 开启了 Spring 的组件扫描和 Spring Boot 自动配置功能。实际上它是一个复合注解，包含 3 个重要的注解@SpringBootConfiguration、@EnableAutoConfiguration、@ComponentScan，其源代码如下：

```
@Target({ElementType.TYPE})
@Retention(RetentionPolicy.RUNTIME)
@Documented
@Inherited
@SpringBootConfiguration
```

```
@EnableAutoConfiguration
@ComponentScan
public @interface SpringBootApplication {

    //省略代码
}
```

- @SpringBootConfiguration 注解：标明该类使用 Spring 基于 Java 的注解。Spring Boot 推荐使用基于 Java 而不是 XML 的配置，所以本书的实战例子都是基于 Java 而不是 XML 的配置。查看@SpringBootConfiguration 源代码，发现它就是对@Configuration 进行简单的"包装"，然后取名为 SpringBootConfiguration：

```
@Target(ElementType.TYPE)
@Retention(RetentionPolicy.RUNTIME)
@Documented
@Configuration
public @interface SpringBootConfiguration {

}
```

我们对@Configuration 注解并不陌生，它就是 JavaConfig 形式的 Spring IoC 容器的配置类使用的@Configuration。

- @EnableAutoConfiguration 注解：可以开启自动配置的功能。@EnableAutoConfiguration 的源代码如下：

```
@Target({ElementType.TYPE})
@Retention(RetentionPolicy.RUNTIME)
@Documented
@Inherited
@AutoConfigurationPackage
@Import({EnableAutoConfigurationImportSelector.class})
public @interface EnableAutoConfiguration {
    //省略代码
}
```

- 从@EnableAutoConfiguration 源代码可以看出，其包含@Import 注解。我们知道，@Import 注解的主要作用就是借助 EnableAutoConfigurationImportSelector 将 Spring Boot 应用所有符合条件的@Configuration 配置都加载到当前 Spring Boot 创建并使用的 IoC 容器中，IoC 容器就是我们所说的 Spring 应用程序上下文 ApplicationContext。学习过 Spring 框架就会知道，Spring 框架提供了很多@Enable 开头的注解定义，比如@EnableScheduling、@EnableCaching 等。这些@Enable 开头的注解都有一个共同的功能，就是借助@Import 的支持，收集和注册特定场景相关的 bean 定义。

- @ComponentScan 注解：启动组件扫描，开发的组件或 bean 定义能被自动发现并注入 Spring 应用程序上下文。比如我们在控制层添加@Controller 注解、服务层添加的@Service 注解和@Component 注解等，这些注解都可以被@ComponentScan 注解扫描到。

在 Spring Boot 早期的版本中，需要在入口类同时添加这 3 个注解。从 Spring Boot 1.2.0 开始，只要在入口类添加@SpringBootApplication 注解即可。

1.6.3　SpringApplication 的 run 方法

除了@SpringBootApplication 注解外，入口类还有一个显眼的地方，就是 SpringApplication.run 方法。在 run 方法中，首先创建一个 SpringApplication 对象实例，然后调用 SpringApplication 的 run 方法。SpringApplication.run 方法的源代码如下：

```java
public class SpringApplication{
    //省略代码
    public ConfigurableApplicationContext run(String... args) {
        StopWatch stopWatch = new StopWatch();
        stopWatch.start();
        ConfigurableApplicationContext context = null;
        FailureAnalyzers analyzers = null;
        configureHeadlessProperty();
        //开启监听器
        SpringApplicationRunListeners listeners = getRunListeners(args);
        listeners.starting();
        try {
            ApplicationArguments applicationArguments =
                        new DefaultApplicationArguments(args);
            ConfigurableEnvironment environment = prepareEnvironment(listeners,
                applicationArguments);
            Banner printedBanner = printBanner(environment);
            //创建应用上下文
            context = createApplicationContext();
            analyzers = new FailureAnalyzers(context);
            //准备上下文
            prepareContext(context, environment, listeners,
                applicationArguments, printedBanner);
            //刷新应用上下文
            refreshContext(context);
            //刷新后操作
            afterRefresh(context, applicationArguments);
            listeners.finished(context, null);
            stopWatch.stop();
            if (this.logStartupInfo) {
                new StartupInfoLogger(this.mainApplicationClass)
                    .logStarted(getApplicationLog(), stopWatch);
            }
            return context;
        }
        catch (Throwable ex) {
            handleRunFailure(context, listeners, analyzers, ex);
            throw new IllegalStateException(ex);
        }
```

```
        }
    }
```

从源代码可以看出，Spring Boot 首先开启了一个 SpringApplicationRunListeners 监听器，然后通过 createApplicationContext、prepareContext 和 refreshContext 方法创建、准备、刷新应用上下文 ConfigurableApplicationContext，通过上下文加载应用所需的类和各种环境配置等，最后启动一个应用实例。

1.6.4　SpringApplicationRunListener 监听器

SpringApplicationRunListener 接口规定了 Spring Boot 的生命周期，在各个生命周期广播相应的事件（ApplicationEvent），实际调用的是 ApplicationListener 类。SpringApplicationRunListener 源代码如下：

```
public interface SpringApplicationRunListener {
    //刚执行 run 方法时触发
    void starting();
    //环境建立好时触发
    void environmentPrepared(ConfigurableEnvironment environment);
    //上下文建立好时触发
    void contextPrepared(ConfigurableApplicationContext context);
    //上下文载入配置时触发
    void contextLoaded(ConfigurableApplicationContext context);
    //上下文刷新完成后，run 方法执行完之前触发
    void finished(ConfigurableApplicationContext context, Throwable exception);
}
```

ApplicationListener 是 Spring 框架对 Java 中实现的监听器模式的一种框架实现，具体源代码如下：

```
public interface ApplicationListener<E extends ApplicationEvent>
        extends EventListener {
    void onApplicationEvent(E var1);
}
```

ApplicationListener 接口只有一个方法 onApplicationEvent，所以自己的类在实现该接口的时候要实现该方法。如果在上下文 ApplicationContext 中部署一个实现了 ApplicationListener 接口的监听器，那么每当 ApplicationEvent 事件发布到 ApplicationContext 时该监听器将会得到通知。如果要为 Spring Boot 应用添加自定义的 ApplicationListener，那么可以通过 SpringApplication.add Listeners() 或者 SpringApplication.setListeners() 方法添加一个或者多个。

1.6.5　ApplicationContextInitializer 接口

Spring Boot 准备上下文 prepareContext 的时候会对 ConfigurableApplicationContext 实例做进一步的设置或者处理。prepareContext 的源代码如下：

```java
    private void prepareContext(ConfigurableApplicationContext context,
            ConfigurableEnvironment environment, SpringApplicationRunListeners listeners,
        ApplicationArguments applicationArguments, Banner printedBanner) {
        context.setEnvironment(environment);
        postProcessApplicationContext(context);
        //对上下文进行设置和处理
        applyInitializers(context);
        listeners.contextPrepared(context);
        if (this.logStartupInfo) {
            logStartupInfo(context.getParent() == null);
            logStartupProfileInfo(context);
        }

        // Add boot specific singleton beans
        context.getBeanFactory().registerSingleton("springApplicationArguments",
            applicationArguments);
        if (printedBanner != null) {
            context.getBeanFactory().
    registerSingleton("springBootBanner", printedBanner);
        }

        // Load the sources
        Set<Object> sources = getSources();
        Assert.notEmpty(sources, "Sources must not be empty");
        load(context, sources.toArray(new Object[sources.size()]));
        listeners.contextLoaded(context);
    }
```

在准备上下文的 prepareContext 方法中，通过 applyInitializers 方法对 context 上下文进行设置和处理。applyInitializers 的源代码如下：

```java
    protected void applyInitializers(ConfigurableApplicationContext context) {
        for (ApplicationContextInitializer initializer : getInitializers()) {
            Class<?> requiredType = GenericTypeResolver.resolveTypeArgument(
                    initializer.getClass(),
ApplicationContextInitializer.class);
            Assert.isInstanceOf(requiredType, context, "Unable to call initializer.");
            initializer.initialize(context);
        }
    }
```

在 applyInitializers 方法中，主要是调用 ApplicationContextInitializer 类的 initialize 方法对应用上下文进行设置和处理。ApplicationContextInitializer 本质上是一个回调接口，用于在 ConfigurableApplicationContext 执行 refresh 操作之前对它进行一些初始化操作。一般情况下，我们不需要自定义 ApplicationContextInitializer，如果需要，可以通过 SpringApplication.addInitializers() 进行设置。

1.6.6 ApplicationRunner 与 CommandLineRunner

ApplicationRunner 与 CommandLineRunner 接口执行点是在容器启动成功后的最后一步回调，我们可以在回调方法 run 中执行相关逻辑。ApplicationRunner 的源代码如下：

```
public interface ApplicationRunner {
    void run(ApplicationArguments args) throws Exception;
}
```

CommandLineRunner 的源代码如下：

```
public interface CommandLineRunner {
    void run(String... args) throws Exception;
}
```

在 ApplicationRunner 或 CommandLineRunner 类中只有一个 run 方法，但是它们的入参不一样，分别是 ApplicationArguments 和可变 String 数组。

如果有多个 ApplicationRunner 或 CommandLineRunner 实现类，并且需要按一定顺序执行，那么可以在实现类上加上 @Order(value=整数值) 注解，Spring Boot 会按照 @Order 中的 value 值从小到大依次执行。

如果想在 Spring Boot 启动的时候运行一些特定的代码，那么可以实现接口 ApplicationRunner 或者 CommandLineRunner。这两个接口的实现方式一样，都只提供一个 run 方法。

例如：

```
/**
 * @author Ay
 * @create 2020/04/08
 **/
public class MyCommandRunner implements CommandLineRunner {

    @Override
    public void run(String... args) throws Exception {
        //do something
    }
}
```

或者：

```
@Bean
public CommandLineRunner init(){
    return (String ... strings)->{

    };
}
```

1.7　SpringApplication 的执行流程

上一节简单学习了 SpringApplication 的 run 方法，本节总结一下 Spring Boot 启动的完整流程（见图 1-26）。

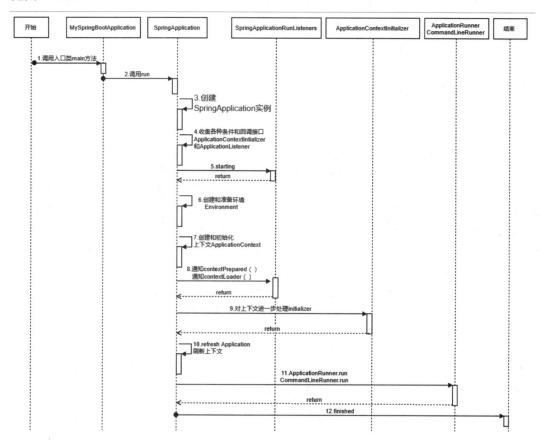

图 1-26　Spring Boot 的启动流程

（1）项目启动时，调用入口类的 main 方法。

（2）入口类的 main 方法会调用 SpringApplication 的静态方法 run。

（3）在 run 方法中首先创建一个 SpringApplication 对象实例，然后调用 SpringApplication 对象实例的 run 方法。

（4）查询和加载所有的 SpringApplicationListener 监听器（通过 SpringFactoriesLoader 加载 META-INF/spring.factories 文件）。

（5）SpringApplicationListener 监听器调用其 starting 方法，Spring Boot 通知这些 SpringApplicationListener 监听器"我马上要开始执行了"。

（6）创建和准备 Spring Boot 应用将要使用的 Environment，包括配置要使用的 PropertySource

以及 Profile。

（7）创建和初始化应用上下文 ApplicationContext。这一步只是准备工作，并未开始正式创建。

（8）这一步是最重要的，Spring Boot 会通过 @EnableAutoConfiguration 获取所有配置以及其他形式的 IoC 容器配置加载到已经准备完毕的 ApplicationContext。

（9）主要是调用 ApplicationContextInitializer 类的 initialize 方法对应用上下文进行设置和处理。

（10）调用 ApplicationContext 上下文的 refresh 方法，使 IoC 容器达到可用状态。

（11）查找当前 ApplicationContext 上下文是否注册有 ApplicationRunner 与 CommandLineRunner，如果有，就循环遍历执行 ApplicationRunner 和 CommandLineRunner 的 run 方法。

（12）执行 SpringApplicationListener 的 finished 方法，Spring Boot 应用启动完毕。

1.7.1　spring-boot-starter 原理

在之前的章节中，pom 文件中引入了很多 spring-boot-starter 依赖，比如 spring-boot-starter-jdbc、spring-boot-starter-jdbc-logging、spring-boot-starter-web 等。这些带有 spring-boot-starter 前缀的依赖都叫作 Spring Boot 起步依赖，它们有助于 Spring Boot 应用程序的构建。

假设要使用 Spring MVC，如果没有起步依赖，那么我们根本记不住 Spring MVC 到底要引入哪些依赖包、到底要使用哪个版本的 Spring MVC、Spring MVC 的 Group 和 Artifact ID 是多少。

Spring Boot 通过提供众多起步依赖降低项目依赖的复杂度。起步依赖本质上是一个 Maven 项目对象模型，定义了对其他库的传递依赖，这些依赖的合集可以对外提供某项功能。起步依赖的命名表明它们提供某种或某类功能。例如，spring-boot-starter-jdbc 表示提供 JDBC 相关的功能，spring-boot-starter-jpa 表示提供 JPA 相关的功能，等等。表 1-5 简单地列举了工作中经常使用的起步依赖。

表 1-5　常用的 spring-boot-starter 起步依赖

名称	描述
spring-boot-starter-logging	提供 logging 相关的日志功能
spring-boot-starter-thymeleaf	使用 Thymeleaf 视图构建 MVC Web 应用程序的启动器
spring-boot-starter-parent	常被作为父依赖，提供智能资源过滤、智能的插件设置、编译级别和通用的测试框架等
spring-boot-starter-web	使用 Spring MVC 构建 Web，包括 RESTful 应用程序。使用 Tomcat 作为默认的嵌入式容器的启动器
spring-boot-starter-test	支持常规的测试依赖，包括 JUnit、Hamcrest、Mockito 以及 spring-test 模块
spring-boot-starter-jdbc	使用 JDBC 与 Tomcat JDBC 连接池的启动器
spring-boot-starter-data-jpa	使用 Spring 数据 JPA 与 Hibernate 的启动器
spring-boot-starter-data-redis	Redis key-value 数据存储与 Spring Data Redis 和 Jedis 客户端启动器
spring-boot-starter-log4j2	提供 log4j2 相关的日志功能
spring-boot-starter-mail	提供邮件相关的功能
spring-boot-starter-activemq	使用 Apache ActiveMQ 的 JMS 启动器
spring-boot-starter-data-mongodb	使用 MongoDB 面向文档的数据库和 Spring Data MongoDB 的启动器

（续表）

名称	描述
spring-boot-starter-actuator	提供应用监控与健康相关的功能
spring-boot-starter-security	使用 Spring security 的启动器
spring-boot-starter-dubbo	提供 dubbo 框架相关的功能

事实上，起步依赖和项目里的其他依赖没有什么区别。引入起步依赖的同时会引入相关的传递依赖，比如 spring-boot-starter-web 起步依赖会引入 spring-webmvc、jackson-databind、spring-boot-starter-tomcat 等传递依赖。如果不想用 spring-boot-starter-web 引入的 spring-webmvc 传递依赖，那么可以使用<exclusions>标签来排除传递依赖，具体代码如下：

```xml
<dependency>
    <groupId>org.springframework.boot</groupId>
    <artifactId>spring-boot-starter-web</artifactId>
    <exclusions>
        <!-- 排查 spring-webmvc -->
        <exclusion>
            <groupId>org.springframework</groupId>
            <artifactId>spring-webmvc</artifactId>
        </exclusion>
    </exclusions>
</dependency>
```

如果 spring-boot-starter-web 引入的传递依赖版本过于低，那么可以在 pom 文件中直接引入所需的版本，告诉 Maven 现在需要这个版本的依赖。

传统的 Spring 应用需要在 application.xml 中配置很多 bean，比如 dataSource 的配置、transactionManager 的配置等。Spring Boot 是如何帮我们完成这些 bean 配置的呢？下面我们来分析这个过程。我们以 MyBatis-spring-boot-starter 依赖为例：

```xml
<dependency>
    <groupId>org.mybatis.spring.boot</groupId>
    <artifactId>mybatis-spring-boot-starter</artifactId>
    <version>2.0.1</version>
</dependency>
```

首先，查看 MyBatis-spring-boot-starter 包下的内容，具体如图 1-27 所示。

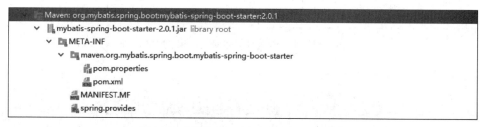

图 1-27　MyBatis-spring-boot-starter 包内容

可以看出在 MyBatis-spring-boot-starter 包中并没有任何源代码，只有一些配置文件，例如 pom.xml 文件，它的作用就是帮我们引入相关的 jar 包。在 pom.xml 中，MyBatis-spring-boot-starter

包引入了 MyBatis-spring-boot-autoconfigure 包，具体代码如下：

```xml
<?xml version="1.0" encoding="UTF-8"?>
<project xmlns="http://maven.apache.org/POM/4.0.0" xmlns:xsi=
"http://www.w3.org/2001/XMLSchema-instance" xsi:schemaLocation=
"http://maven.apache.org/POM/4.0.0 http://maven.apache.org/xsd/maven-4.0.0.xsd">
    <modelVersion>4.0.0</modelVersion>
    <parent>
      <groupId>org.mybatis.spring.boot</groupId>
      <artifactId>mybatis-spring-boot</artifactId>
      <version>2.0.1</version>
    </parent>
    <artifactId>mybatis-spring-boot-starter</artifactId>
    <name>mybatis-spring-boot-starter</name>
    <properties>
      <module.name>org.mybatis.spring.boot.starter</module.name>
    </properties>
    <dependencies>

      <groupId>org.mybatis.spring.boot</groupId>
      <artifactId>mybatis-spring-boot-autoconfigure</artifactId>
    </dependency>

    //省略代码
    </dependencies>
</project>
```

查看 MyBatis-spring-boot-autoconfigure 包下的内容，如图 1-28 所示。

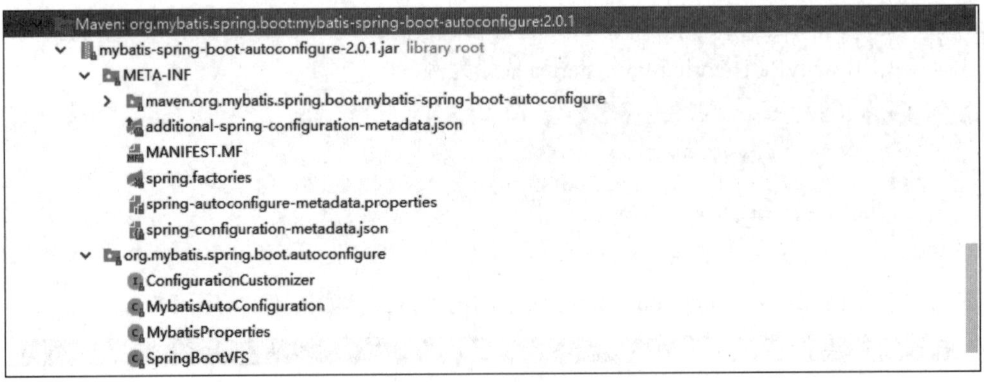

图 1-28 MyBatis-spring-boot-autoconfigure 包内容

再查看 MyBatisAutoConfiguration 源代码：

```java
@Configuration
@ConditionalOnClass({SqlSessionFactory.class, SqlSessionFactoryBean.class})
@ConditionalOnSingleCandidate(DataSource.class)
@EnableConfigurationProperties({MybatisProperties.class})
@AutoConfigureAfter({DataSourceAutoConfiguration.class})
public class MybatisAutoConfiguration implements InitializingBean {
    private static final Logger logger = LoggerFactory.getLogger
```

```
(MybatisAutoConfiguration.class);
    private final MybatisProperties properties;
    private final Interceptor[] interceptors;
    private final ResourceLoader resourceLoader;
    private final DatabaseIdProvider databaseIdProvider;
    private final List<ConfigurationCustomizer> configurationCustomizers;

    @Bean
    @ConditionalOnMissingBean
    public SqlSessionFactory sqlSessionFactory(DataSource dataSource) throws Exception

        //省略代码
    }
```

- @Configuration、@Bean：这两个注解一起使用就可以创建一个基于 Java 代码的配置类。可以把 MyBatisAutoConfiguration 类想象成一份 XML 配置文件。@Configuration 注解的类可以看作 Bean 实例的工厂，能生产让 Spring IoC 容器管理的 Bean。@Bean 注解告诉 Spring，一个带有@Bean 的注解方法将返回一个对象，该对象应该被注册到 Spring 容器中。MyBatisAutoConfiguration 类能自动生成 SqlSessionFactory、SqlSessionTemplate 等 MyBatis 的重要实例并交给 Spring 容器管理，从而完成 Bean 的自动注册。
- @ConditionalOnClass：某个 class 位于类路径上，才会实例化这个 Bean。
- @ConditionalOnBean：仅在当前上下文中存在某个 bean 时才会实例化这个 Bean。
- @ConditionalOnSingleCandidate：类似于@ConditionalOnBean。
- @ConditionalOnExpression：当表达式为 true 时才会实例化这个 Bean。
- @ConditionalOnMissingBean：仅在当前上下文中不存在某个 Bean 时才会实例化这个 Bean。
- @ConditionalOnMissingClass：某个 class 在类路径上不存在时才会实例化这个 Bean。
- @ConditionalOnNotWebApplication：不是 Web 应用时才会实例化这个 Bean。
- @AutoConfigureAfter：在某个 Bean 完成自动配置后实例化这个 Bean。
- @AutoConfigureBefore：在某个 Bean 完成自动配置前实例化这个 Bean。

要完成 MyBatis 的自动配置，需要在类路径中存在 SqlSessionFactory.class、SqlSessionFactoryBean.class 这两个类，需要存在 DataSourceBean 且已完成自动注册。

进入 DataSourceAutoConfiguration 类,可以看到该类属于 spring-boot-autoconfigure 自动配置包，自动配置包帮我们引入了 jdbc、kafka、logging、mail、mongo 等包。很多包需要在引入相应的 jar 后自动配置才生效。

1.7.2 Bean 参数获取

虽然了解 Bean 的配置过程，但是还没有看到 Spring Boot 是如何读取 yml 或者 properties 配置文件的属性来创建数据源的。在 DataSourceAutoConfiguration 类里面使用了 EnableConfigurationProperties 注解，参见如下代码：

```
@Configuration
@ConditionalOnClass({DataSource.class, EmbeddedDatabaseType.class})
@EnableConfigurationProperties({DataSourceProperties.class})
@Import({DataSourcePoolMetadataProvidersConfiguration.class,
DataSourceInitializationConfiguration.class})
public class DataSourceAutoConfiguration {
    public DataSourceAutoConfiguration() {
    }

    //省略代码
}
```

@EnableConfigurationProperties 注解的作用是使@ConfigurationProperties 注解生效。

DataSourceProperties 中封装了数据源的各个属性，且使用注解@ConfigurationProperties 指定了配置文件的前缀，参见如下代码：

```
@ConfigurationProperties(
    prefix = "spring.datasource"
)
public class DataSourceProperties implements BeanClassLoaderAware,
InitializingBean {
    private ClassLoader classLoader;
    private String name;
    private boolean generateUniqueName;
    private Class<? extends DataSource> type;
    private String driverClassName;
    private String url;
    private String username;
    private String password;
    private String jndiName;
    private DataSourceInitializationMode initializationMode;
    private String platform;
    private List<String> schema;
    private String schemaUsername;
    private String schemaPassword;
    private List<String> data;
    private String dataUsername;
    private String dataPassword;
    private boolean continueOnError;

//省略代码
}
```

@ConfigurationProperties 注解的作用是把 yml 或者 properties 配置文件转化为 Bean，通过这种方式把 yml 或者 properties 配置参数转化为 Bean。

1.7.3　Bean 的发现与加载

Spring Boot 默认扫描启动类所在的包下的主类与子类的所有组件，但并没有包括依赖包中的类，那么依赖包中的 Bean 是如何被发现和加载的呢？

我们通常在启动类中加@SpringBootApplication 注解，查看如下源代码：

```
@Target({ElementType.TYPE})
@Retention(RetentionPolicy.RUNTIME)
@Documented
@Inherited
@SpringBootConfiguration
@EnableAutoConfiguration
@ComponentScan(
    excludeFilters = {@Filter(
    type = FilterType.CUSTOM,
    classes = {TypeExcludeFilter.class}
), @Filter(
    type = FilterType.CUSTOM,
    classes = {AutoConfigurationExcludeFilter.class}
)}
)
public @interface SpringBootApplication {

    //省略代码
}
```

@SpringBootConfiguration 是进入@ SpringBootConfiguration 注解的源代码，你会发现它其实和@Configuration 注解的功能是一样的，只是换了一个名字而已。@ SpringBootConfiguration 的源代码如下：

```
@Target({ElementType.TYPE})
@Retention(RetentionPolicy.RUNTIME)
@Documented
@Configuration
public @interface SpringBootConfiguration {

}
```

- @EnableAutoConfiguration：这个注解的功能非常重要，它用于借助@Import 的支持，收集和注册依赖包中相关的 Bean 定义。
- @ComponentScan：该注解的作用是自动扫描并加载符合条件的组件，比如@Component 和 @Repository 等，最终将这些 Bean 定义加载到 Spring 容器中。

@EnableAutoConfiguration 的源代码如下：

```
@Target({ElementType.TYPE})
@Retention(RetentionPolicy.RUNTIME)
@Documented
@Inherited
@AutoConfigurationPackage
//重要
@Import({AutoConfigurationImportSelector.class})
```

```
public @interface EnableAutoConfiguration {

}
```

@EnableAutoConfiguration 注解引入了@AutoConfigurationPackage 和@Import 这两个注解。@AutoConfigurationPackage 的作用是自动配置包，@Import 则是导入需要自动配置的组件。

进入@AutoConfigurationPackage，发现其也引入了@Import 注解，见下述代码：

```
@Target({ElementType.TYPE})
@Retention(RetentionPolicy.RUNTIME)
@Documented
@Inherited
//重要
@Import({Registrar.class})
public @interface AutoConfigurationPackage {

}
```

查看@Import 注解中的 Registrar 类源代码：

```
static class Registrar implements ImportBeanDefinitionRegistrar, DeterminableImports {
        Registrar() {
        }

        public void registerBeanDefinitions(AnnotationMetadata metadata, BeanDefinitionRegistry registry) {
            AutoConfigurationPackages.register(registry, (new AutoConfigurationPackages.PackageImport(metadata)).getPackageName());
        }

        public Set<Object> determineImports(AnnotationMetadata metadata) {
            return Collections.singleton(new AutoConfigurationPackages.PackageImport(metadata));
        }
    }
```

new AutoConfigurationPackages.PackageImport(metadata)).getPackageName()和new AutoConfigurationPackages.PackageImport(metadata)的作用就是加载启动类所在的包下的主类与子类的所有组件注册到 Spring 容器。Spring Boot 默认扫描启动类所在包下的主类与子类的所有组件。

继续查看 AutoConfigurationImportSelector 类的源代码：

```
public class AutoConfigurationImportSelector implements DeferredImportSelector, BeanClassLoaderAware, ResourceLoaderAware, BeanFactoryAware, EnvironmentAware, Ordered {
    private static final AutoConfigurationImportSelector.AutoConfigurationEntry EMPTY_ENTRY = new AutoConfigurationImportSelector.AutoConfigurationEntry();
    private static final String[] NO_IMPORTS = new String[0];
    private static final Log logger = LogFactory.getLog(AutoConfigurationImportSelector.class);
    private static final String PROPERTY_NAME_AUTOCONFIGURE_EXCLUDE =
```

```java
"spring.autoconfigure.exclude";
    private ConfigurableListableBeanFactory beanFactory;
    private Environment environment;
    private ClassLoader beanClassLoader;
    private ResourceLoader resourceLoader;

    public AutoConfigurationImportSelector() {
    }

    protected List<String> getCandidateConfigurations(AnnotationMetadata metadata, AnnotationAttributes attributes) {
        //重要
        List<String> configurations = SpringFactoriesLoader.loadFactoryNames(this.getSpringFactoriesLoaderFactoryClass(), this.getBeanClassLoader());
        Assert.notEmpty(configurations, "No auto configuration classes found in META-INF/spring.factories. If you are using a custom packaging, make sure that file is correct.");
        return configurations;
    }

    //省略代码
}
```

SpringFactoriesLoader.loadFactoryNames 方法调用 loadSpringFactories 方法从所有的 jar 包中读取 META-INF/spring.factories 文件信息。loadSpringFactories 的源代码如下:

```java
    public static List<String> loadFactoryNames(Class<?> factoryClass, @Nullable ClassLoader classLoader) {
        String factoryClassName = factoryClass.getName();
        return (List)loadSpringFactories(classLoader).getOrDefault(factoryClassName, Collections.emptyList());
    }

    private static Map<String, List<String>> loadSpringFactories(@Nullable ClassLoader classLoader) {
        MultiValueMap<String, String> result = (MultiValueMap)cache.get(classLoader);
        if (result != null) {
            return result;
        } else {
            try {
                Enumeration<URL> urls = classLoader != null ? classLoader.getResources("META-INF/spring.factories")
          //加载"META-INF/spring.factories"配置文件中的内容
          : ClassLoader.getSystemResources("META-INF/spring.factories");
                LinkedMultiValueMap result = new LinkedMultiValueMap();

                while(urls.hasMoreElements()) {
                    URL url = (URL)urls.nextElement();
                    UrlResource resource = new UrlResource(url);
```

```
                Properties properties = PropertiesLoaderUtils.
loadProperties(resource);
                Iterator var6 = properties.entrySet().iterator();

                while(var6.hasNext()) {
                    Entry<?, ?> entry = (Entry)var6.next();
                    String factoryClassName = ((String)entry.getKey()).trim();
                    String[] var9 =
StringUtils.commaDelimitedListToStringArray((String)entry.getValue());
                    int var10 = var9.length;

                    for(int var11 = 0; var11 < var10; ++var11) {
                        String factoryName = var9[var11];
                        result.add(factoryClassName, factoryName.trim());
                    }
                }
            }

            cache.put(classLoader, result);
            return result;
        } catch (IOException var13) {
            throw new IllegalArgumentException("Unable to load factories from
location [META-INF/spring.factories]", var13);
        }
    }
}
```

下面是 spring-boot-autoconfigure 的 jar 中 spring.factories 文件的部分内容，其中有一个 key 为 org.springframework.boot.autoconfigure.EnableAutoConfiguration 的值定义了需要自动配置的 Bean，通过读取这个配置获取一组@Configuration 类。

```
# Auto Configure
org.springframework.boot.autoconfigure.EnableAutoConfiguration=\
org.springframework.boot.autoconfigure.admin.SpringApplicationAdminJmxAutoCo
nfiguration,\
org.springframework.boot.autoconfigure.aop.AopAutoConfiguration,\
org.springframework.boot.autoconfigure.amqp.RabbitAutoConfiguration,\
org.springframework.boot.autoconfigure.batch.BatchAutoConfiguration,\
org.springframework.boot.autoconfigure.cache.CacheAutoConfiguration,\
org.springframework.boot.autoconfigure.cassandra.CassandraAutoConfiguration,\
org.springframework.boot.autoconfigure.cloud.CloudServiceConnectorsAutoConfi
guration,\
org.springframework.boot.autoconfigure.context.ConfigurationPropertiesAutoCo
nfiguration,\
org.springframework.boot.autoconfigure.context.MessageSourceAutoConfiguration,\
org.springframework.boot.autoconfigure.context.PropertyPlaceholderAutoConfig
uration,\
org.springframework.boot.autoconfigure.couchbase.CouchbaseAutoConfiguration,\
org.springframework.boot.autoconfigure.dao.PersistenceExceptionTranslationAu
toConfiguration,\
org.springframework.boot.autoconfigure.data.cassandra.CassandraDataAutoConfi
```

```
guration,\
    org.springframework.boot.autoconfigure.data.cassandra.CassandraReactiveDataA
utoConfiguration,\
```

每个 xxxAutoConfiguration 都是一个基于 Java 的 Bean 配置类。实际上，不是所有的 xxxAutoConfiguration 都会被加载，而是要根据 xxxAutoConfiguration 上的@ConditionalOnClass 等条件判断是否加载。通过反射机制将 spring.factories 中的@Configuration 类实例化为对应的 Java 实例。

至此，我们已经知道了怎么发现自动配置的 Bean。最后一步就是怎样将这些 Bean 加载到 Spring 容器中。

将普通类交给 Spring 容器管理通常有以下方法：

（1）使用 @Configuration 与@Bean 注解。

（2）使用@Controller、@Service、@Repository、@Component 注解标注该类，然后启用@ComponentScan 自动扫描。

（3）使用@Import 方法。

Spring Boot 中采取第 3 种方法，@EnableAutoConfiguration 注解中使用@Import({AutoConfigurationImportSelector.class})注解，AutoConfigurationImportSelector 实现了 DeferredImportSelector 接口，DeferredImportSelector 接口继承了 ImportSelector 接口，ImportSelector 接口只有一个 selectImports 方法。

AutoConfigurationImportSelector 的源代码如下：

```java
public class AutoConfigurationImportSelector implements DeferredImportSelector,
BeanClassLoaderAware, ResourceLoaderAware, BeanFactoryAware, EnvironmentAware,
Ordered {
    private static final AutoConfigurationImportSelector.AutoConfigurationEntry
EMPTY_ENTRY = new AutoConfigurationImportSelector.AutoConfigurationEntry();
    private static final String[] NO_IMPORTS = new String[0];
    private static final Log logger =
LogFactory.getLog(AutoConfigurationImportSelector.class);
    private static final String PROPERTY_NAME_AUTOCONFIGURE_EXCLUDE =
"spring.autoconfigure.exclude";
    private ConfigurableListableBeanFactory beanFactory;
    private Environment environment;
    private ClassLoader beanClassLoader;
    private ResourceLoader resourceLoader;

    public String[] selectImports(AnnotationMetadata annotationMetadata) {
        if (!this.isEnabled(annotationMetadata)) {
            return NO_IMPORTS;
        } else {
            AutoConfigurationMetadata autoConfigurationMetadata =
AutoConfigurationMetadataLoader.loadMetadata(this.beanClassLoader);
            AutoConfigurationImportSelector.AutoConfigurationEntry
autoConfigurationEntry = this.getAutoConfigurationEntry(autoConfigurationMetadata,
annotationMetadata);
```

```
                return StringUtils.toStringArray(autoConfigurationEntry.
getConfigurations());
        }
    }
}
```

DeferredImportSelector 与 ImportSelector 的源代码如下：

```
//DeferredImportSelector 源代码
public interface DeferredImportSelector extends ImportSelector {
    @Nullable
    default Class<? extends DeferredImportSelector.Group> getImportGroup() {
        return null;
    }
    //省略代码
}

//ImportSelector 源代码
public interface ImportSelector {
    String[] selectImports(AnnotationMetadata var1);
```

1.7.4 自定义 starter

Spring Boot 提供的 starter 都是以 spring-boot-starter-xxx 的方式命名的，针对自定义的 starter，官方建议以 xxx-spring-boot-starter 命名。

首先，创建一个 Maven 项目，具体步骤如下：

步骤 01 单击菜单栏上的 File→New→Project 命令，如图 1-29 所示。

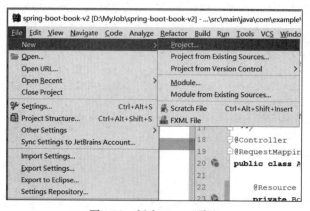

图 1-29 创建 Maven 项目

步骤 02 在打开的 New Project 对话框中选择 Maven，勾选 Create from archetype 复选框，选择 maven-archetype-quickstart 选项，如图 1-30 所示。

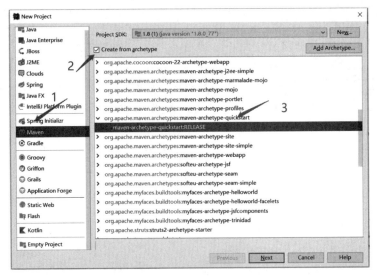

图 1-30　New Project 对话框

步骤 03　填写 GroupId 和 ArtifactId，例如在 GroupId 输入框中填写"org.springframework.boot"、在 ArtifactId 输入框中填写"ay-spring-boot-starter"，如图 1-31 所示。

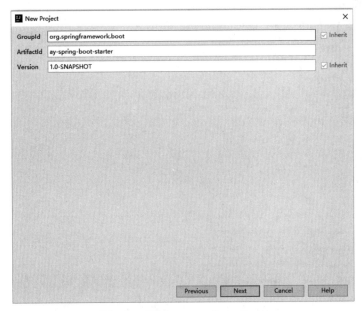

图 1-31　填写 GroupId 和 ArtifactId

步骤 04　选择 Maven 的配置，这里笔者选择自己下载的 Maven，也可以使用默认选项，如图 1-32 所示。

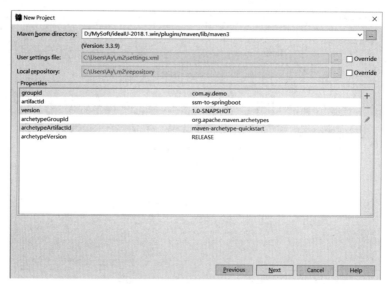

图 1-32　选择 Maven home 地址

步骤 05 填写项目名称和项目存放目录，最后单击 Finish 按钮，如图 1-33 所示。

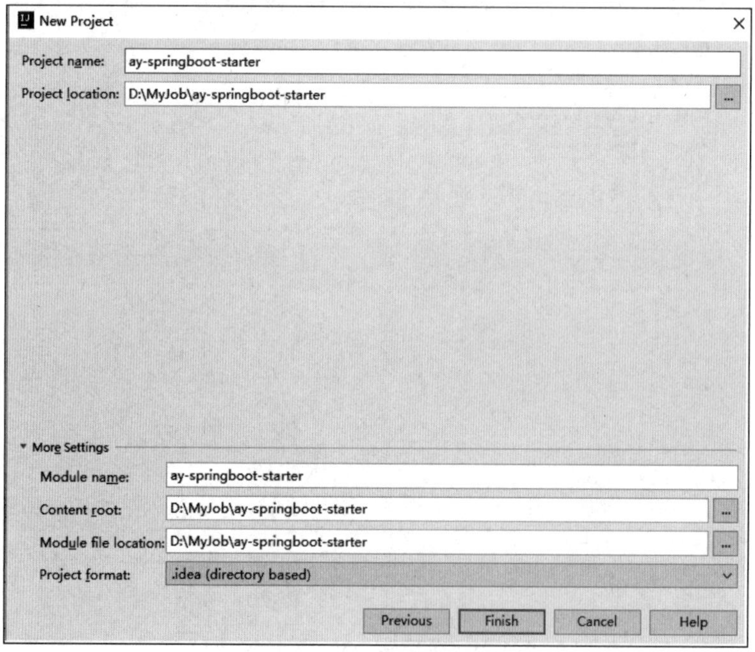

图 1-33　填写项目名称和地址

Maven 项目创建完成后，在 pom.xml 文件中添加依赖，具体代码如下：

```
<?xml version="1.0" encoding="UTF-8"?>
<project xmlns="http://maven.apache.org/POM/4.0.0"
     xmlns:xsi="http://www.w3.org/2001/XMLSchema-instance"
     xsi:schemaLocation="http://maven.apache.org/POM/4.0.0
http://maven.apache.org/xsd/maven-4.0.0.xsd">
   <modelVersion>4.0.0</modelVersion>
```

```xml
    <groupId>org.springframework.boot</groupId>
    <artifactId>ay-spring-boot-starter</artifactId>
    <version>1.0-SNAPSHOT</version>

    <dependencies>
        <dependency>
            <groupId>org.springframework.boot</groupId>
            <artifactId>spring-boot-autoconfigure</artifactId>
        </dependency>
    </dependencies>

    <dependencyManagement>
        <dependencies>
            <dependency>
                <!-- Import dependency management from Spring Boot -->
                <groupId>org.springframework.boot</groupId>
                <artifactId>spring-boot-dependencies</artifactId>
                <version>2.1.6.RELEASE</version>
                <type>pom</type>
                <scope>import</scope>
            </dependency>
        </dependencies>
    </dependencyManagement>
</project>
```

在 resources 下新建包 META-INF，并新增文件 spring.factories，内容如下：

```
org.springframework.boot.autoconfigure.EnableAutoConfiguration=\
  com.ay.config.AyStarterEnableAutoConfiguration
```

其中，AyStarterEnableAutoConfiguration 为自动配置的核心类，Spring Boot 会扫描该文件作为配置类。

在 src\main\java 目录下创建 com.ay.config 包，并在包下创建 AyStarterEnableAutoConfiguration.java 类，具体代码如下：

```java
/**
 * @author Ay
 * @create 2019/09/08
 **/

@Configuration
@ConditionalOnClass(HelloService.class)
@EnableConfigurationProperties(HelloServiceProperties.class)
public class AyStarterEnableAutoConfiguration {

    private final HelloServiceProperties helloServiceProperties;

    @Autowired
    public AyStarterEnableAutoConfiguration(HelloServiceProperties
```

```java
helloServiceProperties) {
        this.helloServiceProperties = helloServiceProperties;
    }

    @Bean
    @ConditionalOnProperty(prefix = "hello.service", name = "enable", havingValue = "true")
    HelloService helloService() {
        return new HelloService(helloServiceProperties.getPrefix(), helloServiceProperties.getSuffix());
    }
}
```

@Configuration、@ConditionalOnClass(HelloService.class)和@EnableConfigurationProperties 注解的含义在之前都提到过，这里不再赘述。

HelloServiceProperties 的代码如下：

```java
/**
 * @author Ay
 * @create 2019/09/08
 **/
@ConfigurationProperties("hello.service")
public class HelloServiceProperties {

    private String prefix;

    private String suffix;
    //省略 set、get 方法
}
```

HelloService 的代码如下：

```java
/**
 * @author Ay
 * @create 2019/09/08
 **/
public class HelloService {
    private String prefix;
    private String suffix;
    public HelloService(String prefix, String suffix) {
        this.prefix = prefix;
        this.suffix = suffix;
    }

    public String say(String text) {
        return String.format("%s, hi, %s, %s", prefix, text, suffix);
    }
}
```

当项目依赖该 starter、配置文件中包含 hello.service 前缀且 hello.service.enable 为 true 时，就会自动生成 HelloService 的 Bean。

最后，我们可以创建一个新的 Spring Boot 项目，并在项目的 pom.xml 文件中引入自定义 starter 依赖：

```xml
<dependency>
    <groupId>org.springframework.boot</groupId>
    <artifactId>ay-spring-boot-starter</artifactId>
    <version>1.0-SNAPSHOT</version>
</dependency>
```

在 application.properties 配置文件中添加如下配置：

```
hello.service.prefix=pre
hello.service.suffix=suf
hello.service.enable=true
```

这样就可以在新的 Spring Boot 项目中使用 HelloService 类并调用 say 方法了。

第 2 章

Spring Cloud/Spring Cloud Alibaba

本章主要介绍 Spring Cloud 的功能特性、模块以及版本信息，Spring Cloud 和 Spring Boot 之间的关系，Spring Cloud Alibaba 简介、主要功能以及相关组件，Netflix、Spring Cloud 以及 Spring Cloud Alibaba 之间的关系。

2.1 Spring Cloud 介绍

Spring Cloud 为开发人员提供了工具，以快速构建分布式系统中的一些常见模式（配置管理、服务发现、断路器、智能路由、微代理、控制总线、令牌 token、全局锁、领导选择、分布式 Session、群集状态等）。使用 Spring Cloud，开发人员可以快速实现这些模式的服务和应用程序。它们可以在任何分布式环境中正常工作，包括开发人员的笔记本电脑、裸机数据中心以及 Cloud Foundry 等托管平台。

2.1.1 Spring Cloud 的特性

Spring Cloud 专注于为典型的用例和可扩展性机制提供良好的开箱即用体验，其主要特性有以下几点：

- 分布式/版本化配置（distributed/versioned configuration）。
- 服务注册和发现（service registration and discovery）。
- 动态路由（routing）。
- 微服务调用（service-to-service calls）。
- 负载均衡（load balancing）。
- 断路器（circuit breakers）。
- 全局锁（global locks）。
- 领导选择和集群状态（leadership selection and cluster state）。

- 分布式消息（distributed messaging）。

Spring Cloud 采用了声明性的方法，通常只需更改类路径或添加一个注释即可获得许多功能，例如（发现客户端的示例应用程序）：

```
@SpringBootApplication
@EnableDiscoveryClient
public class Application {
    public static void main(String[] args) {
        SpringApplication.run(Application.class, args);
    }
}
```

2.1.2　Spring Cloud 的模块

Spring Cloud 包含了很多子项目，具体如下：

- Spring Cloud Config：由 git 存储库支持的集中式外部配置管理。配置资源直接映射到 Spring Environment，如果需要，可以由非 Spring 应用程序使用。
- Spring Cloud Netflix：与各种 Netflix OSS 组件集成（Eureka、Hystrix、Zuul、Archaius 等）。
- Spring Cloud Bus：事件总线，用分布式消息将服务和服务实例链接在一起。对于在整个集群中传播状态更改（例如配置更改事件）很有用。
- Spring Cloud Cloudfoundry：将应用程序与 Pivotal Cloud Foundry 集成，提供服务发现实现，还可以轻松实现 SSO 和 OAuth 2 保护的资源。
- Spring Cloud Open Service Broker：为构建实现 Open Service Broker API 的服务代理提供起点。
- Spring Cloud Cluster：Zookeeper、Redis、Hazelcast 和 Consul 的领导层选举和通用状态模式抽象和实现。
- Spring Cloud Consul：使用 Hashicorp Consul 进行服务发现和配置管理。
- Spring Cloud Security：提供对负载平衡的 OAuth 2 客户端和身份验证标头中继支持。
- Spring Cloud Sleuth：用于 Spring Cloud 应用程序的分布式跟踪，与 Zipkin、HTrace 和基于日志跟踪兼容（例如 ELK）。
- Spring Cloud Data Flow：用于构建数据集成和实时数据处理管道（Pipelines）的工具包。
- Spring Cloud Stream：轻量级的事件驱动型微服务框架，用于快速构建可以连接到外部系统的应用程序。在 Spring Boot 应用程序之间使用 Apache Kafka 或 RabbitMQ 发送和接收消息的简单声明性模型。
- Spring Cloud Stream App Starters：基于 Spring Boot 的 Spring Integration 应用程序，可与外部系统集成。
- Spring Cloud Task：目标是为 Spring Boot 应用程序提供创建短运行期微服务的功能。
- Spring Cloud Task App Starters：是 Spring Boot 应用程序，可以是任何进程，包括不会永远运行的 Spring Batch 作业，它们在有限的数据处理周期后结束/停止。
- Spring Cloud Zookeeper：使用 Apache Zookeeper 进行服务发现和配置管理。
- Spring Cloud CLI：Spring Boot CLI 插件，用于在 Groovy 中快速创建 Spring Cloud 组件应用

程序。
- Spring Cloud Contract：为通过 CDC（Customer Driven Contracts）开发基于 JVM 的应用提供支持，为 TDD（测试驱动开发）提供一种新的测试方式，即基于接口。
- Spring Cloud Gateway：Spring Cloud Gateway 是基于 Project Reactor 的智能可编程路由器。
- Spring Cloud OpenFeign：将 OpenFeign 集成到 Spring Boot 应用中的方式，为微服务架构下服务之间的调用提供了解决方案。
- Spring Cloud Pipelines：提供一个可靠的部署管道，其中包含一些步骤，以确保应用程序可以零停机时间进行部署，并且可以轻松回滚某些错误。
- Spring Cloud Function：通过功能促进业务逻辑的实现，支持无服务器提供商之间的统一编程模型以及独立运行（本地或 PaaS）的能力。

Netflix 是一家做视频的网站，访问量非常大。也正是如此，Netflix 开始把整体的系统往微服务上迁移。Netflix 微服务大规模的应用，在技术上毫无保留地把一整套微服务架构核心技术栈开源出来，叫作 Netflix OSS，也正是如此，在技术上依靠开源社区的力量不断壮大。在 Netflix 开源的一整套核心技术产品线的基础上，做了一系列的封装，就变成了 Spring Cloud。

2.1.3 Spring Cloud 版本介绍

Spring Cloud 是一个由众多独立子项目组成的大型综合项目，每个子项目都有不同的发行节奏，并维护着自己的发布版本号。Spring Cloud 通过一个资源清单 BOM（Bill of Materials）来管理每个版本的子项目清单。为避免与子项目的发布号混淆，没有采用版本号的方式，而是采用命名的方式。

这些版本名称的命名方式采用了伦敦地铁站的名称，同时根据字母表的顺序来对应版本时间顺序，比如最早的 Release 版本 Angel，第二个 Release 版本 Brixton，然后是 Camden、Dalston、Edgware、Finchley、Greenwich，以及 Hoxton。

当一个版本的 Spring Cloud 项目的发布内容积累到临界点或者解决了一个严重 bug 后，就会发布一个 service releases 版本，简称 SRX 版本。其中，X 是一个递增数字。写本书时官方网站上最新的稳定版本是 Hoxton SR3。

Greenwich 可以构建并与 Spring Boot 2.1.x 一起使用，不能与 Spring Boot 1.5.x 一起使用。Dalston 于 2018 年 12 月到期，Edgware 将遵循 Spring Boot 1.5.x 的生命周期。Dalston 和 Edgware 建立在 Spring Boot 1.5.x 上，并且不能与 Spring Boot 2.0.x 一起使用。Spring Cloud 兼容的 Spring Boot 版本如表 2-1 所示。

表 2-1 Spring Cloud 兼容的 Spring Boot 版本

Release Train	Spring Boot Version
Hoxton	2.2.x
Greenwich	2.1.x
Finchley	2.0.x
Edgware	1.5.x
Dalston	1.5.x

Camden 版本的寿命已结束，Camden 发布系列基于 Spring Boot 1.4.x 构建，但也已在 1.5.x 上

进行了测试。

2017 年 7 月，Brixton and Angel 被标记为废弃（EOL）。Brixton 发布系列基于 Spring Boot 1.3.x 构建，已在 1.4.x 上进行了测试。

Angel 系列基于 Spring Boot 1.2.x 构建，并且在某些方面与 Spring Boot 1.3.x 不兼容。Brixton 建立在 Spring Boot 1.3.x 之上，并且与 1.2.x 不兼容。一些库和大多数基于 Angel 构建的应用程序都可以在 Brixton 上正常运行，但是在使用 spring-cloud-security 1.0.x 的 OAuth 2 功能的任何地方都需要进行更改（它们大多在 1.3.0 中移至 Spring Boot）。

2.1.4　Spring Cloud 与 Spring Boot 的关系

Spring Cloud 与 Spring Boot 的关系是：Spring Boot 是 Spring 的一套快速配置脚手架，可以基于 Spring Boot 快速开发单个微服务，Spring Cloud 是一个基于 Spring Boot 实现的云应用开发工具。Spring Boot 专注于快速、方便集成的单个微服务个体，Spring Cloud 关注全局的服务治理框架。Spring Boot 使用了默认大于配置的理念，很多集成方案已经选好，能不配置就不配置，Spring Cloud 很大的一部分是基于 Spring Boot 来实现的。

Spring Boot 可以离开 Spring Cloud 独立使用开发项目；但是 Spring Cloud 离不开 Spring Boot，属于依赖的关系，具体如图 2-1 所示。

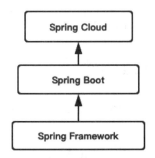

图 2-1　Spring Boot 与 Spring Cloud 的依赖关系

2.2　Spring Cloud Alibaba 简介

Spring Cloud Alibaba 致力于提供微服务开发的一站式解决方案，包含开发分布式应用微服务的必需组件，依托 Spring Cloud Alibaba，只需要添加一些注解和少量配置，就可以将 Spring Cloud 应用接入阿里微服务解决方案，通过阿里中间件来迅速搭建分布式应用系统。

2.2.1　Spring Cloud Alibaba 的主要功能

Spring Cloud Alibaba 项目由两部分组成，包含了阿里巴巴的开源组件和阿里云的产品。Spring Cloud Alibaba 不是一个简单的组件，而是一个综合套件。其中涵盖了非常多的内容，包括服务限流降级、服务注册与发现、分布式配置管理以及阿里开源生态支持的诸多组件。以下内容均来自阿里巴巴官方文档：

- 服务限流降级：默认支持 WebServlet、WebFlux、OpenFeign、RestTemplate、Spring Cloud Gateway、Zuul、Dubbo 和 RocketMQ 限流降级功能的接入，可以在运行时通过控制台实时修改限流降级规则，还支持查看限流降级 Metrics 监控。
- 服务注册与发现：适配 Spring Cloud 服务注册与发现标准，默认集成了 Ribbon 的支持。
- 分布式配置管理：支持分布式系统中的外部化配置，配置更改时自动刷新。
- 消息驱动能力：基于 Spring Cloud Stream 为微服务应用构建消息驱动能力。
- 分布式事务：使用 @GlobalTransactional 注解，高效并且对业务零侵入地解决分布式事务问题。
- 阿里云对象存储：阿里云提供的海量、安全、低成本、高可靠的云存储服务，支持在任何应用、任何时间、任何地点存储和访问任意类型的数据。
- 分布式任务调度：提供秒级、精准、高可靠、高可用的定时（基于 Cron 表达式）任务调度服务。同时提供分布式的任务执行模型，如网格任务。网格任务支持海量子任务均匀分配到所有 Worker（schedulerx-client）上执行。
- 阿里云短信服务：覆盖全球的短信服务，友好、高效、智能的互联化通信能力，帮助企业迅速搭建客户触达通道。

更多内容，请参考阿里巴巴官方文档（https://github.com/alibaba/spring-cloud-alibaba）。

2.2.2　Spring Cloud Alibaba 组件

Spring Cloud Alibaba 提供丰富的组件，有些组件是开源的，有些组件是商业化的。以下内容来自阿里巴巴官方文档：

- Sentinel：阿里巴巴开源产品，把流量作为切入点，从流量控制、熔断降级、系统负载保护等多个维度保护服务的稳定性。
- Nacos：阿里巴巴开源产品，一个更易于构建云原生应用的动态服务发现、配置管理和服务管理平台。
- RocketMQ：一款开源的分布式消息系统，基于高可用分布式集群技术，提供低延时、高可靠的消息发布与订阅服务。
- Dubbo：Apache Dubbo 是一款高性能 Java RPC 框架。
- Seata：阿里巴巴开源产品，一个易于使用的高性能微服务分布式事务解决方案。
- Alibaba Cloud ACM：一款在分布式架构环境中对应用配置进行集中管理和推送的应用配置中心产品。
- Alibaba Cloud OSS：阿里云对象存储服务（Object Storage Service，OSS），是阿里云提供的海量、安全、低成本、高可靠的云存储服务，可以在任何应用、任何时间、任何地点存储和访问任意类型的数据。
- Alibaba Cloud SchedulerX：阿里中间件团队开发的一款分布式任务调度产品，提供秒级、精准、高可靠、高可用的定时（基于 Cron 表达式）任务调度服务。
- Alibaba Cloud SMS：覆盖全球的短信服务，友好、高效、智能的互联化通信能力，帮助企业迅速搭建客户触达通道。

2.2.3 Spring Cloud Alibaba 版本简介

本书使用 Spring Cloud Alibaba 2.2.0.RELEASE 版本进行讲解，组件版本关系如表 2-2 所示。

表 2-2　Spring Cloud Alibaba 组件版本关系（来自阿里巴巴官网）

Spring Cloud Alibaba Version	Sentinel Version	Nacos Version	RocketMQ Version	Dubbo Version	Seata Version
（毕业版本）2.2.0.RELEASE	1.7.1	1.1.4	4.4.0	2.7.4.1	1.0.0
（毕业版本）2.1.1.RELEASE or 2.0.1.RELEASE or 1.5.1.RELEASE	1.7.0	1.1.4	4.4.0	2.7.3	0.9.0
（毕业版本）2.1.0.RELEASE or 2.0.0.RELEASE or 1.5.0.RELEASE	1.6.3	1.1.1	4.4.0	2.7.3	0.7.1
（孵化器版本）0.9.0.RELEASE or 0.2.2.RELEASE or 0.1.2.RELEASE	1.5.2	1.0.0	4.4.0	2.7.1	0.4.2
（孵化器版本）0.2.1.RELEASE or 0.1.1.RELEASE	1.4.0	0.6.2	4.3.1	X	X
（孵化器版本）0.2.0.RELEASE or 0.1.0.RELEASE	1.3.0-GA	0.3.0	X	X	X

Spring Cloud Alibaba 与 Spring Boot、Spring Cloud 版本的兼容关系如表 2-3 所示。

表 2-3　Spring Cloud Alibaba 组件版本关系（来自阿里巴巴官网）

Spring Cloud Version	Spring Cloud Alibaba Version	Spring Boot Version
Spring Cloud Hoxton	2.2.0.RELEASE	2.2.X.RELEASE
Spring Cloud Greenwich	2.1.1.RELEASE	2.1.X.RELEASE
Spring Cloud Finchley	2.0.0.RELEASE	2.0.X.RELEASE
Spring Cloud Edgware	1.5.1.RELEASE	1.5.X.RELEASE

2.4　Netflix/Spring Cloud/Spring Cloud Alibaba 的关系

Spring Cloud Alibaba 的组件孵化自阿里巴巴内部自用的中间件产品，这些中间件经历过多次淘宝双 11 的考验，具备超强的抗压能力。Spring Boot 是对 Spring Framework 的补充，它让框架的集成变得更简单，致力于快速开发独立的 Spring 应用。Spring Cloud 是基于 Spring Boot 设计的一套微服务规范，并增强了应用上下文。Spring Cloud 基于 Spring Boot 构建，而 Spring Cloud Alibaba 采用阿里中间件作为原料，实现了 Spring Cloud 的微服务规范。

Spring Cloud 规范除了 Netflix 提供的实现方案还有很多，不过 Netflix 是最成熟的。

大多数 Spring Cloud 用户很难体会到原生实现的局限性，无论是服务发现、分布式配置还是服务调用和熔断，都不太适合大规模集群场景。Spring Cloud Alibaba 的出现弥补了 Spring Cloud 原生实现在大规模集群场景上的局限性，未来替代 Spring Cloud Netflix 也是非常有可能的。它们之间的简单关系如图 2-2 所示。

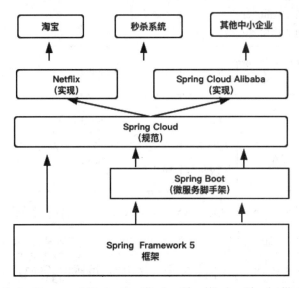

图 2-2　Spring Framework/Spring Boot/Spring Cloud/Spring Cloud Alibaba 的关系

第 3 章

注册中心/配置管理

本章主要介绍 Spring Cloud Alibaba 的服务注册和配置中心组件 Nacos、Nacos 单机模式/集群模式以及 Nacos+Nginx 集群模式搭建，Spring Boot 如何注册到 Nacos 以及如何将配置文件抽到 Nacos 配置中心，Nacos 配置中心和服务发现原理分享，Eureka 简介以及如何通过 Eureka 搭建注册中心集群，Spring Cloud Consul 简介、安装与启动，Spring Cloud Config 简介和原理等内容。

3.1 Nacos 简介

Nacos 是一个易于使用的动态服务发现、配置和服务管理平台，用于构建云原生的应用程序。

服务发现是微服务体系结构中的关键组件之一。在这样的体系结构中，手动为每个客户端配置服务列表可能是一项艰巨的任务，并且使动态扩展极为困难。Nacos Discovery 帮助你自动将服务注册到 Nacos 服务器，并且 Nacos 服务器会跟踪服务并动态刷新服务列表。另外，Nacos Discovery 注册服务实例的某些元数据，例如主机、端口、健康检查 URL。

Nacos 的关键特性包括以下几项。

（1）服务发现和服务健康监测

Nacos 支持基于 DNS 和基于 RPC 的服务发现。服务提供者使用原生 SDK、OpenAPI 等注册服务后，服务消费者可以使用 HTTP&API 查找和发现服务。Nacos 提供对服务的实时健康检查，阻止向不健康的主机或服务实例发送请求。Nacos 支持传输层（PING 或 TCP）和应用层（如 HTTP、MySQL、用户自定义）的健康检查。

（2）动态配置服务

动态配置服务可以让你以中心化、外部化和动态化的方式管理所有环境的应用配置和服务配置。动态配置消除了配置变更时重新部署应用和服务的需要，让配置管理变得更加高效和敏捷。Nacos 提供了一个简洁易用的后台，帮助管理所有服务和应用的配置。

（3）动态 DNS 服务

动态 DNS 服务支持权重路由，让你更容易实现中间层负载均衡、更灵活的路由策略、流量控制以及数据中心内网的简单 DNS 解析服务。

（4）服务及其元数据管理

Nacos 从微服务平台建设的视角管理数据中心的所有服务及元数据，包括管理服务的描述、生命周期、服务的静态依赖分析、服务的健康状态、服务的流量管理、路由及安全策略、服务的 SLA 以及首要的 metrics 统计数据。

更多内容，可参考 Nacos 官方网站。

3.2 Nacos 快速开始

3.2.1 Nacos Server 单机模式

在使用 Nacos 之前，需要先下载 Nacos 并启动 Nacos Server。Nacos Server 有两种运行模式：standalone（单机）和 cluster（集群）。Nacos 快速开始的步骤如下所示。

步骤 01 在 Nacos 官方网站根据不同的操作系统选择不同的安装包类型，下载最新的 release 包（1.2.1 版本），具体如图 3-1 所示。

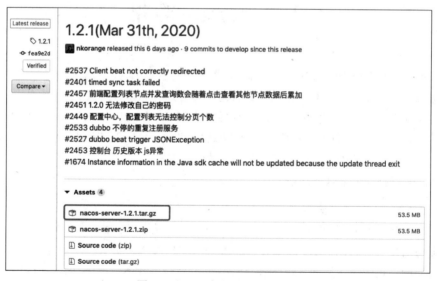

图 3-1 Nacos 官方网站下载界面

步骤 02 解压压缩包，在/Nacos/bin 目录下执行 shell 命令 "sh startup.sh -m standalone"，如图 3-2 所示。

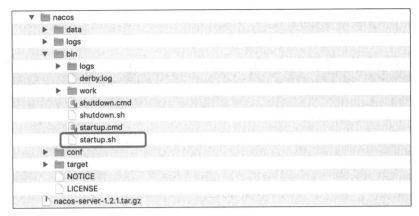

图 3-2　Nacos 目录

步骤 03　在浏览器中访问 http://localhost:8848/Nacos/index.html，输入用户名和密码（Nacos/Nacos），便可进入 Nacos 首页，如图 3-3 所示。

图 3-3　Nacos 首页

步骤 04　通过 curl 命令验证服务注册：

```
### 通过curl命令向nacos注册服务nacos.naming.ay，ip为20.18.7.10，端口为8080
> curl -X POST 'http://127.0.0.1:8848/nacos/v1/ns/instance?serviceName=nacos.naming.ay&ip=20.18.7.10&port=8080'
### 通过curl命令查询服务nacos.naming.ay信息
> curl -X GET 'http://127.0.0.1:8848/nacos/v1/ns/instance/list?serviceName=nacos.naming.ay'
    {"metadata":{},"dom":"nacos.naming.ay","cacheMillis":3000,"useSpecifiedURL":false,"hosts":[{"valid":true,"marked":false,"metadata":{},"instanceId":"20.18.7.10#8080#DEFAULT#DEFAULT_GROUP@@nacos.naming.ay","port":8080,"healthy":true,"ip":"20.18.7.10","clusterName":"DEFAULT","weight":1.0,"ephemeral":true,"serviceName":"nacos.naming.ay","enabled":true}],"name":"DEFAULT_GROUP@@nacos.naming.ay","checksum":"e2bafa7a38a030e36bef6188043c3c2f","lastRefTime":1586169983028,"env":"","clusters":""}
```

在 Nacos 管理后台的"服务管理"→"服务列表"中查看服务注册情况（命令执行完需要马上查看，否则 Nacos 会根据心跳机制判断该服务不健康，将其移除），如图 3-4 所示。

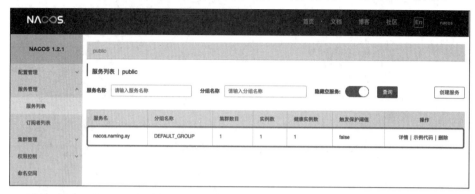

图 3-4　在服务管理中查看服务列表

步骤 05　通过 curl 命令验证服务配置管理：

```
### 通过curl命令将配置"HelloWorld"保存到nacos.cfg.dataId中
> curl -X POST "http://127.0.0.1:8848/nacos/v1/cs/configs?dataId=nacos.cfg.dataId&group=test&content=HelloWorld"
### 通过curl命令获取dataId为nacos.cfg.dataId、group为test的配置
> curl -X GET "http://127.0.0.1:8848/nacos/v1/cs/configs?dataId=nacos.cfg.dataId&group=test"
HelloWorld
```

运行结果如图 3-5 所示。

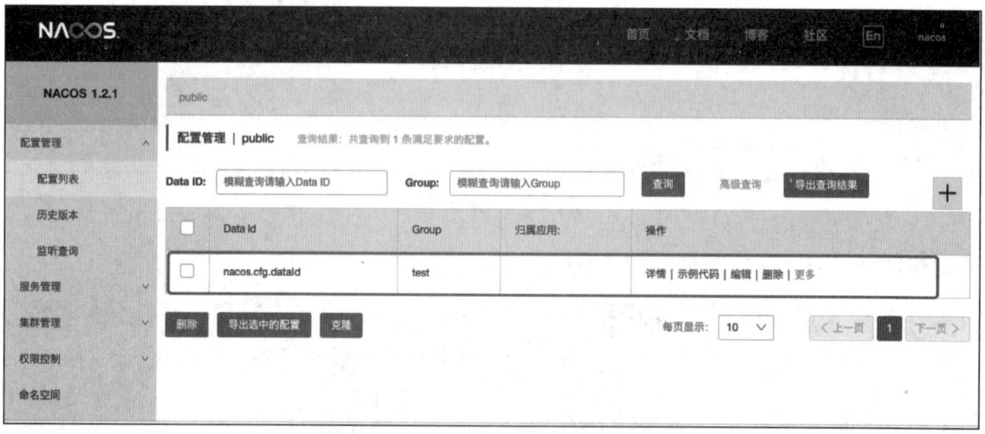

图 3-5　在服务管理中查看配置列表

步骤 06　关闭服务器，在 /Nacos/bin 目录下执行 shell 命令"sh shutdown.sh"。

0.7 版本之前的 Nacos 在单机模式时使用嵌入式数据库实现数据的存储，不方便观察数据存储的基本情况。0.7 版本之后支持 MySQL 数据源能力，具体的操作步骤如下：

步骤 01　安装 MySQL 数据库（版本要求 5.6.5+）。

步骤 02　创建 Nacos_devtest 数据库并初始化表，初始化文件 Nacos-mysql.sql，如图 3-6 所示。

第 3 章 注册中心/配置管理 | 63

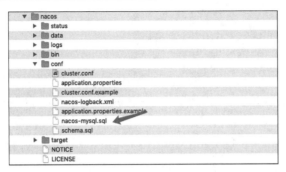

图 3-6 初始化脚本 Nacos-mysql.sql 位置

步骤 03 修改 conf/application.properties 文件，增加支持 MySQL 数据源配置（目前只支持 MySQL），添加 MySQL 数据源的 url、用户名和密码。

```
spring.datasource.platform=mysql
//数据库数量
db.num=1
//数据库连接地址/端口
db.url.0=jdbc:mysql://127.0.0.1:3306/nacos_devtest?characterEncoding=utf8&co
nnectTimeout=1000&socketTimeout=3000&autoReconnect=true
//用户名
db.user=nacos_devtest
//密码
db.password=youdontknow
```

步骤 04 以单机的模式启动 Nacos，通过 http://localhost:8848/Nacos/index.html 地址进入界面，之后 Nacos 所有的数据都会保存到 MySQL。

3.2.2　Nacos Server 集群模式

Nacos 单机模式仅仅适用于测试和单机适用，生产环境大多适用集群模式以确保高可用。如果有多数据中心场景，那么 Nacos 还支持多集群模式。

接下来讲解如何搭建 Nacso 集群环境，具体步骤如下：

步骤 01 将 3.2.1 小节下载的 nacos 安装包复制 2 份，分别取名为 nacos-02、nacos-03，如图 3-7 所示。

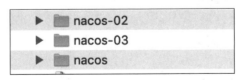

图 3-7　3 个 nacos 实例

步骤 02 修改 nacos、nacos-02、nacos-03 的配置文件 application.properties，将服务启动端口分别改为 8849 和 8850：

```
//nacos 实例的 application.properties 配置文件
```

```
#************** Spring Boot Related Configurations **************#
### Default web context path:
server.servlet.contextPath=/nacos
### Default web server port:
server.port=8848
//省略部分代码

//nacos-02 实例的 application.properties 配置文件
#************** Spring Boot Related Configurations **************#
### Default web context path:
server.servlet.contextPath=/nacos
### Default web server port:
server.port=8849
//省略部分代码

//nacos-03 实例的 application.properties 配置文件
#************** Spring Boot Related Configurations **************#
### Default web context path:
server.servlet.contextPath=/nacos
### Default web server port:
server.port=8850
//省略部分代码
```

步骤 03 在 nacos、nacos-02 和 nacos-03 实例的 conf 目录下添加 cluster.conf 文件（集群配置文件），文件内容如下所示：

```
### 这里简单将 3 个 nacos 实例部署在同一个机器下（同一个 Mac 系统下）
### 192.168.0.100 是本地机器的 ip 地址
192.168.0.100:8848
192.168.0.100:8849
192.168.0.100:8850
```

> **注 意**
>
> 最好用实际 IP 地址，而非 127.0.0.1 或者 localhost，否则会出现问题。

步骤 04 3 个 nacos 实例的数据库连接信息保持不变，分别进入 nacos、nacos-02 和 nacos-03 实例的 bin 目录下，执行 shell 命令：

```
### 在没有参数模式，是集群模式
sh startup.sh
```

步骤 05 访问如下地址来查看 3 个实例的启动情况：

```
### nacos 实例
http://localhost:8848/nacos/index.html
### nacos-02 实例
http://localhost:8849/nacos/index.html
### nacos-02 实例
http://localhost:8850/nacos/index.html
```

步骤 06 在 Nacos 实例后台管理的 "集群管理" → "节点列表" 中查看 Nacos 集群信息，如图 3-8 所示。

图 3-8　在 Nacos 后台管理中查看集群信息

Nacos 实例在启动过程中出现错误时，可在 /logs/start.out 目录下查看启动的日志信息，具体内容如下所示：

```
//省略部分代码
         , --.
       , --.'|
     , --,: : |
   ,`--.'`| ' :                    , ---.              Nacos 1.2.1
   |   :  : | |                                        Running in cluster mode, All function modules
   |   |  \ : |                          ', ,'\   .--.--.    Port: 8850
   :   :  |  , --.--.    , ---.    /   | / /   Pid: 17964
   |   :  ';  | /    \  /   \.  ;, . :|  : /`./  Console: http://192.168.0.100:8850/nacos/index.html
   '   ' ;. ; .--. .-. | /    / '' | |: :|  :  ;_
   |   | | \ | \__\/: . ..   ' /  | .; : \  \`.       https://nacos.io
   '   : |  ; .', '.--.; |' ;:__| : |    `----.  \
   |   | '`--' /  /   ,. |'.'.|   '.'|\   \  / /`--'
   '   : |   ; :   .'   \ :   : :   `----'   '--'
   ;   |.'   |   ,.-./\ \  /                  '--'----'
   '---'     `--`----'  `----'

 2020-04-13 18:29:28,805 INFO The server IP list of Nacos is [127.0.0.1:8848, 127.0.0.1:8849, 127.0.0.1:8850]

 2020-04-13 18:29:29,811 INFO Nacos is starting...

 2020-04-13 18:29:30,817 INFO Nacos is starting...

 2020-04-13 18:29:31,817 INFO Nacos is starting...

 2020-04-13 18:29:32,818 INFO Nacos is starting...
// 省略部分代码
```

3.2.3　Nacos+Nginx 集群模式

前面部署的 Nacos Server 集群一般不适用于生产，本小节详细探讨如何搭建一个生产可用的 Nacos 集群，部署架构如图 3-9 所示。

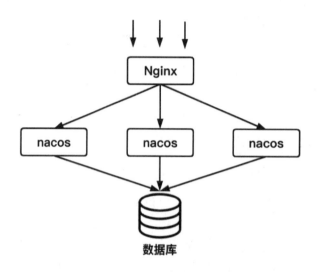

图 3-9　Nacos + Nginx 集群部署

部署图 3-9 所示架构的具体步骤如下：

步骤 01 安装 Nginx（推荐使用 brew install nginx 命令进行安装）：

```
### 使用 brew 安装 Nginx
> brew install nginx
```

步骤 02 在/usr/local/etc/nginx/nginx.conf 目录下找到 Nginx 核心配置文件 nginx.conf, 在文件中添加如下配置：

```
server {
  ###监听端口
  listen 8847;
  ###域名可以配置多个
  server_name nacos.ay.com;
  ###将所有请求/nacos/全部转发到upstream中定义的目标服务器中
  location /nacos/ {
    proxy_pass http://nacos-server/nacos/;
  }
}
###作负载均衡，此配置需要轮询的服务器地址和端口号
upstream nacos-server {
  server 192.168.38.188:8848;
  server 192.168.38.188:8849;
  server 192.168.38.188:8850;
}
```

在系统的 hosts 文件中添加域名与 IP 映射，具体如下：

```
# Nginx
127.0.0.1 nacos.ay.com
```

步骤 03 找到 Nginx 安装目录，执行如下命令：

```
### 启动 nginx，/usr/local/etc/nginx 为 Nginx 安装目录
>sudo nginx -c /usr/local/etc/nginx/nginx.conf
```

如果需要重启 Nginx，就执行如下命令：

```
### 启动 Nginx
>sudo nginx -s reload
```

步骤 04 在浏览器中访问 http://Nacos.ay.com:8847/Nacos/，输入用户名/密码后即可登录 Nacos。

至此，我们完成了 Nacos + Nginx 集群的搭建。

3.3 Spring Boot 注册到 Nacos

本节主要讲解如何通过 Nacos 来实现分布式环境下的配置管理和服务发现。

3.3.1 Nacos 配置管理

通过 Nacos 实现配置管理的主要步骤如下：

步骤 01 参考 1.3.1 小节的内容，快速创建 Spring Boot 项目，项目名称为 Nacos-springboot-config。

步骤 02 参考 3.2.1 和 3.2.2 小节的内容，安装并启动 Nacos 服务。

步骤 03 在 Nacos-springboot-config 项目的 pom.xml 文件中添加如下依赖：

```xml
<dependency>
    <groupId>com.alibaba.boot</groupId>
    <artifactId>nacos-config-spring-boot-starter</artifactId>
    <version>0.2.7</version>
</dependency>
```

其中，Nacos-config-spring-boot-starter（见图 3-10）没有任何实现，只是用来管理依赖，pring.providers 文件声明依赖由 provides: Nacos-config-spring-boot-autoconfigure 提供。

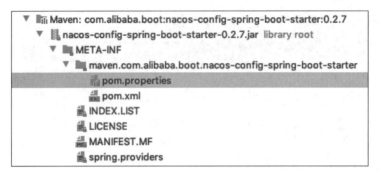

图 3-10　Nacos-config-spring-boot-starter 依赖

步骤 04 在 application.properties 中配置 Nacos Server 的地址:

```
### 192.168.0.100是本地机器的IP地址
nacos.config.server-addr=192.168.0.100:8848
```

步骤 05 在启动类中使用@NacosPropertySource 加载 dataId 为 config-test 的配置源，并开启自动更新:

```
@SpringBootApplication
@NacosPropertySource(dataId = "config-test", autoRefreshed = true)
public class DemoApplication {
    public static void main(String[] args) {
        SpringApplication.run(DemoApplication.class, args);
    }
}
```

步骤 06 通过 Nacos 的@NacosValue 注解设置属性值:

```
@Controller
@RequestMapping("config")
public class ConfigController {
    //设置属性
    @NacosValue(value = "${name:ay}", autoRefreshed = true)
    private String name;

    @RequestMapping(value = "/get", method = GET)
    @ResponseBody
    public String get() {
        return name;
    }
}
```

步骤 07 在 Nacos 配置列表中添加 Data Id: config-test 的配置文件，文件格式为 YAML，文件内容如下:

```
name: ay
```

步骤 08 启动 Nacos-springboot-config 项目，在命令行中执行命令:

```
### 执行curl命令，调用ConfigController类的get方法
```

```
> curl http://localhost:8080/config/get
### 返回 ay 字符串
> ay
```

步骤 09 在 Nacos 配置列表中，修改 config-test 配置文件的 name 属性为 name: al。

3.3.2 Nacos 服务注册

通过 Nacos 实现服务注册的主要步骤如下：

步骤 01 快速创建 Spring Boot 项目，项目名称为 Nacos-springboot-discovery。

步骤 02 安装并启动 Nacos 服务。

步骤 03 在 Nacos-springboot-discovery 项目的 pom.xml 文件中添加如下依赖：

```xml
<dependency>
    <groupId>com.alibaba.boot</groupId>
    <artifactId>nacos-discovery-spring-boot-starter</artifactId>
    <version>0.2.7</version>
</dependency>
```

步骤 04 在 application.properties 中配置 Nacos Server 的地址：

```
### 服务名称
spring.application.name=example
### 服务端口
server.port=18080
### 服务地址
server.address=192.168.0.100
### Nacos 地址
nacos.discovery.server-addr=192.168.0.100:8848
```

步骤 05 修改项目的入口类，具体如下：

```java
/**
 * 入口类
 * @author ay
 * @since 2020-05-17
 */
@SpringBootApplication
public class DemoApplication implements CommandLineRunner {

    @NacosInjected
    private NamingService namingService;

    @Value("${spring.application.name}")
    private String applicationName;

    @Value("${server.port}")
    private Integer serverPort;
```

```
    @Value("${server.address}")
    private String address;

    public static void main(String[] args) {
        SpringApplication.run(DemoApplication.class, args);
    }

    @Override
    public void run(String... args) throws Exception {
        //应用启动时，将服务注册到Nacos
        namingService.registerInstance(applicationName, address, serverPort);
    }
}
```

- CommandLineRunner：接口执行点是在容器启动成功后的最后一步回调的，可以在回调方法 run 中执行相关逻辑。
- NamingService：使用@NacosInjected 注入 Nacos 的 NamingService 实例。NamingService 是 Nacos 对外提供给使用者的接口，提供了以下方法：
    - registerInstance：注册实例。
    - deregisterInstance：注销实例。
    - getAllInstances：获取某一服务的所有实例。
    - selectInstances：获取某一服务中健康或不健康的实例。
    - selectOneHealthyInstance：根据权重选择一个健康的实例。
    - getServerStatus：检测服务端的健康状态。
    - subscribe：注册对某个服务的监听。
    - unsubscribe：注销对某个服务的监听。
    - getSubscribeServices：获取被监听的服务。

步骤06 启动 Nacos-springboot-discovery 项目，可以看到服务已经注入到 Nacos 中。

3.4 Nacos Spring Cloud

本节主要讲解 Spring Cloud 的使用者如何通过 Nacos 实现配置管理和服务发现。

3.4.1 Nacos 配置管理

步骤01 快速创建 Spring Boot 项目，项目名称为 Nacos-springcloud-config。

步骤02 安装并启动 Nacos 服务。

步骤03 在 Nacos-springcloud-config 项目的 pom.xml 文件中添加如下依赖：

```
<properties>
    <java.version>1.8</java.version>
```

```xml
    <spring-cloud-alibaba.version>2.2.1.RELEASE</spring-cloud-alibaba.version>
</properties>

<dependencies>
    <dependency>
        <groupId>org.springframework.boot</groupId>
        <artifactId>spring-boot-starter-web</artifactId>
    </dependency>
    <!-- 配置管理所需依赖 -->
    <dependency>
        <groupId>com.alibaba.cloud</groupId>
        <artifactId>spring-cloud-starter-alibaba-nacos-config</artifactId>
    </dependency>
</dependencies>
<!-- spring cloud alibaba 依赖 -->
<dependencyManagement>
    <dependencies>
        <dependency>
            <groupId>com.alibaba.cloud</groupId>
            <artifactId>spring-cloud-alibaba-dependencies</artifactId>
            <version>${spring-cloud-alibaba.version}</version>
            <type>pom</type>
            <scope>import</scope>
        </dependency>
    </dependencies>
</dependencyManagement>
```

步骤 04 在 application.properties 中配置 Nacos Server 的地址：

```
### Nacos 地址，192.168.1.9 为本机的 IP 地址
spring.cloud.nacos.config.server-addr=192.168.1.9:8848
### 应用名称
spring.application.name=example
```

spring.application.name 之后会构成 Nacos 配置管理 dataId 字段的一部分。在 Nacos Spring Cloud 中，dataId 的完整格式如下：

```
${prefix}-${spring.profile.active}.${file-extension}
```

- ${prefix}：默认为 spring.application.name 的值，也可以通过 spring.cloud.Nacos.config.prefix 配置项来配置。
- ${spring.profile.active}：当前环境对应的 profile。当 spring.profile.active 为空时，对应的连接符 "-" 也将不存在，dataId 的拼接格式变成 ${prefix}.${file-extension}。
- ${file-extension}：配置内容的数据格式，可以通过 spring.cloud.Nacos.config.file-extension 来配置。目前只支持 properties 和 yaml 类型。

步骤 05 开发控制层类 ConfigController，具体代码如下：

```
import org.springframework.cloud.context.config.annotation.RefreshScope;
import org.springframework.web.bind.annotation.RequestMapping;
```

```
import org.springframework.web.bind.annotation.RestController;

@RestController
@RequestMapping("/config")
@RefreshScope
public class ConfigController {

    @Value("${author}")
    private String name;

    @RequestMapping("/get")
    public String get() {
        return name;
    }

}
```

步骤 06 在 Nacos 配置列表中创建 example.properties 配置文件,配置内容如下:

```
author=ay
```

步骤 07 在浏览器中输入访问地址 "http://localhost:8080/config/get" 便可以返回 ay 字符串。

> **提 示**
>
> 如果出现 java.lang.IllegalArgumentException: Could not resolve placeholder 'author' in value "${author}",就检查 Nacos 配置、application.properties 配置并尝试重启 Nacos 服务器。

3.4.2 Nacos 服务注册

下面将演示如何在 Spring Cloud 项目中启用 Nacos 的服务发现功能,具体步骤如下:

步骤 01 快速创建 Spring Boot 项目,项目名称为 Nacos-springcloud-discovery-01。

步骤 02 安装并启动 Nacos 服务。

步骤 03 在 application.properties 配置文件中添加如下配置:

```
### 服务端口
server.port=8070
### 服务名称 service-provider
spring.application.name=service-provider
### Nacos 地址
spring.cloud.nacos.discovery.server-addr=192.168.43.27:8848
```

步骤 04 在 pom.xml 配置文件中添加如下配置:

```xml
<properties>
    <java.version>1.8</java.version>
    <spring-cloud-alibaba.version>2.2.1.RELEASE</spring-cloud-alibaba.version>
</properties>
    <dependency>
```

```xml
        <groupId>com.alibaba.cloud</groupId>
        <artifactId>spring-cloud-starter-alibaba-nacos-discovery</artifactId>
</dependency>
<dependencyManagement>
    <dependencies>
        <dependency>
            <groupId>com.alibaba.cloud</groupId>
            <artifactId>spring-cloud-alibaba-dependencies</artifactId>
            <version>${spring-cloud-alibaba.version}</version>
            <type>pom</type>
            <scope>import</scope>
        </dependency>
    </dependencies>
</dependencyManagement>
```

spring-cloud-starter-alibaba-Nacos-discovery 依赖是服务注册所需依赖，版本 2.1.x.RELEASE 对应的是 Spring Boot 2.1.x 版本。版本 2.0.x.RELEASE 对应的是 Spring Boot 2.0.x 版本，版本 1.5.x.RELEASE 对应的是 Spring Boot 1.5.x 版本。

步骤 05 在启动类上添加@EnableDiscoveryClient 注解。@EnableDiscoveryClient 注解可以开启服务注册发现功能。

步骤 06 创建 HelloController 控制层，具体代码如下：

```java
/**
 * @author ay
 * @since 2020-06-14
 */
@RestController
@RequestMapping("/test")
public class HelloController {

    @RequestMapping("/hello")
    public String hello(){
        return "ay";
    }
}
```

步骤 07 运行启动类，在浏览器中输入访问地址 "http://localhost:8070/test/hello"，页面将显示 ay 字符串。

> **提 示**
>
> 如果不想手动在 pom.xml 文件中添加依赖包，那么可以使用 Intellij IDEA 开发工具。在创建 Spring Boot 项目的时候，勾选 Spring Cloud Alibaba 相关的组件包，如图 3-11 所示。

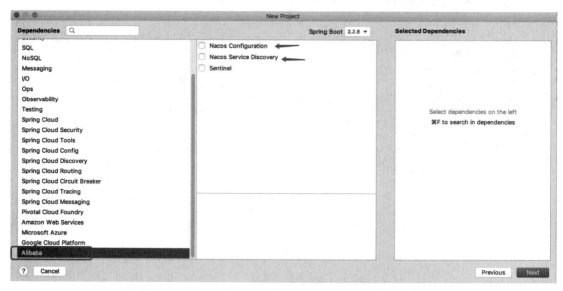

图 3-11　在 Intellij IDEA 中选择 Spring Cloud Alibaba 组件

至此，服务提供者开发完成。接下来开发服务消费者，具体步骤如下：

步骤 01 快速创建 Spring Boot 项目，项目名称为 Nacos-springcloud-discovery-02。

步骤 02 在 application.properties 配置文件中添加如下配置：

```
### 服务端口
server.port=8080
### 服务名称service-consumer
spring.application.name=service-provider
### Nacos 地址
spring.cloud.nacos.discovery.server-addr=192.168.43.27:8848
```

步骤 03 在启动类上添加@EnableDiscoveryClient 注解，@EnableDiscoveryClient 注解可以开启服务注册发现功能。

```
@SpringBootApplication
@EnableDiscoveryClient
public class DemoApplication {

    @LoadBalanced
    @Bean
    public RestTemplate restTemplate() {
        return new RestTemplate();
    }

    public static void main(String[] args) {
        SpringApplication.run(DemoApplication.class, args);
    }
}
```

- RestTemplate：向服务的某个具体实例发起 HTTP 请求，具体的请求路径是通过拼接完成的。

- @LoadBalanced：添加@LoadBalanced 注解，使 RestTemplate 具有负载均衡的功能。

步骤 04 开发 EchoController 控制层类，具体代码如下：

```
@RestController
@RequestMapping("/echo")
public class EchoController {

    @Resource
    private RestTemplate restTemplate;

    @RequestMapping("/hello")
    public String echo(){
        return restTemplate.getForObject("http://service-provider/test/hello",
String.class);
    }
}
```

在真正调用服务接口的时候，原来 host 部分是通过手工拼接 IP 和端口的，现在直接采用服务名作为请求路径。在真正调用的时候，Spring Cloud 会将请求拦截下来，然后通过负载均衡器选出节点，并替换服务名部分为具体的 IP 和端口，从而实现基于服务名的负载均衡调用。

步骤 05 运行启动类，在浏览器中输入请求地址"http://localhost:8080/echo/hello"，调用 service-consumer 服务。service-consumer 服务通过 RestTemplate 类向 service-provider 服务发起请求，Spring Cloud 将请求拦截下来，通过负载均衡器选出节点进行调用，最后将请求结果返回。

3.5　Nacos 原理解析

3.5.1　Nacos 配置中心原理分析

　　动态配置管理是 Nacos 的重要功能之一，通过动态配置服务可以以集中和动态的方式管理所有应用程序或服务的配置信息。

　　动态配置中心可以实现配置更新时无须重新部署应用程序，极大地增加了系统的运维能力。接下来将和大家一起了解 Nacos 动态配置的能力，分析 Nacos 是如何以简单、优雅、高效的方式管理配置以及实现配置的动态变更的。

　　首先，在 Nacos 的控制台中添加配置，如图 3-12 所示。

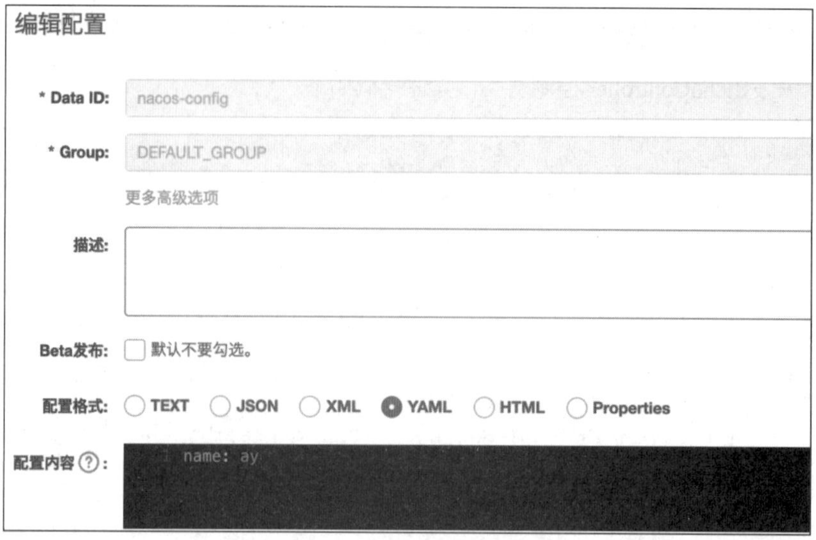

图 3-12 创建 Nacos-config 配置

配置文件添加完成后，可单击图 3-13 中的示例代码按钮，具体代码如下：

```java
/**
 * @author ay
 * @since 2020-06-17
 */
public class ConfigExample {

    public static void main(String[] args) throws NacosException,
InterruptedException, IOException {
        //服务地址
        String serverAddr = "localhost";
        String dataId = "nacos-config";
        String group = "DEFAULT_GROUP";
        Properties properties = new Properties();
        properties.put(PropertyKeyConst.SERVER_ADDR, serverAddr);
        ConfigService configService =
NacosFactory.createConfigService(properties);
        String content = configService.getConfig(dataId, group, 5000);
        System.out.println(content);
        //监听器
        configService.addListener(dataId, group, new Listener() {
            @Override
            public void receiveConfigInfo(String configInfo) {
                System.out.println("recieve:" + configInfo);
            }
            @Override
            public Executor getExecutor() {
                return null;
            }
        });
        //推送配置到nacos
```

```
            boolean isPublishOk =
            configService.publishConfig(dataId, group, "content");
            System.out.println(isPublishOk);
            Thread.sleep(3000);
            //从nacos中获取数据
            content = configService.getConfig(dataId, group, 5000);
            System.out.println(content);
            //删除配置
            boolean isRemoveOk = configService.removeConfig(dataId, group);
            System.out.println(isRemoveOk);
            Thread.sleep(3000);
            //获取配置，检查配置是否仍存在
            content = configService.getConfig(dataId, group, 5000);
            System.out.println(content);
            Thread.sleep(300000);
    }
}
```

图 3-13　获取示例代码

Nacos 客户端连接到服务端之后，根据 Data ID、Group 可以获取到具体的配置信息，当服务端的配置发生变更时，客户端会收到通知。

在示例代码中，首先创建 ConfigService 类。ConfigService 是通过 ConfigFactory 类创建的，具体源码如下：

```
    public static ConfigService createConfigService(Properties properties) throws NacosException {
        try {
            Class<?> driverImplClass = Class.forName("com.alibaba.nacos.client.config.NacosConfigService");
            Constructor constructor = driverImplClass.getConstructor(Properties.class);
            //每次调用都会创建ConfigService的实例
            ConfigService vendorImpl = (ConfigService)constructor.newInstance(properties);
            return vendorImpl;
        } catch (Throwable var4) {
            throw new NacosException(-400, var4);
        }
    }
```

这里通过反射调用了 NacosConfigService 构造方法来创建 ConfigService，而且 NacosConfigService 的构造方法有一个 Properties 参数。NacosConfigService 的源码如下所示：

```java
/**
 * http agent
 */
private HttpAgent agent;
/**
 * longpolling
 */
private ClientWorker worker;
public NacosConfigService(Properties properties) throws NacosException {
    String encodeTmp = properties.getProperty(PropertyKeyConst.ENCODE);
    if (StringUtils.isBlank(encodeTmp)) {
        encode = Constants.ENCODE;
    } else {
        encode = encodeTmp.trim();
    }
    initNamespace(properties);
    agent = new MetricsHttpAgent(new ServerHttpAgent(properties));
    agent.start();
    worker = new ClientWorker(agent, configFilterChainManager, properties);
}
```

实例化时主要是初始化 HttpAgent 和 ClientWorker 对象。HttpAgent 是接口类，ServerHttpAgent 和 MetricsHttpAgent 都是其子类。agent 会作为 ClientWorker 初始化的参数。ClientWorker 源码如下所示：

```java
//省略部分代码
//ClientWorker 构造方法
@SuppressWarnings("PMD.ThreadPoolCreationRule")
public ClientWorker(final HttpAgent agent, final ConfigFilterChainManager configFilterChainManager, final Properties properties) {
    this.agent = agent;
    this.configFilterChainManager = configFilterChainManager;

    // Initialize the timeout parameter

    init(properties);
    //第一个线程池
    executor = Executors.newScheduledThreadPool(1, new ThreadFactory() {
        @Override
        public Thread newThread(Runnable r) {
            Thread t = new Thread(r);
            t.setName("com.alibaba.nacos.client.Worker." + agent.getName());
            t.setDaemon(true);
            return t;
        }
    });
    //第二个线程池
```

```
        executorService =
Executors.newScheduledThreadPool(Runtime.getRuntime().availableProcessors(), new
ThreadFactory() {
            @Override
            public Thread newThread(Runnable r) {
                Thread t = new Thread(r);
                //长轮询
                t.setName("com.alibaba.nacos.client.Worker.longPolling." +
agent.getName());
                t.setDaemon(true);
                return t;
            }
        });

        executor.scheduleWithFixedDelay(new Runnable() {
            @Override
            public void run() {
                try {
                    checkConfigInfo();
                } catch (Throwable e) {
                    LOGGER.error("[" + agent.getName() + "] [sub-check] rotate check
error", e);
                }
            }
        }, 1L, 10L, TimeUnit.MILLISECONDS);
    }
```

在 ClientWorker 构造方法中创造两个线程池：第一个线程池只拥有一个线程，用来执行定时任务的 executor，这里 executor 每隔 10ms 就会执行一次 checkConfigInfo()方法（根据方法名得知每 10 ms 检查一次配置信息）；第二个线程池是一个普通的线程池，用来做长轮询。

checkConfigInfo 源码如下所示：

```
    public void checkConfigInfo() {
        // 分任务
        int listenerSize = cacheMap.get().size();
        // 向上取整为批数
        int longingTaskCount = (int) Math.ceil(listenerSize /
ParamUtil.getPerTaskConfigSize());
        if (longingTaskCount > currentLongingTaskCount) {
          for (int i = (int) currentLongingTaskCount; i < longingTaskCount; i++) {
            // 如何判断任务是否在执行，需要好好想想。任务列表现在是无序的，变化过程可能有问题
              executorService.execute(new LongPollingRunnable(i));
          }
          currentLongingTaskCount = longingTaskCount;
        }
    }
```

checkConfigInfo 方法取出一批任务，提交给 executorService 线程池去执行，执行的任务是 LongPollingRunnable。LongPollingRunnable 主要分为两部分：第一部分是检查本地的配置信息；第二部分是先获取服务端的配置信息，然后更新到本地。

```java
class LongPollingRunnable implements Runnable {
    private int taskId;

    public LongPollingRunnable(int taskId) {
        this.taskId = taskId;
    }

    @Override
    public void run() {

        List<CacheData> cacheDatas = new ArrayList<CacheData>();
        List<String> inInitializingCacheList = new ArrayList<String>();
        try {
            // check failover config
            for (CacheData cacheData : cacheMap.get().values()) {
                if (cacheData.getTaskId() == taskId) {
                    cacheDatas.add(cacheData);
                    try {
                        //检查本地配置
                        checkLocalConfig(cacheData);
                        if (cacheData.isUseLocalConfigInfo()) {
                            cacheData.checkListenerMd5();
                        }
                    } catch (Exception e) {
                        LOGGER.error("get local config info error", e);
                    }
                }
            }

            // 检查服务配置 (check server config)
            List<String> changedGroupKeys =
                    checkUpdateDataIds(cacheDatas, inInitializingCacheList);
            LOGGER.info("get changedGroupKeys:" + changedGroupKeys);

            for (String groupKey : changedGroupKeys) {
                String[] key = GroupKey.parseKey(groupKey);
                String dataId = key[0];
                String group = key[1];
                String tenant = null;
                //省略代码
            }
            for (CacheData cacheData : cacheDatas) {
                if (!cacheData.isInitializing() || inInitializingCacheList
                        .contains(GroupKey.getKeyTenant(cacheData.dataId,
                                cacheData.group, cacheData.tenant))) {
                    //重点关注的源码
                    cacheData.checkListenerMd5();
                    cacheData.setInitializing(false);
                }
            }
        }
```

```
    } catch (Throwable e) {
        //重新通过executorService提交本任务
        LOGGER.error("longPolling error : ", e);
        executorService.schedule(this, taskPenaltyTime,
        TimeUnit.MILLISECONDS);
    }
}
```

取出与该 taskId 相关的缓存对象 CacheData，然后对 CacheData 进行检查，包括本地配置检查和监听器的 Md5 检查。本地检查主要是做一个故障容错，当服务端挂掉后，Nacos 客户端可以从本地的文件系统中获取相关的配置信息。

查看 checkLocalConfig 源码：

```
private void checkLocalConfig(CacheData cacheData) {
    final String dataId = cacheData.dataId;
    final String group = cacheData.group;
    final String tenant = cacheData.tenant;
    //nacos 配置文件保存路径，通过查看getFailoverFile源码即可知道该路径如何生成
    File path = LocalConfigInfoProcessor.getFailoverFile(agent.getName(),
dataId, group, tenant);
    //省略部分代码


}
```

由源码可知，Nacos 将配置信息保存在如下目录：

```
###nacos 配置文件保存目录
~/nacos/config/fixed-{address}_8848_nacos/snapshot/DEFAULT_GROUP/{dataId}
```

通过 checkUpdateDataIds() 方法从服务端获取发生变化的 dataId 列表的值。在 getServerConfig 方法中，根据 dataId 获取最新的配置信息，将最新的配置信息保存到 CacheData 中。最后调用 CacheData 的 checkListenerMd5 方法。

ConfigExample 类的具体代码如下：

```
//省略部分代码
ConfigService configService =
    NacosFactory.createConfigService(properties);
String content = configService.getConfig(dataId, group, 5000);
System.out.println(content);
//监听器
configService.addListener(dataId, group, new Listener() {
    @Override
    public void receiveConfigInfo(String configInfo) {
        System.out.println("recieve:" + configInfo);
    }
    @Override
    public Executor getExecutor() {
        return null;
    }
});
//省略部分代码
```

在 ConfigExample 类中，通过 addListener 方法给 configService 添加监听器，具体代码如下：

```java
@Override
public void addListener(String dataId, String group, Listener listener) throws NacosException {
    worker.addTenantListeners(dataId, group, Arrays.asList(listener));
}
```

addTenantListeners 源码如下所示：

```java
public void addTenantListeners(String dataId, String group, List<? extends Listener> listeners) throws NacosException {
    group = null2defaultGroup(group);
    String tenant = agent.getTenant();
    CacheData cache = addCacheDataIfAbsent(dataId, group, tenant);
    for (Listener listener : listeners) {
        cache.addListener(listener);
    }
}
```

根据 dataId、group 和当前的租户获取一个 CacheData 对象，将当前的 listener 对象添加到 CacheData，即 listener 最终被 CacheData 所持有，listener 的回调方法 receiveConfigInfo 最终是在 CacheData 中触发的。CacheData 类出现的频率非常高，在 LongPollingRunnable 的任务中，几乎所有的方法都围绕着 CacheData 类，添加 Listener 时，实际上 Listener 会委托给 CacheData 类。CacheData 源码如下：

```java
public class CacheData {
    private final String name;
    private final ConfigFilterChainManager configFilterChainManager;
    public final String dataId;
    public final String group;
    public final String tenant;
    private final CopyOnWriteArrayList<ManagerListenerWrap> listeners;

    private volatile String md5;
    /**
     * whether use local config
     */
    private volatile boolean isUseLocalConfig = false;
    /**
     * last modify time
     */
    private volatile long localConfigLastModified;
    private volatile String content;
    private int taskId;
    private volatile boolean isInitializing = true;
    private String type;
}
```

在 CacheData 类中，包含 dataId、group、content、taskId、listeners、md5 等与配置相关的属性，

属性 md5 的值是根据 content 属性计算出来的。listeners 是包装后的 ManagerListenerWrap 对象，该对象除了持有 Listener 对象，还持有 lastCallMd5 属性。

ConfigService 类的 Listener 在什么时候会触发回调方法 receiveConfigInfo 呢？

在 ClientWorker 定时任务中，启动长轮询的任务 LongPollingRunnable，该任务多次执行 cacheData.checkListenerMd5()方法。checkListenerMd5() 方法的源码如下所示：

```
void checkListenerMd5() {
    for (ManagerListenerWrap wrap : listeners) {
        if (!md5.equals(wrap.lastCallMd5)) {
            safeNotifyListener(dataId, group, content, type, md5, wrap);
        }
    }
}
```

checkListenerMd5 方法会检查 CacheData 当前的 md5 值与 CacheData 持有的所有 Listener 中保存的 md5 值是否一致，如果不一致，就执行安全的监听器的通知方法 safeNotifyListener，通知 Listener 使用者该 Listener 所关注的配置信息已经发生改变。safeNotifyListener 的具体源码如下所示：

```
private void safeNotifyListener(final String dataId...省略代码) {
    final Listener listener = listenerWrap.listener;

    Runnable job = new Runnable() {
        @Override
        public void run() {
            ClassLoader myClassLoader =
                    Thread.currentThread().getContextClassLoader();
            ClassLoader appClassLoader =
                    listener.getClass().getClassLoader();
            try {
                if (listener instanceof AbstractSharedListener) {
                    AbstractSharedListener adapter = (AbstractSharedListener) listener;
                    adapter.fillContext(dataId, group);
                    LOGGER.info("[{}] [notify-context] dataId={}, group={}, md5={}",
name, dataId, group, md5);
                }
                //执行回调之前先将线程classloader 设置为具体webapp 的classloader,
                // 以免回调方法中调用 spi 接口时出现异常或错用 (多应用部署才会有该问题)
                Thread.currentThread().setContextClassLoader(appClassLoader);
                ConfigResponse cr = new ConfigResponse();
                cr.setDataId(dataId);
                cr.setGroup(group);
                cr.setContent(content);
                configFilterChainManager.doFilter(null, cr);
                //获取最新配置信息
                String contentTmp = cr.getContent();
                //调用 Listener 回调方法
                listener.receiveConfigInfo(contentTmp);
```

```
            //省略部分代码
            //更新listenerWrap的md5值
            listenerWrap.lastCallMd5 = md5;
    }
```

在 safeNotifyListener 方法中，获取最新的配置信息，调用 Listener 的回调方法，将最新的配置信息作为参数传入，这样 Listener 的使用者就能接收到变更后的配置信息，最后更新 ListenerWrap 的 md5 值。

CacheData 的 md5 值又是何时发生改变的呢？

LongPollingRunnable 在执行任务时，当服务端配置信息发生变更时，将最新的 content 数据写入 CacheData 中，同时更新该 CacheData 的 md5 值，所以下次执行 checkListenerMd5 方法时会发现当前 listener 所持有的 md5 值和 CacheData 的 md5 值不一致，也就意味着服务端的配置信息发生了改变，这时需要将最新的数据通知给 Listener 的持有者。

最终，我们发现 Nacos 并不是通过推的方式将服务端最新的配置信息发送给客户端的，而是客户端维护长轮询的任务定时去拉取发生变更的配置信息，然后将最新的数据推送给 Listener 的持有者。

3.5.2 Nacos 服务发现原理分析

服务注册与发现是 Nacos 的一个重要功能。说到这个功能，一定要了解阿里 Dubbo 框架，如图 3-14 所示。

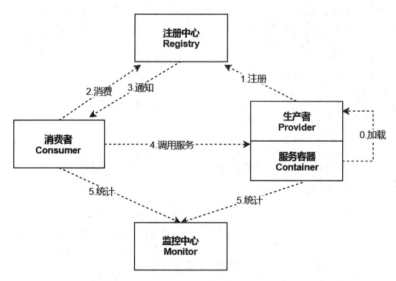

图 3-14 Dubbo 底层框架原理

Registry 是服务注册与发现的注册中心，Provider 是暴露服务的服务提供方，Consumer 是调用远程服务的服务消费方，Monitor 是统计服务的调用次数和调用时间的监控中心，Container 服务运行容器。Dubbo 简单的调用关系如下：

（0）服务容器（Container）负责启动、加载、运行服务提供者（Provider）。

（1）服务提供者（Provider）在启动时向注册中心（Registry）注册自己提供的服务。

（2）服务消费者（Consumer）在启动时向注册中心（Registry）订阅自己所需的服务。

（3）注册中心（Registry）返回服务提供者地址列表给消费者（Consumer），如果有变更，注册中心（Registry）将基于长连接推送变更数据给消费者（Consumer）。

（4）服务消费者（Consumer）从提供者地址列表中基于软负载均衡算法选一台提供者进行调用，如果调用失败，就再选一台调用。

（5）服务消费者（Consumer）和提供者（Provider）在内存中累计调用次数和调用时间，定时每分钟发送一次统计数据到监控中心（Monitor）。

现在我们开始深入了解 Nacos 是如何实现服务注册与发现的。首先在 Nacos 后台管理中单击"服务列表"，查看示例代码，如图 3-15 所示。

图 3-15　服务列表示例代码

示例代码如下所示：

```
/**
 * @author nkorange
 */
public class NamingExample {
    public static void main(String[] args) throws NacosException {
        Properties properties = new Properties();
        properties.setProperty("serverAddr",
        System.getProperty("serverAddr"));
        properties.setProperty("namespace", System.getProperty("namespace"));

        NamingService naming = NamingFactory.createNamingService(properties);
        //通过 registerInstance 注册服务
        naming.registerInstance("service-consumer", "11.11.11.11", 8888, "TEST1");

        naming.registerInstance("service-consumer", "2.2.2.2", 9999,"DEFAULT");
        //通过 getAllInstances 方法拉取所有可用的实例
        System.out.println(naming.getAllInstances("service-consumer"));
        //注销服务注册
        naming.deregisterInstance("service-consumer", "2.2.2.2", 9999,"DEFAULT");
```

```
            System.out.println(naming.getAllInstances("service-consumer"));
            //
            naming.subscribe("service-consumer", new EventListener() {
                @Override
                public void onEvent(Event event) {
                    System.out.println(((NamingEvent)event).getServiceName());
                    System.out.println(((NamingEvent)event).getInstances());
                }
            });
            try{
                //阻止主线程退出
                System.in.read();
            }catch (IOException e){
                e.printStackTrace();
            }
        }
    }
```

通过 NamingService 接口的 registerInstance 方法可以注册服务，注册完成后，通过调用 getAllInstances 方法立即获取所有可用的实例。

通过调用 subscribe 方法，服务的消费者向注册中心订阅某个服务，并提交一个监听器，当注册中心中的服务发生变更时，监听器会收到通知，这时消费者更新本地的服务实例列表，以保证所有的服务均是可用的。当我们运行上面的实例时，onEvent 事件会多次触发，因为 11.11.11.11:8888 和 2.2.2.2:9999 这种有问题的 IP 地址的服务会被下线。

Nacos 客户端获取到服务的完整实例列表后，在客户端通过负载均衡算法（轮询法、随机法）获取可用的实例。

Nacos 客户端进行服务注册由两个部分组成：

- 将服务信息注册到服务端。
- 向服务端发送心跳包。

这两个操作都是通过 NamingProxy 和服务端进行数据交互的。

Nacos 客户端进行服务订阅时也由两部分组成：

- 不断从服务端查询可用服务实例的定时任务。
- 不断从已变服务队列中取出服务并通知 EventListener 持有者的定时任务。

3.6　Eureka 服务发现

3.6.1　Eureka 简介

Eureka 是 Netflix 公司开发的服务发现框架，Spring Cloud 对它提供了支持，将它集成在自己的 spring-cloud-netflix 子项目中。

Eureka 是基于 REST（Representational State Transfer）服务的，主要以 AWS 云服务为支撑，提供服务发现并实现负载均衡和故障转移。Eureka 提供了 Java 客户端组件 Eureka Client，方便与服务端的交互。客户端内置了基于 round-robin 实现的简单负载均衡，在 Netflix 中为 Eureka 提供更为复杂的负载均衡方案进行封装，以实现高可用，包括基于流量、资源利用率以及请求返回状态的加权负载均衡。

Eureka 高级架构如图 3-16 所示。其中，Application Server 表示服务提供方，Application Client 表示服务消费方，Make Remote Call 表示远程调用。服务在 Eureka 上注册，然后每隔 30 秒发送心跳来更新它们的租约。如果客户端不能多次续订租约，就将在大约 90 秒内从服务器注册表中剔除。注册信息和更新被复制到集群中的所有 Eureka 节点。来自任何区域的客户端都可以查找注册表信息（每 30 秒发生一次）来定位它们的服务（可能在任何区域）并进行远程调用。

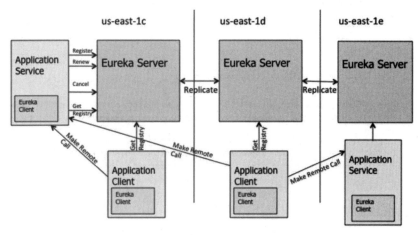

图 3-16　Eureka 高级架构

服务发现有两种模式：一种是客户端发现模式，一种是服务端发现模式。Eureka 采用的是客户端发现模式。Eureka 采用 C-S 的设计架构，如图 3-17 所示。

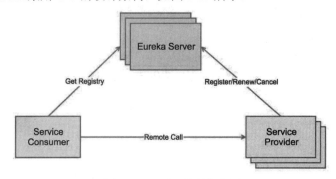

图 3-17　Eureka 简单架构图

- Register（注册）：Eureka 客户端将关于运行实例的信息注册到 Eureka 服务器。注册发生在第一次心跳。
- Renew（更新 / 续借）：Eureka 客户端需要更新最新注册信息（续借），每 30 秒发送一次心跳。更新通知是为了告诉 Eureka 服务器实例仍然存活。如果服务器在 90 秒内没有看到更新,就会将实例从注册表中删除。建议不要更改更新间隔,因为服务器使用该信息来确定客户机与服务器之间的通信是否存在广泛传播的问题。

- Fetch Registry(抓取注册信息):Eureka 客户端从服务器获取注册表信息并在本地缓存。之后,客户端使用这些信息来查找其他服务。通过在上一个获取周期和当前获取周期之间获取增量更新,这些信息会定期更新(每 30 秒更新一次),获取的时候可能返回相同的实例。Eureka 客户端自动处理重复信息。
- Cancel(取消):Eureka 客户端在关机时向 Eureka 服务器发送一个取消请求。这将从服务器的实例注册表中删除实例,从而有效地将实例从流量中取出。

3.6.2 如何看待 Eureka 停产

Eureka 2.0 停产说明如图 3-18 所示。

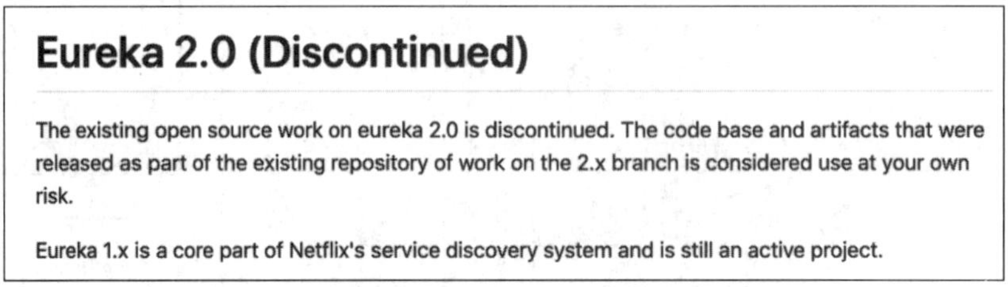

Eureka 2.0 (Discontinued)

The existing open source work on eureka 2.0 is discontinued. The code base and artifacts that were released as part of the existing repository of work on the 2.x branch is considered use at your own risk.

Eureka 1.x is a core part of Netflix's service discovery system and is still an active project.

图 3-18　Eureka 2.0 停产说明

官方只是说 Eureka 2.0 的开发被停止了,如果你将 Eureka 2.0 分支用在生产中,那么后果自负。事实上,Eureka 2.x 从来就没有正式发布过,Eureka 2.0 最终没有孵化出来,但是绝不代表 Eureka 的闭源!官方依然在积极地维护 Eureka 1.x。

国内使用 Eureka 的用户群体还是比较多的,可能考虑到会有团队将 Eureka 2.x 用于线上,甚至基于 2.x 开发,所以官方友情提示一下。大部分用户都是因为 Spring Cloud 才接触到 Eureka,Spring Cloud 使用的是 Eureka 1.x,所以 Eureka 2.x 版本的开源停产对于目前的架构没有多大影响。一般架构师都不会使用一个非正式发布的版本。

Spring Cloud 支持使用 Eureka、Zookeeper、Consul 实现服务发现的能力,即使 Eureka 闭源,也可以从 Eureka 切换成 Zookeeper,只需要改一个依赖、加两行配置就可以了。

3.6.3 搭建 Eureka 注册中心

搭建 Eureka 注册中心的具体步骤如下:

步骤 01 创建 netflix-eureka-server 项目,依赖选择 Eureka Server,如图 3-19 所示。

第 3 章 注册中心/配置管理

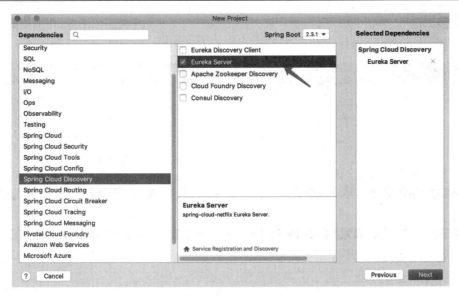

图 3-19 创建 netflix-eureka-server 项目

步骤 02 在启动类上添加 @EnableEurekaServer 注解，启用 Euerka 注册中心功能，具体代码如下：

```
//启用 Euerka 注册中心功能
@EnableEurekaServer
@SpringBootApplication
public class DemoApplication {
    public static void main(String[] args) {
        SpringApplication.run(DemoApplication.class, args);
    }
}
```

步骤 03 在配置文件 application.yml 中添加 Eureka 注册中心的配置，具体代码如下：

```yaml
# 指定运行端口
server:
  port: 8001

# 指定服务名称
spring:
  application:
    name: eureka-server

# 指定主机地址
eureka:
  instance:
    hostname: localhost
  client:
    # 指定是否从注册中心获取服务(注册中心不需要开启)
    fetch-registry: false
    # 指定是否将服务注册到注册中心(注册中心不需要开启)
    register-with-eureka: false
```

步骤04 运行 main 方法启动服务，在浏览器中访问"http://localhost:8001/"便可以看到 Eureka 注册中心的界面，如图 3-20 所示。

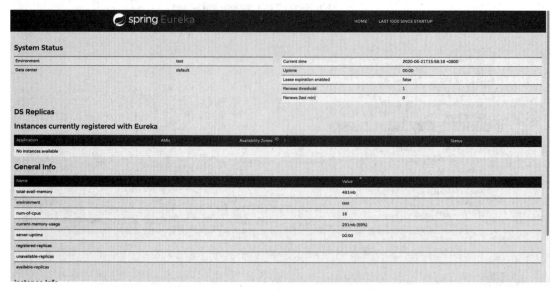

图 3-20　Eureka 注册中心首页

Eureka 注册中心服务端搭建成功后，接下来搭建客户端，具体步骤如下：

步骤01 创建 netflix-eureka-client 项目，依赖选择 Spring Web 以及 Eureka Discovery Client，如图 3-21 和图 3-22 所示。

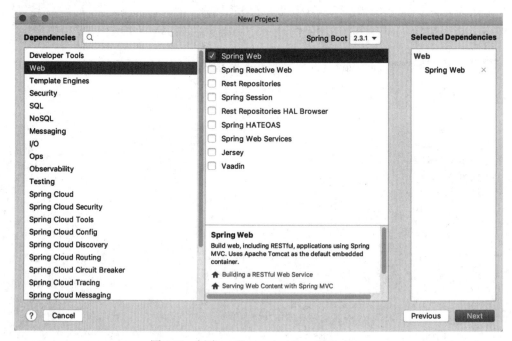

图 3-21　创建 netflix-eureka-client 项目（1）

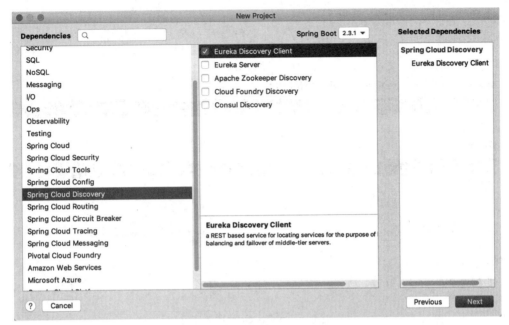

图 3-22　创建 netflix-eureka-client 项目（2）

步骤02 在启动类上添加@EnableDiscoveryClient 注解，表明是一个 Eureka 客户端。

步骤03 在配置文件 application.yml 中添加 Eureka 客户端的配置，具体代码如下：

```yaml
# 指定运行端口
server:
  port: 8101

# 指定服务名称
spring:
  application:
    name: eureka-client

eureka:
  client:
    # 注册到 Eureka 的注册中心
    register-with-eureka: true
    # 获取注册实例列表
    fetch-registry: true
    service-url:
      # 配置注册中心地址
      defaultZone: http://192.168.49.79:8001/eureka
```

步骤04 运行 main 方法，启动 netflix-eureka-client 项目，刷新 http://localhost:8001 页面，即可看到 eureka-client 已经注入 Eureka 服务，如图 3-23 所示。

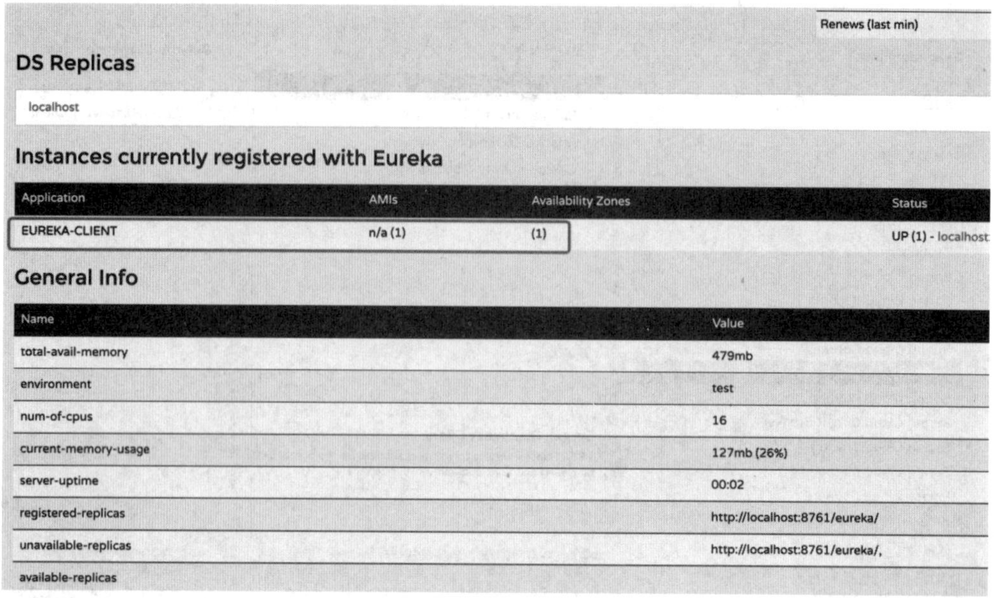

图 3-23　eureka-client 服务注册

至此，我们简单搭建了 Eureka 注册中心。

3.6.4　搭建 Eureka 注册中心集群

由于所有服务都会注册到注册中心，服务之间的调用都是通过从注册中心获取服务列表来调用的，注册中心一旦宕机，所有服务调用都会出现问题，因此需要多个注册中心组成集群来提供服务。下面将搭建一个双节点的注册中心集群，具体步骤如下所示。

步骤 01 在 netflix-eureka-server 项目中添加配置文件 application-replica1.yml，配置第一个注册中心，具体代码如下：

```yml
# 指定运行端口
server:
  port: 8002

# 指定服务名称
spring:
  application:
    name: eureka-server

# 指定主机地址
eureka:
  instance:
    hostname: replica1
  client:
    fetch-registry: true
    register-with-eureka: true
    service-url:
```

```
      # 注册到另一个 Eureka 注册中心
      defaultZone: http://replica2:8003/eureka/
```

步骤02 继续在 netflix-eureka-server 项目中添加配置文件 application-replica1.yml，配置第二个注册中心，具体代码如下：

```
# 指定运行端口
server:
  port: 8003

# 指定服务名称
spring:
  application:
    name: eureka-server

# 指定主机地址
eureka:
  instance:
    hostname: replica2
  client:
    fetch-registry: true
    register-with-eureka: true
    service-url:
      # 注册到另一个 Eureka 注册中心
      defaultZone: http://replica1:8002/eureka/
```

通过两个注册中心互相注册，搭建注册中心的双节点集群。由于 defaultZone 使用了域名，因此需在本机的 host 文件中配置，具体代码如下：

```
# Eureka
127.0.0.1 replica1
127.0.0.1 replica2
```

步骤03 在 IDEA 中，可以通过使用不同的配置文件启动同一个 Spring Boot 应用，从原启动配置中复制一份，修改启动配置文件的 Name 和 Active profiles，如图 3-24 和图 3-25 所示。

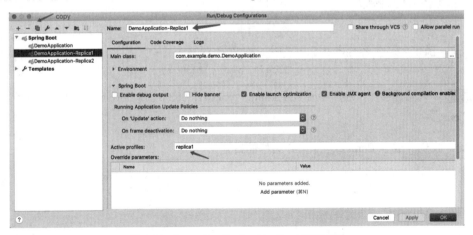

图 3-24　DemoApplication-Replica1 配置文件

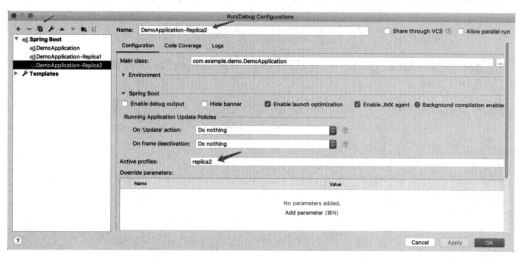

图 3-25 DemoApplication-Replica2 配置文件

步骤 04 启动两个 eureka-server 服务,访问其中一个注册中心 http://replica1:8002/,发现另一个已经成为其备份,修改并启动 netflix-eureka-client 项目,让其连接到集群,netflix-eureka-client 配置修改为:

```yaml
# 指定运行端口
server:
  port: 8101

# 指定服务名称
spring:
  application:
    name: eureka-client

eureka:
  client:
    # 注册到Eureka的注册中心
    register-with-eureka: true
    # 获取注册实例列表
    fetch-registry: true
    service-url:
      # 随便注册到一个注册中心
      defaultZone: http://replica1:8002/eureka
```

步骤 05 访问任意一个注册中心节点(http://replica1:8002 或者 http://replica1:8003)都可以看到 eureka-client 服务。

至此,我们完成了 Eureka 注册中心集群的搭建。

3.7　Spring Cloud Consul

3.7.1　Consul 简介

Consul 是 HashiCorp 公司推出的开源工具，用于实现分布式系统的服务发现与配置。与其他分布式服务注册与发现的方案相比，Consul 的方案更具"一站式"的特点，其可实现以下功能：

①服务注册与发现。
②Key/Value 存储。
③健康检查：支持 HTTP 接口、脚本、TCP 等形式定时任务检测。
④支持多数据中心。
⑤可视化界面。
⑥分布一致性协议实现：raft 算法。

3.7.2　Consul 安装与启动

Spring Cloud Consul 安装非常简单，对于 Mac OS 操作系统，可通过 Homebrew 进行安装，具体操作如下：

```
$ brew install consul
### 以下是命令行窗口打印的信息
Updating Homebrew...
==> Auto-updated Homebrew!
Updated 2 taps (homebrew/cask-versions and homebrew/cask).
### 省略代码
To have launchd start consul now and restart at login:
  brew services start consul
Or, if you don't want/need a background service you can just run:
  ### 可通过如下命令启动 Consul 软件
  consul agent -dev -advertise 127.0.0.1
==> Summary
/usr/local/Cellar/consul/1.6.1: 8 files, 94.0MB
```

执行完 brew install consul 命令后，从命令行打印的信息可知，执行如下命令可启动 Consul 软件：

```
### 启动 Consul 软件
$ consul agent -dev -advertise 127.0.0.1
```

Consul 启动完成后，在浏览器中访问 http://localhost:8500/便可以看到 Consul 首页，如图 3-26 所示。

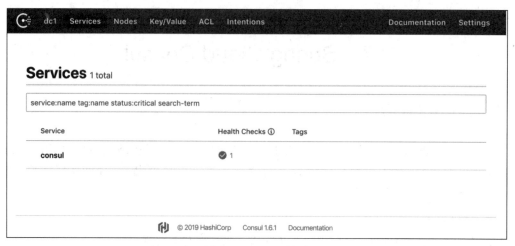

图 3-26　Consul 首页

提　示
其他操作系统的安装步骤可参考 https://learn.hashicorp.com/consul/getting-started/install.html 中的文档。

3.7.3　Consul 服务注册与发现

使用 Spring Cloud Consul 实现服务注册与发现，具体步骤如下：

步骤 01　使用 Intellij IDEA，创建 spring-cloud-consul-producer 项目，具体如图 3-27、图 3-28、图 3-29 所示。

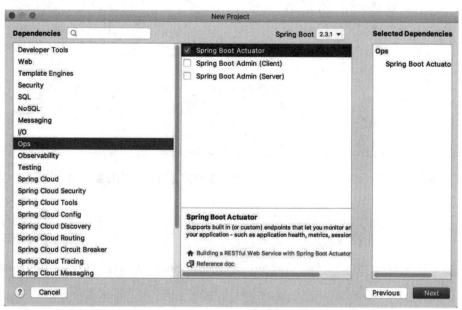

图 3-27　Spring Boot Actuator 依赖

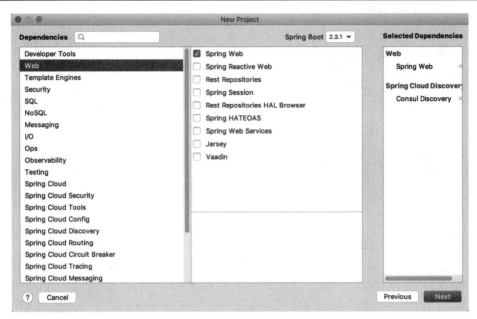

图 3-28　Spring Web 依赖

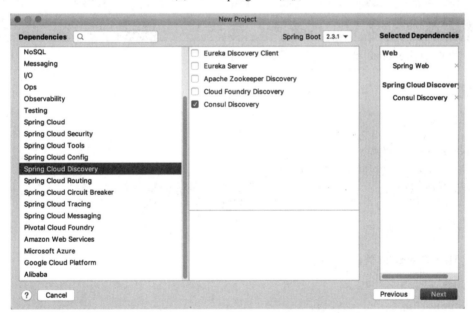

图 3-29　Consul Discovery 依赖

其中，Spring Boot Actuator 是健康检查需要依赖的包。

步骤02 在 application.yml 配置文件中添加如下配置：

```yaml
server:
  port: 18080
spring:
  application:
    name: consul-producer
```

```yaml
    cloud:
      consul:
        host: localhost
        port: 8500
        discovery:
          instance-id: ${spring.application.name}:${server.port}
```

在 resource 目录下创建 application-02.yml 文件，添加如下配置：

```yaml
server:
  ### 端口为 18081
  port: 18081
spring:
  application:
    name: consul-producer
  cloud:
    consul:
      host: localhost
      port: 8500
      discovery:
        instance-id: ${spring.application.name}:${server.port}
```

> **提 示**
>
> 只要保证 service-name 相同，就表示这是同一个服务。

步骤 03 在启动类上添加注解@EnableDiscoveryClient，具体代码如下：

```java
@SpringBootApplication
//Spring Boot 2.0 之后可不用添加
@EnableDiscoveryClient
public class DemoApplication {
    public static void main(String[] args) {
        SpringApplication.run(DemoApplication.class, args);
    }
}
```

开发 HelloController 类，具体代码如下：

```java
/**
 * @author ay
 * @since 2020-06-27
 */
@RestController
@RequestMapping("/test")
public class HelloController {
    @RequestMapping("/hello")
    public String hello(){
        return "hello";
    }
}
```

步骤 04 使用 Intellij IDEA 开发工具，通过一份代码，两份不同配置，启动两个应用，具体可参考 3.6.4 小节内容。在 Consul 控制台中可看到如图 3-30 所示的内容。

图 3-30　将 consul-producer 注册到 consul

接下来创建服务消费者，具体步骤如下：

步骤 01 创建项目 spring-cloud-consul-consumer，参考 spring-cloud-consul-producer 项目创建，步骤基本一样。

步骤 02 修改服务端口为 18082。

```
server:
  ### 端口为18082
  port: 18082
spring:
  application:
    name: consul-consumer
  cloud:
    consul:
      host: localhost
      port: 8500
      discovery:
        instance-id: ${spring.application.name}:${server.port}
```

步骤 03 开发 TestController 类，具体代码如下：

```
/**
 * @author ay
 * @since 2020-06-27
 */
@RestController
public class TestController {
    @Resource
    private LoadBalancerClient loadBalancer;
    @Resource
    private DiscoveryClient discoveryClient;
    //获取所有服务
```

```java
@RequestMapping("/services")
public Object services() {
    return discoveryClient.getInstances("consul-producer");
}
//从所有服务中选择一个服务（轮询）
@RequestMapping("/discover")
public Object discover() {
    return loadBalancer.choose("consul-producer").getUri().toString();
}
//服务调用
@RequestMapping("/call")
public String call() {
    ServiceInstance serviceInstance =
                loadBalancer.choose("consul-producer");
    System.out.println("服务地址: " + serviceInstance.getUri());
    System.out.println("服务名称: " + serviceInstance.getServiceId());
    //使用RestTemplate进行远程调用
    String callServiceResult =
        new RestTemplate().getForObject(serviceInstance.getUri()
            .toString() + "/test/hello", String.class);
    System.out.println(callServiceResult);
    return callServiceResult;
}
```

步骤 04 在浏览器中访问 http://localhost:18082/services，返回 consul-producer 实例信息，具体信息如下：

```
[{"instanceId":"consul-producer-18080","serviceId":"consul-producer","host":"192.168.49.79","port":18080,"secure":false,"metadata":{"secure":"false"},"uri":"http://192.168.49.79:18080","scheme":null},{"instanceId":"consul-producer-18081","serviceId":"consul-producer","host":"192.168.49.79","port":18081,"secure":false,"metadata":{"secure":"false"},"uri":"http://192.168.49.79:18081","scheme":null}]
```

在浏览器中多次访问 http://localhost:18082/discover，将会交替返回下面的信息：

```
http://192.168.49.79:18080
http://192.168.49.79:18081
```

在浏览器中访问 http://localhost:18082/call，页面会返回 hello 字符串。

3.7.4 Consul 配置中心

将 Spring Cloud Consul 作为配置中心，具体步骤如下：

步骤 01 参考 3.7.2 小节的内容，创建 spring-cloud-consul-config 项目。在使用 Intellij IDEA 创建项目的过程中，除了选择 Spring Web、Spring Boot Actuator、Consul Discovery 依赖外，还需要选择 Consul Configuration 依赖，如图 3-31 所示。

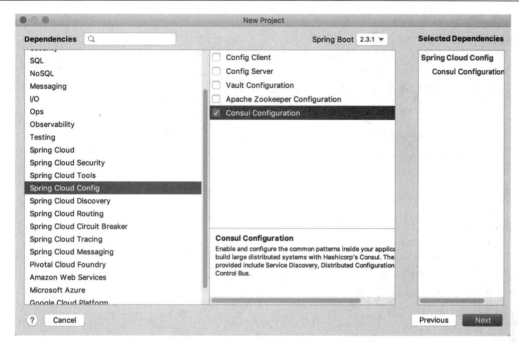

图 3-31 Consul Configuration 依赖

步骤 02 在 application.yml 配置文件中，添加如下配置：

```yaml
spring:
  application:
    name: consul-config
server:
  port: 18084
```

创建 bootstrap.yml 配置文件，具体内容如下：

```yaml
spring:
  cloud:
    consul:
      host: localhost
      port: 8500
      config:
        #false 禁用 Consul 配置，默认为 true
        enabled: true
        # 表示 consul 上面文件的格式，有四种：YAML、PROPERTIES、KEY-VALUE、FILES
        format: YAML
        #表示 consul 上面的 KEY 值(或者说文件的名字) 默认是 data
        defaultContext: consul_config    # 可以理解为 mysql_config 的上级目录
        data-key: data
        # 可以理解为配置文件所在的最外层目录
        prefix: config
```

步骤 03 定义 MySqlComplexConfig 配置类。

```
@Component
```

```
@ConfigurationProperties(prefix = "mysql")
public class MySqlComplexConfig {
    public static class UserInfo{
        private String username;
        private String password;
    }
    private String host;
    private UserInfo user;
    //省略 set、get 方法
}
```

步骤 04 启动类添加@EnableConfigurationProperties 注解。

```
@SpringBootApplication
@EnableDiscoveryClient
@EnableConfigurationProperties({MySqlComplexConfig.class})
public class DemoApplication {
    public static void main(String[] args) {
        SpringApplication.run(DemoApplication.class, args);
    }
}
```

@EnableConfigurationProperties：启用配置属性类，当 Spring Boot 程序启动时会立即加载 @EnableConfigurationProperties 注解中指定的类对象。

@ConfigurationProperties：被标注的类会使用外部文件给 bean 注入属性。

步骤 05 开发 ConfigController 控制层类:

```
/**
 * @author ay
 * @since 2020-06-27
 */
@RestController
public class ConfigController {
    @Resource
    private MySqlComplexConfig mySqlComplexConfig;

    @GetMapping(value = "/getHost")
    public String getMysqlHost(){
        return mySqlComplexConfig.getHost();
    }

    @GetMapping(value = "/getUser")
    public String getMysqlUser(){
        MySqlComplexConfig.UserInfo userInfo = mySqlComplexConfig.getUser();
        return userInfo.toString();
    }
}
```

步骤 06 在浏览器中分别请求 http://localhost:18084/getUser 和 http://localhost:18084/getHost，服务会从 Consul 中获取配置，并返回字符串 UserInfo{username='ay', password='123456'}和 localhost。

3.7.5　Consul 简单架构

Consul 官方架构如图 3-32 所示。

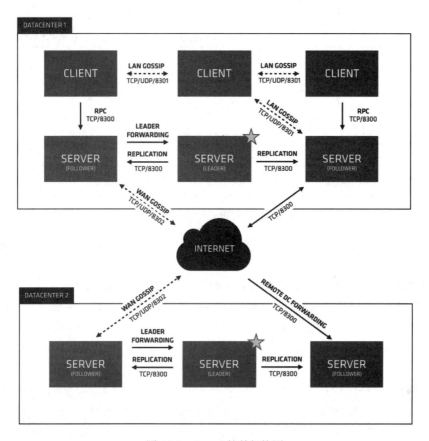

图 3-32　Consul 简单架构图

Consul 支持多数据中心，在图 3-32 中有两个数据中心，数据中心之间通过 Internet 互联，为了提高通信效率，只有 Server 节点才能加入跨数据中心的通信。

在单个数据中心中，Consul 分为 Client 和 Server 两种节点（所有的节点被称为 Agent）。Server 节点保存数据，推荐数量是 3 个或者 5 个；Client 节点负责健康检查及转发数据请求到 Server。

Server 节点包含一个 Leader 和多个 Follower，Leader 节点会将数据同步到 Follower，在 Leader 挂掉的时候会启动选举机制产生一个新的 Leader。

集群内的 Consul 节点通过 gossip 协议（流言协议）维护成员关系，也就是说某个节点了解集群内现在还有哪些节点，这些节点是 Client 还是 Server。单个数据中心的流言协议同时使用 TCP 和 UDP 通信，并且都使用 8301 端口。跨数据中心的流言协议也同时使用 TCP 和 UDP 通信，端口使用 8302。集群内数据的读写请求既可以直接发到 Server，也可以通过 Client 使用 RPC 转发到 Server，请求最终会到达 Leader 节点，集群内数据的读写和复制都是通过 TCP 的 8300 端口完成的。

下面简单描述服务发现的完整流程，具体如图 3-33 所示。

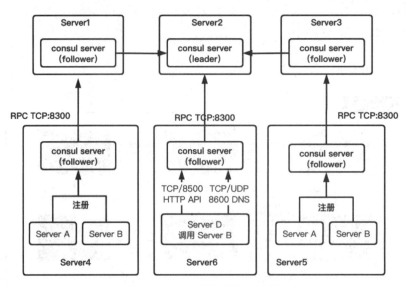

图 3-33　Consul 服务发现完整流程

服务器 Server1、Server2、Server3 部署 Consul Server 集群，Server2 上的 Consul Server 节点为 Leader。

Server4 和 Server5 上通过 Consul Client 分别注册 Service A、B 应用服务，每个应用服务分别部署在两个服务器上，这样可以避免应用服务单点问题。服务注册到 Consul 既可以通过 HTTP API（8500 端口）的方式，也可以通过 Consul 配置文件的方式。Consul Client 是无状态的，它将注册信息通过 RPC 转发到 Consul Server，服务信息保存在 Server 的各个节点中，并且通过 Raft 实现了强一致性。

假设服务器 Server6 中的 Server D 需要访问 Service B，就先访问本机 Consul Client 提供的 HTTP API，本机 Client 会将请求转发到 Consul Server。Consul Server 查询到 Service B 当前的信息，最终 Server D 拿到 Service B 所有部署的 IP 和端口，选择 Service B 中的一个应用服务并向其发起请求。

如果服务发现采用的是 DNS 方式，则 Service D 中直接使用 Service B 的服务发现域名，域名解析请求首先到达本机 DNS 代理，然后转发到本机 Consul Client，本机 Client 会将请求转发到 Consul Server。

Consul Server 查询到 Service B 当前的信息返回，最终 Server D 拿到 Service B 某个部署的 IP 和端口。

3.8　Spring Cloud Config

3.8.1　Spring Cloud Config 简介

Spring Cloud Config 项目是一个解决分布式系统的配置管理方案。它包含 Client 和 Server 两部分，Server 提供配置文件的存储，以接口的形式将配置文件的内容提供出去；Client 通过接口获取数据，并依据此数据初始化自己的应用。Spring Cloud Config 使用 Git 或 SVN 存放配置文件，默认情况下使用 Git。

Spring Cloud Config 支持以下功能：

- 提供服务端和客户端支持。
- 集中管理各环境的配置文件。
- 配置文件修改之后，可以快速生效。
- 可以进行版本管理。
- 支持大的并发查询。
- 支持各种语言。

3.8.2 Spring Cloud Config 快速入门

本节介绍如何构建一个基于 Git 存储的分布式配置中心，具体步骤如下：

1. 服务端创建

步骤 01 在 GitHub 上创建 config-repo 项目（https://github.com/huangwenyi10/config-repo），同时创建开发环境、预发环境以及正式环境的 3 个配置文件（ay-dev.properties、ay-pre.properties、ay-release.properties），并分别在 3 个配置文件中添加如下配置信息：

```
### 在 ay-dev.properties 配置文件中添加如下内容
config.name=dev_ay
### 在 ay-pre.properties 配置文件中添加如下内容
config.name=pre_ay
### 在 ay-release.properties 配置文件中添加如下内容
config.name=release_ay
```

步骤 02 使用 Intellij IDEA 创建 Spring Boot 项目，项目名为 spring-cloud-config-server，除了勾选 Spring Web 依赖外，还需要勾选 Config Server 依赖，具体如图 3-34 所示。

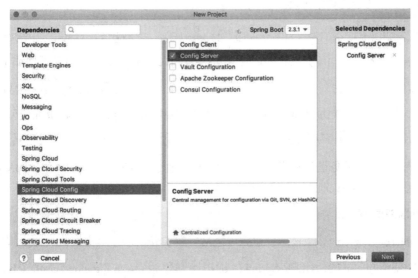

图 3-34　Config Server 依赖

步骤03 在启动类上添加@EnableConfigServer注解，开启Spring Cloud Config的服务端功能。

```
@SpringBootApplication
@EnableConfigServer
public class DemoApplication {
    public static void main(String[] args) {
        SpringApplication.run(DemoApplication.class, args);
    }
}
```

步骤04 在application.properties中添加配置服务的基本信息以及Git仓库的相关信息，具体代码如下：

```
###服务名
spring.application.name=config-server
### 端口
server.port=8888
#配置git仓库地址
spring.cloud.config.server.git.uri=
https://github.com/huangwenyi10/config-repo.git
#主分支
spring.cloud.config.label=master
#访问git仓库的用户名，config-repo是public，无须用户名
spring.cloud.config.server.git.username=
#访问git仓库的用户密码，config-repo是public，无须密码
spring.cloud.config.server.git.password=
```

在application.properties配置文件中配置了服务器名、服务端口和最重要的一个配置：spring.cloud.config.server.git.uri，它指向Git仓库地址。config-server会从该地址中查找配置文件，如果Git仓库访问权限是公有的，就无须配置用户名和密码。

步骤05 启动config-server服务，在浏览器中输入访问地址"http://localhost:8888/ay-dev.properties"，页面返回如下内容：

```
config.name: dev_ay
```

在浏览器中输入访问地址"http://localhost:8888/ay-pre.properties"，页面返回如下内容：

```
config.name: pre_ay
```

在浏览器中输入访问地址"http://localhost:8888/ay-release.properties"，页面返回如下内容：

```
config.name: release_ay
```

通过该请求，配置服务已经从指定的Git仓库中找到了ay-dev.properties和ay-pre.properties配置文件。至此，config-server服务创建成功。

2. 客户端创建

配置服务的服务端config-server创建完成之后，我们开始创建客户端，具体步骤如下：

步骤01 使用Intellij IDEA创建Spring Boot项目，项目名为spring-cloud-config-client，除了勾选Spring Web依赖外，还需要勾选Config Client依赖，具体如图3-35所示。

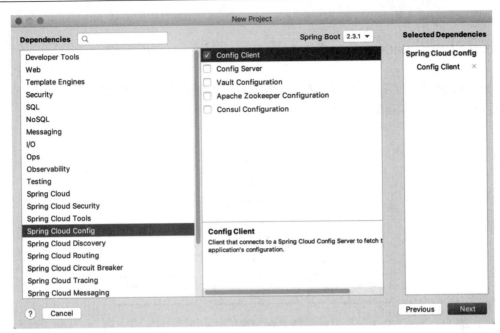

图 3-35　Config Client 依赖

步骤 02　在配置文件 application.properties 中添加如下配置：

```
spring.application.name=config-client
```

步骤 03　在项目的 resources 目录下创建配置文件 bootstrap.properties，具体内容如下：

```
#指定使用 git 主分支
spring.cloud.config.label=master
#指定 dev 开发环境配置文件
spring.cloud.config.profile=dev
#指定配置文件名字 ay，此名字对应 git 中的配置文件名字
spring.cloud.config.name=ay
#指定配置服务地址
spring.cloud.config.uri=http://localhost:8888/
```

在 resources 目录下创建了两个配置文件：application.properties（默认的）和 bootstrap.properties。这两个文件都可以用于 Spring Boot 的配置，只是 bootstrap.properties 中的配置优先级高于前者，而 bootstrap.properties 一般用来加载外部配置。

步骤 04　开发 HelloController 类，该类会读取 config.name 配置（在 Git 中配置过）。

步骤 05　依次启动 config-server 和 config-client 服务，在浏览器中输入访问地址 "http://localhost:8080/hello" 便可以看到 dev_ay 信息，说明确实从 Git 中获取了配置参数。修改 bootstrap.properties 配置文件中的 spring.cloud.config.profile=pre 即可完成环境的切换，使用 ay-pre.properties 作为配置。

至此，配置服务的客户端开发完成。

3.8.3 Spring Cloud Config 配置中心原理

上一小节演示了如何使用 Spring Cloud Config 搭建配置中心、获取远程的配置（基于 Git），获取远程配置的具体原理如图 3-36 所示。

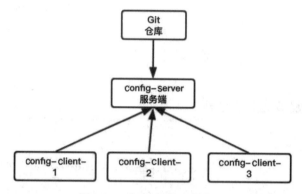

图 3-36 基于 Git 的配置中心

这样仍然不够，因为还不能做到自动刷新配置文件。例如，在 Git 上更改了配置文件，还需要重启服务才能够读取最新的配置，因此有了图 3-37 所示的解决方案。

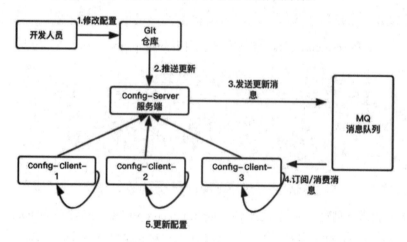

图 3-37 自动更新配置简单原理

Spring Cloud Bus 会对外提供 HTTP 接口，将这个接口配置到远程 Git 上。当开发人员修改项目配置、Git 上的文件内容发生变动时会触发 WebHook。WebHook 会自动调用 HTTP 接口通知 Config-Server。Config-Server 会发布更新消息到消息队列 Spring Cloud Bus 中，其他客户端服务订阅到该消息，向 Config-Server 请求配置，Config-Server 拉取配置仓库中的最新配置文件并转成相应的 json 格式回传给 Config-Client，随后 Config-Client 将内容更新到上下文中。

Spring Cloud Bus 相关的内容将会在后面章节详细描述。

第 4 章

微服务网关

本章首先介绍 Zuul 网关、快速搭建 Zuul 网关、Zuul 网关路由配置/过滤器/管理端点；接着介绍 Spring Cloud Gateway 相关内容，包括如何快速入门 Spring Cloud Gateway、Gateway 路由断言工厂、全局过滤器、跨域、HTTP 超时配置、TLS/SSL 配置、Gateway 底层原理等；最后对比 Gateway 和 Zuul 网关区别。

4.1 Zuul 网关

4.1.1 Zuul 概述

Spring Cloud Zuul 是 Spring Cloud Netflix 子项目的核心组件之一，可以作为微服务架构中的 API 网关使用，支持动态路由与过滤功能等。Netflix 主要将 Zuul 用于以下用途：

- 鉴权：对于访问每个服务的请求进行鉴权，拒绝鉴权失败的请求。
- 监控：可以对整个系统的请求进行监控，记录详细的请求响应日志，可以实时统计出当前系统的访问量以及监控状态。
- 压力测试：帮助对整个集群进行可控的压力测试。
- 灰度测试：灰度发布可以保证整体系统的稳定，在初始灰度的时候就可以发现、调整问题，以保证其影响度。
- 动态路由：基于请求路径，将请求可控的分发到指定的客户端。
- 负载控制：统一控制各客户端请求压力，超过压力的请求直接拒绝。
- 静态响应处理：在边缘位置直接建立部分响应，从而避免其流入内部集群。

4.1.2 Zuul 快速入门

本节介绍如何通过 Zuul 快速构建网关，具体步骤如下：

步骤01 使用 Intellij IDEA 快速创建 Spring Boot 项目，项目名称为 spring-cloud-zuul-service，同时勾选 Zuul 依赖，如图 4-1 所示。

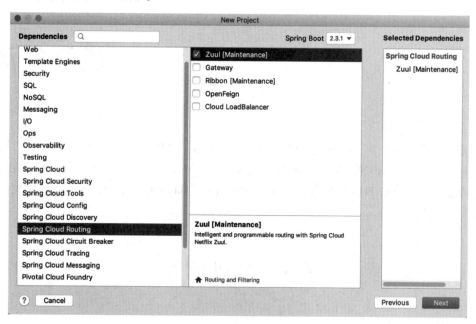

图 4-1 勾选 Zuul 依赖

勾选 Zuul 依赖相当于在 pom.xml 文件中添加如下配置：

```xml
<dependency>
    <groupId>org.springframework.cloud</groupId>
    <artifactId>spring-cloud-starter-netflix-zuul</artifactId>
</dependency>
```

步骤02 在 application.yml 配置文件中添加如下配置：

```yaml
spring:
  application:
    name: zuul-service

server:
  port: 9080

zuul:
  routes:
    blog:
      path: /baidu/**
      ### url用于配置符合path的请求路径路由到的服务地址
```

```
    url: https://www.baidu.com
```

步骤 03 在启动类中添加@EnableZuulProxy 注解，具体代码如下：

```
@SpringBootApplication
//开启 zuul 网关
@EnableZuulProxy
public class DemoApplication {
    public static void main(String[] args) {
        SpringApplication.run(DemoApplication.class, args);
    }
}
```

Spring Cloud Netflix 安装了许多过滤器，具体取决于用于启用 Zuul 的注解。@EnableZuulProxy 是@EnableZuulServer 的超集。换句话说，@ EnableZuulProxy 包含@EnableZuulServer 安装的所有过滤器。@EnableZuulServer 安装的过滤器在后面章节会讲。

步骤 04 启动项目，在浏览器中输入访问地址"http://localhost:9080/baidu"，发现请求被路由到百度界面，Zuul 服务搭建成功。

4.1.3 Zuul 路由配置

上一节，我们使用路径的方式匹配路由规则，key 的结构如下：

```
### 其中 customName 为用户自定义名称
zuul:
  routes:
    customName:
      path: xxx
```

可使用的通配符有以下几种：

- ?：单个字符。
- *：任意多个字符，不包含多级路径。
- **：任意多个字符，包含多级路径。

除了使用以上的 URL 路径匹配，还可以使用服务名称匹配，例如：

```
zuul:
  routes:
    ### users 为用户自定义名称
    users:
      path: /myusers/**
      ### serviceId 用于配置符合 path 的请求路径路由到的服务名称
      serviceId: users_service
```

服务名称匹配也可以使用简化的配置：

```
zuul:
  routes:
    service-provider:
```

```
      path: /**
```

如果只配置 path 不配置 serviceId，则 customName 相当于服务名称，即 service-provider 会被当作服务名称。

如果想排查配置，可以使用 ignored-services，例如：

```
zuul:
  ignoredServices: '*'
  routes:
    users: /myusers/**
```

ignored-services 可以配置不被 zuul 管理的服务列表，多个服务名称使用逗号分隔，配置的服务将不被 Zuul 代理。在上面的实例中，除了用户服务外，所有的服务均被忽略。

可以通过 zuul.prefix 配置路由前缀，例如：

```
zuul:
  routes:
    users:
      path:/myusers/**
  prefix: /api
```

配置请求路径前缀，所有基于此前缀的请求都由 Zuul 网关提供代理。

4.1.4　Zuul 过滤器

Zuul 定义过滤器用来过滤代理请求，提供额外功能逻辑，如权限验证、日志记录等。Zuul 提供的过滤器父类是 ZuulFilter，通过父类中定义的抽象方法 filterType 来决定当前的 Filter 种类是什么。

Zuul 内部有一套完整的机制，可以动态读取编译运行 filter 机制，filter 与 filter 之间不直接通信，在请求线程中会通过 RequestContext 来共享状态，它内部是用 ThreadLocal 实现的，例如 HttpServletRequest、HttpServletResponse、异常信息等。

Zuul 过滤器分为前置过滤、路由后过滤、后置过滤以及异常过滤。

（1）前置过滤：在请求进入 Zuul 后，立刻执行的过滤逻辑。

（2）路由后过滤：在请求进入 Zuul 后，Zuul 实现请求路由，并在远程服务调用之前过滤的逻辑。

（3）后置过滤：远程服务调用结束后执行的过滤逻辑。

（4）异常过滤：任意一个过滤器发生异常或远程服务调用无结果反馈时执行的过滤逻辑。无结果反馈是指远程服务调用超时。

Zuul 提供的四种过滤器之间的关系如图 4-2 所示。

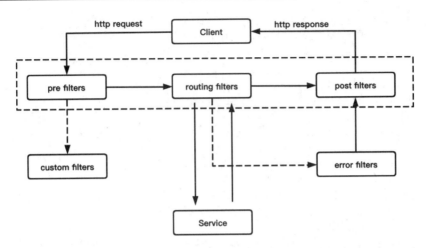

图 4-2 四种过滤器之间的关系

ZuulFilter 类及其父类 IZuulFilter 共提供了四种抽象方法，分别是 filterType、filterOrder、shouldFilter 以及 run 方法。

（1）filterType：返回字符串数据，代表当前过滤器的类型，可选值有 pre、route、post 以及 error。

- pre：前置过滤器，在请求被路由前执行，通常用于处理身份认证、日志记录等。
- route：在路由执行后、服务调用前被调用。
- error：任意一个 filter 发生异常或远程服务调用没有反馈时执行（超时），通常用于处理异常。
- post：在 route 或 error 执行后被调用，一般用于收集服务信息、统计服务性能指标等，也可以对 response 结果做特殊处理。

（2）filterOrder：返回 int 数据，用于为同一种 filterType 的多个过滤器定制执行顺序，返回值越小，执行顺序越优先。

（3）shouldFilter：返回 boolean 数据，代表当前 filter 是否生效。

（4）run：具体的过滤执行逻辑。例如，pre 类型的过滤器可以通过对请求的验证来决定是否将请求路由到服务上，post 类型的过滤器可以对服务响应结果做加工处理（如为每个响应增加 footer 数据）。

接下来，我们看一个具体实例。

```
/**
 * 自定义过滤器
 *
 * @author ay
 * @since 2020-07-03
 */
@Component
public class TokenFilter extends ZuulFilter {
```

```java
    public boolean shouldFilter() {
        //判断过滤器是否生效,true 代表生效
        return true;
    }
    //具体的过滤执行逻辑
    public Object run() throws ZuulException {
        //获取上下文
        RequestContext currentContext = RequestContext.getCurrentContext();
        //获取 request 对象
        HttpServletRequest request = currentContext.getRequest();
        //从请求头获取 token 的参数
        String userToken = request.getParameter("token");
        if (StringUtils.isEmpty(userToken)) {
            //返回错误提示
            //false:表示不会继续往下执行,不会调用服务接口,直接响应给客户
            currentContext.setSendZuulResponse(false);
            currentContext.setResponseBody("token is null!!!");
            currentContext.setResponseStatusCode(401);
            return null;
        }
        //否则正常执行,调用服务接口...
        return null;
    }

    @Override
    public String filterType() {
        //定义过滤器的类型,pre 表示在请求被路由前执行
        return "pre";
    }

    @Override
    public int filterOrder() {
        //返回 int 数据,用于为同一种 filterType 的多个过滤器定制执行顺序,
        //返回值越小,执行顺序越优先
        return 0;
    }
}
```

TokenFilter 类继承 ZuulFilter 类,实现 ZuulFilter 类的抽象方法,在 run 方法中开发相关的业务逻辑,当请求中没有携带 token 参数时返回 401 状态码,并提示 "token is null!!!" 信息。

4.1.5 管理端点

默认情况下,如果将@EnableZuulProxy 与 Spring Boot Actuator 结合使用,则会启用两个附加端点:Routes 和 Filters。

提示,zuul 服务需要添加 spring-boot-starter-actuator 依赖:

```xml
<dependency>
    <groupId>org.springframework.boot</groupId>
```

```
        <artifactId>spring-boot-starter-actuator</artifactId>
</dependency>
```

application.properties 配置文件需要添加如下配置：

```
management:
  endpoint:
    health:
      show-details: always
  endpoints:
    web:
      exposure:
        include: "*"
```

请求 http://ip:port/actuator/routes 地址，返回如下自定义的路由信息：

```
{
    "/baidu/**":"https://www.baidu.com"
}
```

可以通过向 /routes 添加 ?format = details 来请求路由详细信息。

请求 http://localhost:9080/actuator/filters 地址，返回 Zuul 过滤器的列表以及它们的详细信息。

4.1.6 禁用 Zuul 过滤器

有关 Spring Cloud Zuul 可启用的过滤器列表，请参见 Zuul 过滤器包。如果要禁用某个过滤器，请设置 zuul.<SimpleClassName>.<filterType>.disable = true。按照惯例，过滤器所在的包是 Zuul 过滤器类型。例如，要禁用 org.springframework.cloud.netflix.zuul.filters.post.SendResponseFilter，请设置 zuul.SendResponseFilter.post.disable = true。

4.1.7 启用 Zuul 跨域请求

默认情况下，Zuul 将所有跨域请求（CORS）路由到服务。如果希望 Zuul 处理这些请求，则可以通过提供自定义的 WebMvcConfigurer 对象来完成：

```
@Bean
public WebMvcConfigurer corsConfigurer() {
    return new WebMvcConfigurer() {
        public void addCorsMappings(CorsRegistry registry) {
            registry.addMapping("/path-1/**")
                    .allowedOrigins("https://allowed-origin.com")
                    .allowedMethods("GET", "POST");
        }
    };
}
```

在上面的示例中，允许 allowed-origin.com 的 GET 和 POST 方法将跨域请求发送到以 /path-1 开头的端点。可以使用 /** 映射将 CORS 配置应用于特定的路径模式或整个应用于可以通过此配置来自定义 allowedOrigins、allowedMethods、allowedHeaders、exposedHeaders、allowCredentials 和 maxAge 等属性。

4.1.8 Eureka 整合 Zuul

本小节主要讲解 Eureka 集成 Zuul 网关，实现简单的服务注册、服务路由以及简单的服务调用，具体步骤如下：

步骤01 参考 3.6.3 小节的内容，使用 Intellij IDEA 搭建 Eureka 注册中心，项目名称为 netflix-eureka-server，服务端口为 8001。

步骤02 参考 4.1.2 小节的内容，使用 Intellij IDEA 搭建 Zuul 网关，项目名称为 spring-cloud-zuul-service-eureka，在创建过程中需要勾选 Eureka Discovery Client 和 Spring Cloud Routing 依赖，如图 4-3 和图 4-4 所示。

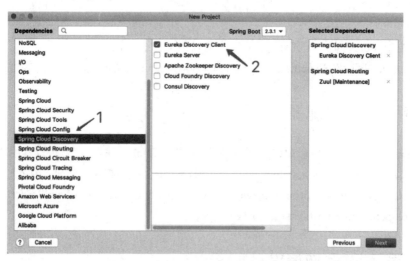

图 4-3 勾选 Eureka Discovery Client 依赖

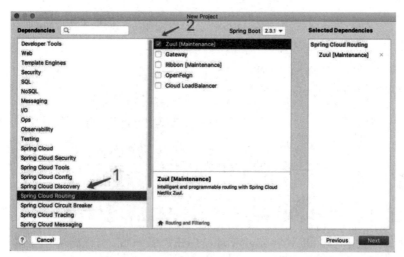

图 4-4 勾选 Spring Cloud Routing 依赖

在入口类中添加@EnableDiscoveryClient 和@EnableZuulProxy 注解，具体代码如下：

```
@SpringBootApplication
@EnableDiscoveryClient
@EnableZuulProxy
public class DemoApplication {
    public static void main(String[] args) {
        SpringApplication.run(DemoApplication.class, args);
    }
}
```

在 application.yml 配置文件中添加如下配置：

```yaml
# 指定运行端口
server:
  port: 18101

# 指定服务名称
spring:
  application:
    name: zuul-service

eureka:
  client:
    # 注册到 Eureka 的注册中心
    register-with-eureka: true
    # 获取注册实例列表
    fetch-registry: true
    service-url:
      # 配置注册中心地址
      defaultZone: http://localhost:8001/eureka
zuul:
  routes:
    service-provider:
      path: /provider/hello
      serviceId: service-provider
    service-consumer:
      path: /consumer/invoke
      serviceId: service-consumer

### ribbon 后续的章节会讲
ribbon:
  eureka:
    enabled: true
```

步骤 03 使用 Intellij IDEA 创建 Spring Boot 微服务，作为服务提供者，项目名称为 spring-cloud-zuul-eureka-provider，同时勾选 Eureka Discovery Client 依赖。如果想自己手动添加依赖包，可以在 pom.xml 文件中添加如下依赖：

```xml
<!-- web 依赖包 -->
<dependency>
```

```xml
        <groupId>org.springframework.boot</groupId>
        <artifactId>spring-boot-starter-web</artifactId>
</dependency>
<!-- eureka 依赖包 -->
<dependency>
        <groupId>org.springframework.cloud</groupId>
        <artifactId>spring-cloud-starter-netflix-eureka-client</artifactId>
</dependency>
```

在 spring-cloud-zuul-eureka-provider 项目的 application.yml 配置文件中添加如下配置：

```yaml
# 指定运行端口
server:
  port: 18102

# 指定服务名称
spring:
  application:
    name: service-provider

eureka:
  client:
    # 注册到 Eureka 的注册中心
    register-with-eureka: true
    # 获取注册实例列表
    fetch-registry: true
    service-url:
      # 配置注册中心地址
      defaultZone: http://localhost:8001/eureka
```

在 spring-cloud-zuul-eureka-provider 项目中开发控制层类 HelloController，具体代码如下：

```java
@RestController
@RequestMapping("provider")
public class HelloController {

    @RequestMapping("/hello")
    public String sayHello(){
        return "Hello, ay";
    }
}
```

在 HelloController 类中只定义简单的 sayHello 方法，供其他服务调用。

步骤 04 使用 Intellij IDEA 创建 Spring Boot 微服务，作为服务消费者，项目名称为 spring-cloud-zuul-eureka-consumer，同时勾选 Eureka Discovery Client 依赖，并在 spring-cloud-zuul-eureka-provider 项目的 application.yml 配置文件中添加如下配置：

```yaml
# 指定运行端口
server:
  port: 18103
```

```yaml
# 指定服务名称
spring:
  application:
    name: service-consumer

eureka:
  client:
    # 注册到 Eureka 的注册中心
    register-with-eureka: true
    # 获取注册实例列表
    fetch-registry: true
    service-url:
      # 配置注册中心地址
      defaultZone: http://localhost:8001/eureka
```

在启动类中，通过@Bean注解注入RestTemplate实体类，通过@LoadBalanced注解让RestTemplate具备负载均衡能力，具体代码如下：

```java
@SpringBootApplication
@EnableEurekaClient
public class DemoApplication {
    public static void main(String[] args) {
        SpringApplication.run(DemoApplication.class, args);
    }
    @Bean
    @LoadBalanced
    RestTemplate restTemplate(){
        return new RestTemplate();
    }
}
```

开发 InvokeController 控制层类，并在 InvokeController 控制层类中通过 LoadBalancerClient 获取服务提供者 service-provider、通过 RestTemplate 类向服务提供者发起请求，具体代码如下：

```java
@RestController
public class InvokeController {

    @Resource
    private RestTemplate restTemplate;
    @Resource
    private LoadBalancerClient loadBalancer;

    @RequestMapping("/consumer/invoke")
    public String invoke(){
        ServiceInstance serviceInstance =
                loadBalancer.choose("service-provider");
        System.out.println("服务地址: " + serviceInstance.getUri());
        System.out.println("服务名称: " + serviceInstance.getServiceId());
        //使用 RestTemplate 进行远程调用
        String callServiceResult = restTemplate.getForObject("http://"+
```

```
serviceInstance.getServiceId()
                .toString() + "/provider/hello", String.class);
        System.out.println(callServiceResult);
        return callServiceResult;
    }
}
```

步骤05 分别启动 Eureka 注册中心、Zuul 网关、服务提供者 service-provider、服务消费者 service-consumer，访问 http://localhost:8001/可看到服务都注入 Eureka 中，具体如图 4-5 所示。

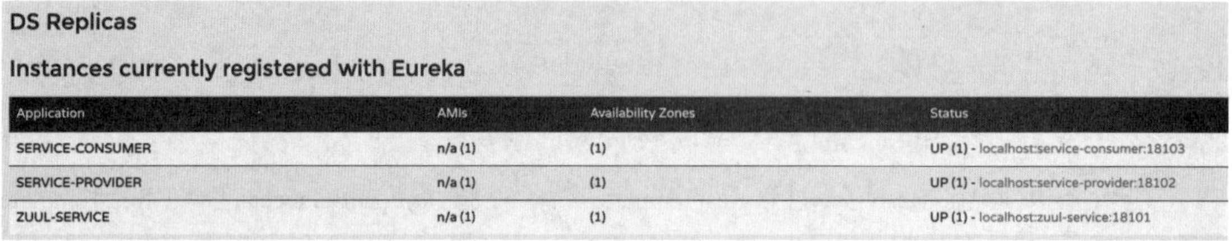

Application	AMIs	Availability Zones	Status
SERVICE-CONSUMER	n/a (1)	(1)	UP (1) - localhost:service-consumer:18103
SERVICE-PROVIDER	n/a (1)	(1)	UP (1) - localhost:service-provider:18102
ZUUL-SERVICE	n/a (1)	(1)	UP (1) - localhost:zuul-service:18101

图 4-5　Eureka 服务注册

在浏览器中访问 http://localhost:18101/consumer/invoke。其中，18101 是 Zuul 网关的端口，Zuul 网关将请求路由到服务消费者 service-consumer，请求服务消费者 service-consumer 的 invoke()方法，在 invoke 方法中通过 RestTemplate.getForObject 调用服务提供者的 sayHello()方法，sayHello()方法最终返回"Hello，ay"字符串到浏览器，请求结束。

4.2　Spring Cloud Gateway

4.2.1　Gateway 简介

Spring Cloud Gateway 基于 Spring 5、Spring Boot 2 和 Project Reactor 等技术，是在 Spring 生态系统之上构建的 API 网关服务。Gateway 旨在提供一种简单而有效的方式来对 API 进行路由（Route），以及提供一些强大的过滤器功能，例如熔断、限流、重试等。

Spring Cloud Gateway 具有如下特性：

- 基于 Spring Framework 5、Project Reactor 以及 Spring Boot 2.0 进行构建。
- 能够匹配任何请求属性。
- 可以对路由指定 Predicate（断言）和 Filter（过滤器）。
- 集成 Hystrix 的断路器功能。
- 集成 Spring Cloud 服务发现功能。
- 易于编写的 Predicate 和 Filter。
- 请求限流功能。
- 路径重写。

其中涉及几个关键术语，具体如下：

- Route（路由）：构建网关的基本模块，由 ID、目标 URI、一系列的断言和过滤器组成，如果断言为 true 就匹配该路由。
- Predicate（断言）：Java 8 的 Function Predicate，输入类型是 Spring 框架中的 ServerWebExchange。这使开发人员可以匹配 HTTP 请求中的所有内容，例如请求头或请求参数。如果请求与断言相匹配，就进行路由。
- Filter（过滤器）：Spring 框架中的 GatewayFilter 实例通过使用过滤器可以在路由前后对请求进行修改。

4.2.2 Gateway 快速入门

本小节介绍如何创建 Spring Cloud Gateway 网关，具体步骤如下：

步骤 01 使用 Intellij IDEA 快速创建 Spring Boot 项目，项目名称为 spring-cloud-gateway-01，同时需要勾选 Gateway 依赖，具体如图 4-6 所示。

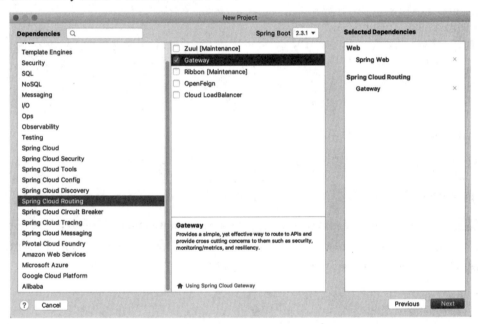

图 4-6　勾选 Spring Cloud Gateway 依赖

步骤 02 在 application.yml 配置文件中添加如下配置：

```
server:
  port: 9201
### 自定义配置属性
service-url:
  user-service: http://localhost:8201
spring:
  cloud:
    gateway:
      routes:
```

```
### 路由 ID
- id: path_route
  ### 匹配后路由地址
  uri: ${service-url.user-service}/user/{id}
  ### 断言，路径相匹配的进行路由
  predicates:
    - Path=/user/{id}
```

Gateway 提供两种不同的方式用于配置路由：一种是通过 yml 文件来配置；另一种是通过 Java Bean 来配置。

步骤 03 使用 Intellij IDEA 快速创建用户服务，项目名称为 spring-cloud-web-user。

步骤 04 在 spring-cloud-web-user 项目中创建 UserController 类，具体代码如下：

```java
/**
 * @author ay
 * @since 2020-06-27
 */
@RestController
public class UserController {
    @GetMapping("/user/{id}")
    public String getUser(@PathParam("id") Integer id){
        return "ay:" + id;
    }
}
```

在 application.yml 配置文件中添加如下配置：

```
### 指定服务端口
server:
  port: 8201
```

步骤 05 启动 spring-cloud-gateway-01 网关和 spring-cloud-web-user 用户服务，在浏览器中输入访问地址"http://localhost:9201/user/1"，发现请求被路由到用户服务的 http://localhost:8201/user/1。

步骤 06 在 spring-cloud-gateway-01 项目下创建网关配置类 GatewayConfig，具体代码如下：

```java
/**
 * Java Bean 配置
 * @author ay
 * @since 2020-06-27
 */
@Configuration
public class GatewayConfig {

    @Bean
    public RouteLocator customRouteLocator(RouteLocatorBuilder builder) {
        return builder.routes()
                .route("path_route2", r -> r.path("/user2/getUser")
                        .uri("http://localhost:8201/user2/getUser"))
                .build();
    }
}
```

步骤 07 在 spring-cloud-web-user 项目的 UserController 类下继续添加 getUser2 方法,代码如下:

```
@RestController
public class UserController {

    @GetMapping("/user2/getUser")
    public String getUser2(){
        return "ay";
    }
}
```

步骤 08 启动 spring-cloud-gateway-01 网关和 spring-cloud-web-user 用户服务,在浏览器中输入访问地址 " http://localhost:9201/user2/getUser ",发现请求被路由到用户服务的 http://localhost:8201/user2/getUser 上。

4.2.3 Gateway 路由断言工厂

Spring Cloud Gateway 包括许多内置的路由断言(Route Predicate)工厂,所有这些 Predicate 都与 HTTP 请求的不同属性匹配。多个 Route Predicate 工厂可以进行组合,下面我们来介绍一些常用的 Route Predicate。

1. After Route Predicate Factory

在指定时间之后的请求会匹配该路由,具体代码如下:

```
spring:
  cloud:
    gateway:
      routes:
      - id: after_route
        uri: https://example.org
        predicates:
        - After=2017-01-20T17:42:47.789-07:00[America/Denver]
```

2. Before Route Predicate Factory

在指定时间之前的请求会匹配该路由,具体代码如下:

```
spring:
  cloud:
    gateway:
      routes:
      - id: before_route
        uri: https://example.org
        predicates:
        - Before=2017-01-20T17:42:47.789-07:00[America/Denver]
```

3. Between Route Predicate Factory

在指定时间区间内的请求会匹配该路由,具体代码如下:

```
spring:
  cloud:
    gateway:
      routes:
      - id: between_route
        uri: https://example.org
        predicates:
        - Between=2017-01-20T17:42:47.789-07:00[America/Denver], 2017-01-21T17:42:47.789-07:00[America/Denver]
```

4. Cookie Route Predicate Factory

带有指定 Cookie 的请求会匹配该路由，具体代码如下：

```
spring:
  cloud:
    gateway:
      routes:
      - id: cookie_route
        uri: https://example.org
        predicates:
        - Cookie=chocolate, ch.p
```

使用 curl 工具发送带有 cookie 为 chocolate=ch.p 的请求可以匹配该路由。

```
curl http://localhost:9201/user/1 --cookie "chocolate=ch.p"
```

5. Header Route Predicate Factory

带有指定请求头的请求会匹配该路由，具体代码如下：

```
spring:
  cloud:
    gateway:
      routes:
      - id: header_route
        uri: https://example.org
        predicates:
        - Header=X-Request-Id, \d+
```

使用 curl 工具发送带有请求头为 X-Request-Id:123 的请求可以匹配该路由：

```
curl http://localhost:localhost/hello -H "X-Request-Id:123"
```

6. Host Route Predicate Factory

带有指定 Host 的请求会匹配该路由，具体代码如下：

```
spring:
  cloud:
    gateway:
      routes:
      - id: host_route
        uri: https://example.org
        predicates:
        - Host=**.somehost.org,**.anotherhost.org
```

使用 curl 工具发送带有请求头为 Host:www.anotherhost.org 的请求可以匹配该路由：

```
curl http://localhost:9201/user/1 -H "Host:www.anotherhost.org"
```

7. Method Route Predicate Factory

发送指定方法的请求会匹配该路由，具体代码如下：

```yaml
spring:
  cloud:
    gateway:
      routes:
      - id: method_route
        uri: https://example.org
        predicates:
        - Method=GET,POST
```

使用 curl 工具发送 GET、POST 请求可以匹配该路由：

```
curl http://localhost:8080/hello
curl -X POST http://localhost:8080/hello
```

8. Path Route Predicate Factory

发送指定路径的请求会匹配该路由，具体代码如下：

```yaml
spring:
  cloud:
    gateway:
      routes:
      - id: path_route
        uri: https://example.org
        predicates:
        - Path=/red/{segment},/blue/{segment}
```

使用 curl 工具发送 /red/1 或者 /blue/1 路径请求可以匹配该路由，具体代码如下：

```
curl http://localhost:8080/red/1
curl http://localhost:8080/blue/1
```

9. Query Route Predicate Factory

带指定查询参数的请求可以匹配该路由，具体代码如下：

```yaml
spring:
  cloud:
    gateway:
      routes:
      - id: query_route
        uri: https://example.org
        predicates:
        - Query=green
```

使用 curl 工具发送带 green=1 查询参数的请求可以匹配该路由：

```
curl http://localhost:9201/hello?green=1
```

10. RemoteAddr Route Predicate Factory

从指定远程地址发起的请求可以匹配该路由,具体代码如下:

```yaml
spring:
  cloud:
    gateway:
      routes:
      - id: remoteaddr_route
        uri: https://example.org
        predicates:
        - RemoteAddr=192.168.1.1/24
```

使用 curl 工具从 192.168.1.1 发起请求可以匹配该路由:

```
curl http://localhost:9201/hello
```

11. Weight Route Predicate Factory

使用权重来路由相应请求,以下代码表示有 80% 的请求会被路由到 weighthigh.org、20% 会被路由到 weightlow.org。

```yaml
spring:
  cloud:
    gateway:
      routes:
      - id: weight_high
        uri: https://weighthigh.org
        predicates:
        - Weight=group1, 8
      - id: weight_low
        uri: https://weightlow.org
        predicates:
        - Weight=group1, 2
```

更多内容请参考官方文档(https://cloud.spring.io/spring-cloud-gateway)。

可以使用 metadata 为每个 route 增加附加属性:

```yaml
spring:
  cloud:
    gateway:
      routes:
      - id: route_with_metadata
        uri: https://example.org
        metadata:
          optionName: "OptionValue"
          compositeObject:
            name: "value"
          iAmNumber: 1
```

还可以从 exchange 获取所有元数据属性:

```
Route route = exchange.getAttribute(GATEWAY_ROUTE_ATTR);
// get all metadata properties
```

```
route.getMetadata();
// get a single metadata property
route.getMetadata(someKey);
```

4.2.4 Gateway 过滤器工厂

路由过滤器可用于修改进入的 HTTP 请求和返回的 HTTP 响应。Spring Cloud Gateway 内置了多种路由过滤器，由 GatewayFilter 的工厂类产生。下面介绍常用路由过滤器的用法。

1. AddRequestParameter GatewayFilter

AddRequestParameter GatewayFilter 是给请求添加参数的过滤器，具体代码如下：

```
spring:
  cloud:
    gateway:
      routes:
      - id: add_request_parameter_route
        uri: http://localhost:8080
        filters:
        - AddRequestParameter=username, ay
        predicates:
        - Method=GET
```

以上配置会对 GET 请求添加 username=ay 的请求参数，通过 curl 工具使用以下命令进行测试：

```
curl http://localhost:9201/test/sayHello
```

相当于发起如下请求：

```
curl http://localhost:8080/test/sayHello?username=ay
```

2. PrefixPath GatewayFilter

PrefixPath GatewayFilter 是对指定数量的路径前缀进行去除的过滤器，具体代码如下：

```
spring:
  cloud:
    gateway:
      routes:
      - id: strip_prefix_route
        uri: http://localhost:8080
        predicates:
        - Path=/user-service/**
        filters:
        - StripPrefix=2
```

以上配置会把以 /user-service/ 开头的请求的路径去除两位，可以利用 curl 工具使用以下命令进行测试：

```
curl http://localhost:9201/user-service/a/user/1
```

相当于发起如下请求：

```
curl http://localhost:8080/user/1
```

3. PrefixPath GatewayFilter

与 StripPrefix 过滤器恰好相反，PrefixPath GatewayFilter 是会对原有路径进行增加操作的过滤器，具体代码如下：

```yaml
spring:
  cloud:
    gateway:
      routes:
      - id: prefix_path_route
        uri: http://localhost:8080
        predicates:
        - Method=GET
        filters:
        - PrefixPath=/user
```

以上配置会对所有 GET 请求添加/user 路径前缀，可以利用 curl 工具使用以下命令进行测试：

```
curl http://localhost:9201/1
```

相当于发起如下请求：

```
curl http://localhost:8080/user/1
```

这里简单列举几种过滤器，更多的过滤器请参考官方文档：https://cloud.spring.io/spring-cloud-gateway。

4.2.5　Gateway 全局过滤器

GlobalFilter 全局过滤器与普通的过滤器 GatewayFilter 具有相同的接口定义。只不过 GlobalFilter 会作用于所有路由。

本小节将详细探讨 Spring Cloud Gateway 内置的全局过滤器，包括 Combined Global Filter and GatewayFilter Ordering、Forward Routing Filter、LoadBalancerClient Filter、ReactiveLoadBalancerClientFilter、Netty Routing Filter、Netty Write Response Filter、RouteToRequestUrl Filter、Websocket Routing Filter、Gateway Metrics Filter、Marking An Exchange As Routed。

1. Combined Global Filter and GatewayFilter Ordering

发起请求时，Filtering Web Handler 处理器会添加所有 GlobalFilter 实例和匹配的 GatewayFilter 实例到过滤器链中。过滤器链会使用@Ordered 注解所指定的顺序进行排序。Spring Cloud Gateway 和 Zuul 网关不同，只有 pre 和 post 阶段，优先级高的过滤器将会在 pre 阶段最先执行，优先级最低的过滤器则在 post 阶段最后执行。记住，@Ordered 注解所指定的顺序是数值越小越靠前执行。例如：

```java
@Bean
public GlobalFilter customFilter() {
    return new CustomGlobalFilter();
}
```

```
//自定义GlobalFilter
public class CustomGlobalFilter implements GlobalFilter, Ordered {

    @Override
    public Mono<Void> filter(ServerWebExchange exchange, GatewayFilterChain chain) {
        log.info("custom global filter");
        return chain.filter(exchange);
    }

    @Override
    public int getOrder() {
        return -1;
    }
}
```

除了以上写法可以定义全局过滤器外,还可以通过更简洁的方式定义:

```
@Bean
@Order(-1)
public GlobalFilter a() {
    return (exchange, chain) -> {
        log.info("first pre filter");
        return chain.filter(exchange).then(Mono.fromRunnable(() -> {
            log.info("third post filter");
        }));
    };
}
```

2. Forward Routing Filter

Forward Routing Filter 用于本地 Forward,也就是将请求在 Gateway 服务内进行转发,而不是转发到下游服务 Forward Routing Filter,大致原理是:ForwardRoutingFilter 会查看 exchange 的属性 ServerWebExchangeUtils.GATEWAY_REQUEST_URL_ATTR 的值(一个 URI),如果该值 l 的 scheme 是 forward(比如 forward://localendpoint),则它会使用 Spring 的 DispatcherHandler 处理该请求。请求 URL 的路径部分会被 forward URL 中的路径覆盖。未修改的原始 URL 会被追加到 ServerWebExchangeUtils.GATEWAY_ORIGINAL_REQUEST_URL_ATTR 属性中。

3. LoadBalancerClient Filter

LoadBalancerClient Filter 用于整合 Ribbon 实现负载均衡,LoadBalancerClientFilter 会查看 exchange 的属性值(一个 URI):

```
ServerWebExchangeUtils.GATEWAY_REQUEST_URL_ATTR
```

如果该值的 scheme 是 lb,比如 lb://myservice,就将会使用 Spring Cloud 的 LoadBalancerClient 将 myservice 解析成实际的 host 和 port,并替换掉 ServerWebExchangeUtils.GATEWAY_REQUEST_URL_ATTR 的内容,原始地址会追加到 ServerWebExchangeUtils.GATEWAY_ORIGINAL_REQUEST_URL_ATTR 中。该过滤器还会查看 ServerWebExchangeUtils.GATEWAY_SCHEME_PREFIX_ATTR 属性,如果发现该属性的值是 lb,就会执行相同的逻辑。例如:

```
spring:
  cloud:
    gateway:
      routes:
      - id: myRoute
        uri: lb://service
        predicates:
        - Path=/service/**
```

默认情况下，如果无法在 LoadBalancer 中找到指定服务的实例，就会返回 503，可使用以下配置让其返回 404：

```
spring.cloud.gateway.loadbalancer.use404=true
```

LoadBalancer 返回 ServiceInstance 的 isSecure 值，会覆盖请求的 scheme。例如，请求打到 Gateway 上使用的是 HTTPS，但 ServiceInstance 的 isSecure 是 false，那么下游收到的是 HTTP 请求，反之亦然。如果该路由指定了 GATEWAY_SCHEME_PREFIX_ATTR 属性，那么前缀将会被剥离，并且路由 URL 中的 scheme 会覆盖 ServiceInstance 的配置。

4. Netty Routing Filter

使用 Netty 的 HttpClient 转发 http、https 请求，如果 ServerWebExchangeUtils.GATEWAY_REQUEST_URL_ATTR 值的 scheme 是 http 或 https，则运行 Netty Routing Filter。它使用 Netty HttpClient 向下游发送代理请求，获得的响应将放在 exchange 的 ServerWebExchangeUtils.CLIENT_RESPONSE_ATTR 属性中，以便在后面的 filter 中使用。

5. Netty Write Response Filter

将代理响应写回网关的客户端侧，如果 exchange 的 ServerWebExchangeUtils.CLIENT_RESPONSE_ATTR 属性中有 HttpClientResponse，则运行 NettyWriteResponseFilter。该过滤器在所有其他过滤器执行完成后执行，并将代理响应写回网关的客户端侧。

6. RouteToRequestUrl Filter

将从 request 里获取的原始 url 转换成 Gateway 进行请求转发时所使用的 url。

7. Websocket Routing Filter

如果 exchange 中 ServerWebExchangeUtils.GATEWAY_REQUEST_URL_ATTR 属性值的 scheme 是 ws 或者 wss，则运行 Websocket Routing Filter。它的底层使用 Spring Web Socket 将 Websocket 请求转发到下游。可为 URI 添加 lb 前缀实现负载均衡，例如 lb:ws://serviceid，具体示例如下：

```
spring:
  cloud:
    gateway:
      routes:
      # Normwal Websocket route
      - id: websocket_route
        uri: ws://localhost:3001
        predicates:
```

```
        - Path=/websocket/**
```

8. Gateway Metrics Filter

整合监控相关，提供监控指标，要启用 Gateway Metrics，需添加 spring-boot-starter-actuator 依赖。只要 spring.cloud.gateway.metrics.enabled 的值不是 false，就会运行 Gateway Metrics Filter。此过滤器添加名为 gateway.requests 的时序度量（timer metric），其中包含以下标记：

- routeId：路由 ID。
- routeUri：API 将路由到的 URI。
- outcome：HttpStatus.Series 分类的结果。
- status：返回给客户端的 Http Status。
- httpStatusCode：返回给客户端的请求的 Http Status。
- httpMethod：请求所使用的 Http 方法。

这些指标可以从/actuator/metrics/gateway.requests 中抓取，并且可以轻松地将它们与 Prometheus 集成以创建 Grafana 仪表板。

> **提　示**
>
> Prometheus 是开源监控告警解决方案。

9. Marking An Exchange As Routed

当一个请求走完整条过滤器链后，负责转发请求到下游的过滤器会在 exchange 中添加一个 gatewayAlreadyRouted 属性，从而将 exchange 标记为 routed（已路由）。一旦请求被标记为 routed，其他路由过滤器将不会再次路由该请求，而是直接跳过。

不同协议的请求会由不同的过滤器转发到下游，所以负责添加这个 gatewayAlreadyRouted 属性的过滤器就是最终负责转发请求的过滤器：

- http、https：请求会由 NettyRoutingFilter 或 WebClientHttpRoutingFilter 添加这个属性。
- forward：请求会由 ForwardRoutingFilter 添加这个属性。
- websocket：请求会由 WebsocketRoutingFilter 添加这个属性。

这些过滤器调用以下方法将 exchange 标记为 routed，或检查 exchange 是否是 routed：

- ServerWebExchangeUtils.isAlreadyRouted：检查 exchange 是否为 routed 状态。
- ServerWebExchangeUtils.setAlreadyRouted：将 exchange 设置为 routed 状态。

4.2.6　Gateway 跨域

Gateway 是支持 CORS 的配置，可以通过不同的 URL 规则匹配不同的 CORS 策略，例如：

```
spring:
  cloud:
    gateway:
      globalcors:
        corsConfigurations:
```

```yaml
      '[/**]':
        allowedOrigins: "https://docs.spring.io"
        allowedMethods:
        - GET
```

在上面的示例中，对于所有 GET 请求，将允许来自 docs.spring.io 的 CORS 请求。除此之外，Gateway 还提供更为详细的配置，例如 yml 配置：

```yaml
spring:
  cloud:
    gateway:
      globalcors:
        cors-configurations:
          '[/**]':
            # 允许携带认证信息
            allow-credentials: true
            # 允许跨域的源(网站域名/ip)，设置*为全部
            allowed-origins:
            - "http://localhost:13009"
            - "http://localhost:13010"
            # 允许跨域请求里的 head 字段，设置*为全部
            allowed-headers: "*"
            # 允许跨域的 method，默认为 GET 和 OPTIONS，设置*为全部
            allowed-methods:
            - OPTIONS
            - GET
            - POST
            # 跨域允许的有效期
            max-age: 3600
            # 允许 response 的 head 信息
            # 默认仅允许如下 6 个：
            #     Cache-Control
            #     Content-Language
            #     Content-Type
            #     Expires
            #     Last-Modified
            #     Pragma
            #exposed-headers:
```

4.2.7　Gateway Actuator API

通过 Actuator 端点/gateway 可以监视 Spring Cloud Gateway 应用程序并与之交互。为了可远程访问，必须在应用程序属性中启用和公开端点，具体操作如下所示。

在 pom.xml 文件中添加如下依赖：

```xml
<dependency>
    <groupId>org.springframework.boot</groupId>
    <artifactId>spring-boot-starter-actuator</artifactId>
</dependency>
```

在 application.properties 中添加如下配置:

```yaml
management:
  endpoint:
    gateway:
      enabled: true
  endpoints:
    web:
      exposure:
        include: "*"
```

访问/actuator/gateway/routes 端点，可查看与每个路由关联的谓词、过滤器以及任何可用的配置:

```json
[
    {
        "predicate":"Paths: [/user2/getUser], match trailing slash: true",
        "route_id":"path_route2",
        "filters":[

        ],
        "uri":"http://localhost:8201/user2/getUser",
        "order":0
    },
    {
        "predicate":"Paths: [/user/{id}], match trailing slash: true",
        "route_id":"path_route",
        "filters":[

        ],
        "uri":"http://localhost:8201/user/%7Bid%7D",
        "order":0
    }
]
```

默认情况下启用此功能。要禁用它，请设置以下属性:

```
spring.cloud.gateway.actuator.verbose.enabled=false
```

> **提示**
>
> 在将来的版本中，该默认值为 true。

要检索应用于所有路由的全局过滤器，可向/actuator/gateway/globalfilters 发出 GET 请求。产生的响应类似于以下内容:

```
{
"org.springframework.cloud.gateway.filter.LoadBalancerClientFilter@77856cc5":
10100,
"org.springframework.cloud.gateway.filter.RouteToRequestUrlFilter@4f6fd101":
10000,
"org.springframework.cloud.gateway.filter.NettyWriteResponseFilter@32d22650":
```

```
    -1,
"org.springframework.cloud.gateway.filter.ForwardRoutingFilter@106459d9":
2147483647,
"org.springframework.cloud.gateway.filter.NettyRoutingFilter@1fbd5e0":
2147483647,
"org.springframework.cloud.gateway.filter.ForwardPathFilter@33a71d23": 0,
"org.springframework.cloud.gateway.filter.AdaptCachedBodyGlobalFilter@135064ea
": 2147483637,
"org.springframework.cloud.gateway.filter.WebsocketRoutingFilter@23c05889":
2147483646
    }
```

该响应包含全局过滤器的详细信息。对于每个全局过滤器（例如 org.springframework.cloud.gateway.filter.LoadBalancerClientFilter@77856cc5:10100），10100 表示过滤器链中的相应顺序。

要清除路由缓存，可向/actuator/gateway/refresh 发出 POST 请求，请求会返回 200。

4.2.8　HTTP 超时配置

1. 全局超时

- connect-timeout：连接超时，单位为毫秒。
- response-timeout：响应超时，单位为 java.time.Duration。

例如：

```
spring:
  cloud:
    gateway:
      httpclient:
        #### 连接超时配置
        connect-timeout: 1000
        ### 响应超时配置
        response-timeout: 5s
```

2. 每个路由配置

- connect-timeout：单位为毫秒。
- response-timeout：单位为毫秒。

```
- id: per_route_timeouts
  uri: https://example.org
  predicates:
    - name: Path
      args:
        pattern: /delay/{timeout}
  metadata:
    response-timeout: 200
    connect-timeout: 200
```

3. Java 配置

```
@Bean
```

```
public RouteLocator customRouteLocator(RouteLocatorBuilder routeBuilder){
    return routeBuilder.routes()
            .route("test1", r -> {
                return r.host("*.somehost.org").and().path("/somepath")
                    .filters(f -> f.addRequestHeader("header1",
"header-value-1"))
                    .uri("http://someuri")
                    .metadata(RESPONSE_TIMEOUT_ATTR, 200)
                    .metadata(CONNECT_TIMEOUT_ATTR, 200);
            })
            .build();
}
```

4.2.9 TLS / SSL 设置

在 Web 服务应用中，为了数据的传输安全，会使用安全证书以及 TLS/SSL 加密。本小节将简单介绍 Spring Cloud Gateway 的 HTTPS 配置。

- TLS：安全传输层协议，用于在两个通信应用程序之间提供保密性和数据完整性。
- SSL：安全套接层协议。TSL/SSL 都属于加密协议，在传输层与应用层之间对网络连接进行加密，用于在网络数据传输中保护隐私和数据的完整性。

网关可以通过遵循常规的 Spring 服务器配置来侦听 HTTPS 上的请求。例如：

```
server:
  ssl:
    //启用 ssl
    enabled: true
    //启用证书
    key-alias: scg
    //证书密码
    key-store-password: scg1234
    //证书地址
    key-store: classpath:scg-keystore.p12
    //证书类型
    key-store-type: PKCS12
```

可以使用以下配置为 Gateway 配置一组可信任的已知证书：

```
spring:
  cloud:
    gateway:
      httpclient:
        ssl:
          trustedX509Certificates:
            - cert1.pem
            - cert2.pem
```

提示

ssl 证书放在 resources 目录下。

如果 Spring Cloud Gateway 未配置可信证书，就会使用默认的安全证书（可以使用系统属性 javax.net.ssl.trustStore 覆盖）。

网关路由后端时，使用客户端连接池。通过 HTTPS 进行通信时，客户端会启动 TLS 握手。握手可能会超时，可以按以下方式配置超时时间：

```
spring:
  cloud:
    gateway:
      httpclient:
        ssl:
          handshake-timeout-millis: 10000
          close-notify-flush-timeout-millis: 3000
          close-notify-read-timeout-millis: 0
```

4.2.10 Gateway 底层原理

Spring Cloud Gateway 的核心处理流程如图 4-7 所示。

Spring Cloud Gateway 的工作机制大体如下：

（1）Gateway 接收客户端请求。

（2）客户端请求与路由信息进行匹配，匹配成功被发往相应的下游服务。

（3）请求经过 Filter 过滤器链，执行 pre（前置过滤器）处理逻辑，如修改请求头信息等。

（4）请求被转发至下游服务并返回响应。

（5）响应经过 Filter 过滤器链，执行 post（后置过滤器）处理逻辑。

（6）向客户端响应应答。

在 Spring MVC 中，通过 HandlerMapping 解析请求链接，根据请求链接找到执行请求 Controller 类。在 Spring Cloud Gateway 中也是使用 HandlerMapping 对请求的链接进行解析，匹配对应的 Route 进行代理转发到对应的服务。

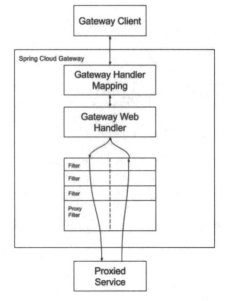

图 4-7　Spring Cloud Gateway 工作原理

Gateway 的客户端向 Spring Cloud Gateway 发起请求，请求被 HttpWebHandlerAdapter 进行提取组装成 Gateway 上下文，Gateway 上下文会传递到 DispatcherHandler。DispatcherHandler 是所有请求的分发处理器，负责分发请求对应的处理器。比如将请求分发到对应的 RoutePredicateHandlerMapping（路由断言处理映射器）。RoutePredicateHandlerMapping 是 Gateway Handler Mapping 其中的一个 HandlerMapping。

路由断言处理映射器主要用于路由的查找，以及找到路由后返回对应的 FilteringWebHandler。

FilteringWebHandler 是一个存放过滤器的 Handler，主要负责组装 Filter 链表并调用 Filter 执行一系列 Filter 处理，然后把请求转到后端对应的代理服务处理，处理完毕后将 Response 返回到 Gateway 客户端。

4.3　Gateway 与 Zuul 的区别

Spring Cloud Gateway 是 Spring 官方基于 Spring 5.0、Spring Boot 2.0 和 Project Reactor 等技术开发的网关，旨在提供一种简单而有效的路由 API 的方法。Spring Cloud Gateway 作为 Spring Cloud 生态系中的网关，目标是替代 Netflix ZUUL，Gateway 与 Zuul 的区别如表 4-1 所示。

表 4-1　Gateway 与 Zuul 的区别

维度	技术		
	Gateway	Zuul 1.0	Zuul 2.0
开源时间	2017 年 7 月	2013 年 6 月	2017 年 11 月
架构	基于 Netty	基于 Servlet	基于 Netty
运行方式	异步非阻塞	同步阻塞	异步非阻塞
Spring Cloud 集成	√	√	×
支持长连接、Web Socket	√	×	√
限流、限速	√	第三方	即将推出
监控	√	√	√
上手难度	中等	简单	复杂

从性能角度进行对比，具体如表 4-2 所示。

表 4-2　Gateway 与 Zuul 的性能对比

组件	平均延迟	RPS（每秒请求数）
gateway	6.61ms	3.24k
Zuul 1.0	12.56ms	2.09k
Zuul 2.0	?	?

从性能测试结果可知，Spring Cloud Gateway 的 RPS 是 Zuul 1.0 的 1.55 倍，平均延迟是 Zuul 1.0 的一半。Spring Cloud Gateway、Zuul 2.0 的实现方式非常接近，最大的不同在于 Spring Cloud 暂时还没有对 Zuul 2.0 的整合计划。从目前来看，Gateway 替代 Zuul 是未来趋势。

第 5 章

Ribbon 负载均衡

本章主要介绍 Ribbon 负载均衡器、常用负载均衡算法、如何自定义负载均衡算法和 Ribbon 客户端，以及 Eureka/Nacos 如何整合 Ribbon 客户端。

5.1 Ribbon 基础知识

5.1.1 Ribbon 简介

Ribbon 是 Netflix 客户端的负载均衡器，可对 HTTP 和 TCP 客户端的行为进行控制。为 Ribbon 配置服务提供者地址后，Ribbon 就可以基于某种负载均衡算法自动帮助服务消费者去请求。Ribbon 默认为我们提供了很多负载均衡算法，例如轮询、随机等。当然，也可以为 Ribbon 实现自定义的负载均衡算法。

要将 Ribbon 包含在项目中，请添加如下依赖：

```xml
<dependency>
    <groupId>org.springframework.cloud</groupId>
    <artifactId>spring-cloud-starter-ribbon</artifactId>
    <version>1.4.7.RELEASE</version>
</dependency>
```

在 Ribbon 中有以下几个重要概念：

（1）Rule：该组件主要决定从候选服务器中返回哪个服务器地址进行远程调用的操作。

（2）Ping：在后台运行的组件，用来确认哪些服务器是存活可用的。

（3）ServerList：当前可以用作 LB 的服务器列表，该列表可以是静态的，也可以是动态的。如果是动态列表（例如从 Eureka 服务器获取），就会有一个后台线程按照时间间隔刷新列表。

Ribbon 提供了以下几种 Rule：

（1）RoundRobinRule：最简单的规则，会在 ServerList 中依次轮询调用。
（2）RandomRule：随机。
（3）AvailabilityFilteringRule：在这种规则下 Ribbon 集成了 Hystrix 的功能，默认情况下调用某个远程方法失败 3 次后断路器的开关会被打开，而之后的请求中 Ribbon 会跳过这个服务器地址，直到 30 秒之后断路器关闭后才会重新加入调用列表。
（4）WeightedResponseTimeRule：将响应时间作为权重的负载规则，某个服务器的响应时间越长，它的权重就越低。具体选择服务器时，结合权重进行随机选择。
（5）RetryRule：按照 RoundRobinRule（轮询）策略获取服务，如果获取服务失败，就在指定时间内重试，获取可用的服务。
（6）BestAvailableRule：先过滤掉由于多次访问故障而处于断路器跳闸状态的服务，然后选择一个并发量最小的服务。
（7）ZoneAvoidanceRule：复合判断 Server 所在区域的性能和 Server 的可用性选择服务器。

Spring Cloud Netflix 默认为 Ribbon 提供的 bean 对象如表 5-1 所示。

表 5-1 Spring Cloud Netflix 默认为 Ribbon 提供的 bean 对象

Bean Type	Bean Name	Class Name
IClientConfig	ribbonClientConfig	DefaultClientConfigImpl
IRule	ribbonRule	ZoneAvoidanceRule
IPing	ribbonPing	DummyPing
ServerList<Server>	ribbonServerList	ConfigurationBasedServerList
ServerListFilter<Server>	ribbonServerListFilter	ZonePreferenceServerListFilter
ILoadBalancer	ribbonLoadBalancer	ZoneAwareLoadBalancer
ServerListUpdater	ribbonServerListUpdater	PollingServerListUpdater

Ribbon 通过 IRule 接口实现负载均衡策略，接口实现结构图如图 5-1 所示。

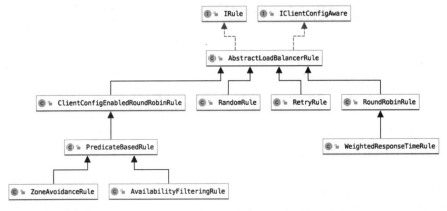

图 5-1 负载均衡规则——类继承关系图

Ribbon 默认使用的是 ZoneAvoidanceRule 规则。

5.1.2 负载均衡算法

服务消费者从服务配置中心获取服务的地址列表后需要选取其中一台发起 RPC/HTTP 调用，这时需要用到具体的负载均衡算法。常用的负载均衡算法有轮询法、加权轮询法、随机法、加权随机法、源地址哈希法、一致性哈希法等。

1. 轮询法

轮询法是指将请求按顺序轮流分配到后端服务器上，均衡地对待后端的每一台服务器，不关心服务器实际的连接数和当前系统负载。轮询法的具体实例如图 5-2 所示。

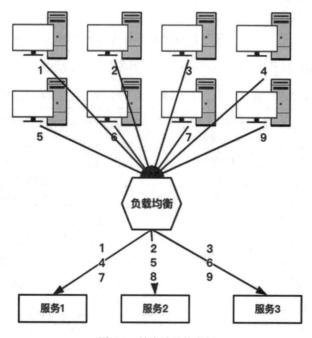

图 5-2 轮询法具体实例

在图 5-2 中，有 9 个客户端请求、3 台后端服务器。当第一个请求到达负载均衡服务器时，负载均衡服务器会将这个请求分派到后端服务器 1；当第二个请求到来时，负载均衡服务器会将这个请求分派到后端服务器 2；当第三个请求到来时，负载均衡服务器会将这个请求分派到后端服务器 3；当第四个请求到来时，负载均衡服务器会将这个请求分派到后端服务器 1，以此类推。

2. 加权轮询法

加权轮询法是指根据真实服务器的不同处理能力来调度访问请求，这样可以保证处理能力强的服务器处理更多的访问流量。简单的轮询法并不考虑后端机器的性能和负载差异。加权轮询法可以很好地处理这一问题，它将按照顺序且按照权重分派给后端服务器：给性能高、负载低的机器配置较高的权重，让其处理较多的请求；给性能低、负载高的机器配置较低的权重，让其处理较少的请求。

假设有 9 个客户端请求、3 台后端服务器如图 5-3 所示，后端服务器 1 被赋予权值 1，后端服务器 2 被赋予权值 2，后端服务器 3 被赋予权值 3。这样一来，客户端请求 1、2、3 都被分派到服务器 3 处理，客户端请求 4、5 被分派到服务器 2 处理，客户端请求 6 被分派到服务器 1 处理，客户端请求 7、8、9 被分派到服务器 3 处理，以此类推。

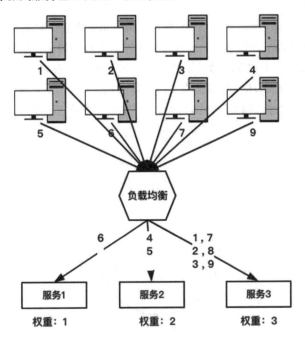

图 5-3　轮询法具体实例

3. 随机法

随机法也很简单，就是随机选择一台后端服务器进行请求处理。由于每次服务器被挑中的概率都一样，因此客户端的请求可以被均匀地分派到所有的后端服务器上。由概率统计理论可以得知，随着调用量的增大，其实际效果越来越接近于平均分配流量到每一台后端服务器，也就是轮询的效果。

4. 加权随机法

加权随机法跟加权轮询法类似，根据后台服务器不同的配置和负载情况配置不同的权重。不同的是，它是按照权重来随机选取服务器的，而非顺序。加权随机算法一般应用的场景是在一个集合 S{A，B，C，D} 中随机抽取一项，但是抽取的概率不同，比如希望抽到 A 的概率是 50%、抽到 B 和 C 的概率是 20%、抽到 D 的概率是 10%。一般来说，我们可以给各项附加一个权重，抽取的概率正比于这个权重，上述集合就成了 {A:5，B:2，C:2，D:1}。扩展这个集合，使每一项出现的次数与其权重正相关，即 {A，A，A，A，A，B，B，C，C，D}，然后就可以用均匀随机算法从中选取了。

5. 源地址哈希法

源地址哈希（Hash）是根据获取客户端的 IP 地址，通过哈希函数计算得到一个数值，用该数值对服务器列表的大小进行取模运算，得到的结果便是客户端要访问服务器的序号。采用源地址哈希法进行负载均衡，当后端服务器列表不变时，同一个 IP 地址的客户端，每次都会映射到同一台后端服务器进行访问。源地址哈希法的缺点是，当后端服务器增加或者减少时，采用简单的哈希取模方法会使命中率大大降低。这个问题可以采用一致性哈希法来解决。

6. 一致性哈希法

一致性哈希（Hash）法解决了分布式环境下机器增加或者减少时简单的取模运算无法获取较高命中率的问题。通过虚拟节点的使用，一致性哈希算法可以均匀分担机器的负载，使得这一算法更具现实意义。正因如此，一致性哈希算法被广泛应用于分布式系统中。

一致性哈希算法通过哈希环的数据结构实现，环的起点是 0，终点是 $2^{32}-1$，并且起点与终点连接。哈希环中间的整数按逆时针分布，故哈希环的整数分布范围是[0, $2^{32}-1$]，具体如图 5-4 所示。

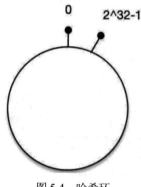

图 5-4　哈希环

在负载均衡中，首先为每一台机器计算一个哈希值，然后把这些哈希值放置在哈希环上。假设有 3 台 Web 服务器 s1、s2、s3，它们计算得到的哈希值分别为 h1、h2、h3，在哈希环上的位置如图 5-5 所示。

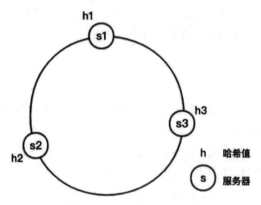

图 5-5　服务器分布哈希环

计算每一个请求 IP 的哈希值：hash("192.168.0.1")，并把这些哈希值放置到哈希环上。假设有 5 个请求，对应的哈希值为 q1、q2、q3、q4、q5，放置到哈希环上的位置如图 5-6 所示。

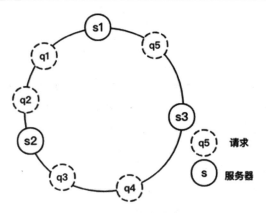

图 5-6 请求分布哈希环

接下来为每一个请求找到对应的机器，在哈希环上顺时针查找距离这个请求的哈希值最近的机器，结果如图 5-7 所示。

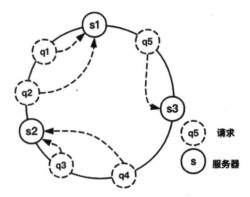

图 5-7 请求寻找最近的服务器

对于线上的业务，增加或者减少一台机器的部署是常有的事情。增加服务器 s4 的部署并将机器 s4 加入哈希环的机器 s3 与 s2 之间。这时，只有机器 s3 与 s4 之间的请求需要重新分配新的机器。只有请求 q4 被重新分配到了 s4 上，其他请求仍在原有机器上，如图 5-8 所示。

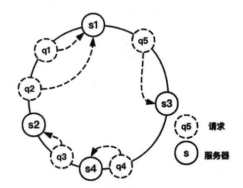

图 5-8 请求寻找最近的服务器（加入服务器 s4）

从图 5-8 的分析可以知道，增减机器只会影响相邻的机器，这就导致了添加机器时只会分担其中一台机器的负载、删除机器时会把负载全部转移到相邻的一台机器上，这并不是我们希望看到的。

我们希望看到的是：增加机器时，新的机器可以合理地分担所有机器的负载；删除机器时，多出来的负载可以均匀地分给剩余的机器。

例如，系统中只有两台服务器，由于某种原因下线 Node B（节点 B），此时必然造成大量数据集中到 Node A（节点 A）上，而只有极少量会定位到 Node B 上。为此，我们引入虚拟节点来解决负载不均衡的问题，即对每一个服务节点计算多个哈希，每个计算结果位置都放置一个服务节点，称为虚拟节点。可以在服务器 IP 或主机名的后面增加编号来实现，例如为每台服务器计算 3 个虚拟节点，分别计算 "Node A#1" "Node A#2" "Node A#3" "Node B#1" "Node B#2" "Node B#3" 的哈希值，形成 6 个虚拟节点。同时数据定位算法不变，只是多了一步虚拟节点到实际节点的映射，例如定位到 "Node A#1" "Node A#2" "Node A#3" 三个虚拟节点的数据均定位到 Node A 上。这样就解决了服务节点少时数据倾斜的问题。在实际应用中，通常将虚拟节点数设置为 32 甚至更大，因此即使很少的服务节点也能做到相对均匀的数据分布。

5.1.3 第一个 Ribbon 程序

本小节介绍如何通过 Ribbon 实现简单的负载均衡，具体步骤如下：

步骤 01 使用 Intellij IDEA 快速创建 Spring Boot 项目，项目名称为 spring-cloud-ribbon-provider，同时需要勾选 Spring Web 依赖。

步骤 02 在 spring-cloud-ribbon-provider 项目中添加配置文件 application-01.properties 和 application-02.properties，具体内容如下：

```
###application-01.properties 配置文件内容
server.port=8080


###application-02.properties 配置文件内容
server.port=8081
```

步骤 03 开发 UserController 类，具体代码如下：

```
@RestController
@RequestMapping("/user")
public class UserController {

    @Resource
    Environment environment;

    public String getPort(){
        return environment.getProperty("local.server.port");
    }

    @RequestMapping("getName")
    public String getUserName(){
        return "hello, ay" + "-" + getPort();
    }
}
```

通过注入 Environment 对象获取服务启动的端口，getUserName 方法用于打印字符串和端口信息。

步骤 04 在 IDEA 中，通过使用不同的配置文件，可以启动多个 Spring Boot 应用，从原启动配置中复制一份，修改启动配置文件的 Name 和 Active profiles，如图 5-9 所示。

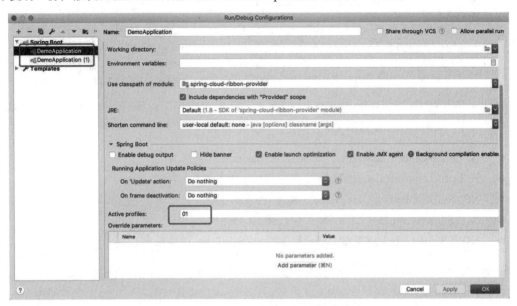

图 5-9　Spring Boot 应用使用不同配置

步骤 05 分别以 DemoApplication 和 DemoApplication(1)配置启动 Spring Boot 应用，相当于启动两个 Spring Boot 应用，端口分别是 8080 和 8081。

步骤 06 使用 Intellij IDEA 快速创建 Spring Boot 项目，项目名称为 spring-cloud-ribbon-consumer，同时需要勾选 Ribbon 依赖，具体如图 5-10 所示。

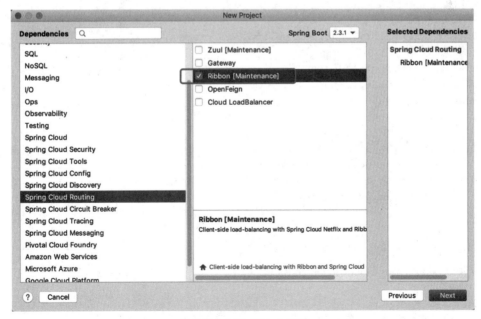

图 5-10　勾选 Ribbon 依赖

步骤07 在 application.yml 配置文件中添加如下配置：

```yaml
### 负载均衡配置
my-client:
  ribbon:
    ### 配置服务列表
    listOfServers: localhost:8080,localhost:8081
    ### 配置负载均衡算法 RoundRobinRule (轮询)
    NFLoadBalancerRuleClassName:com.netflix.loadbalancer.RoundRobinRule
```

Ribbon 的配置格式是 "<clientName> : ribbon :需要配置的属性"。其中，<clientName>是 Ribbon 的客户端名称，如果省略就配置所有客户端。配置的属性有以下几种：

- NFLoadBalancerClassName：配置 ILoadBalancer 的实现类。
- NFLoadBalancerRuleClassName：配置 IRule 的实现类。
- NFLoadBalancerPingClassName：配置 IPing 的实现类。
- NIWSServerListClassName：配置 ServerList 的实现类。
- NIWSServerListFilterClassName：配置 ServerListFilter 的实现类。

例如，配置 life-base 这个 Ribbon Client 的负载均衡规则，在 yml 文件中可以配置为：

```yaml
life-base:
  ribbon:
    NFLoadBalancerRuleClassName: com.netflix.loadbalancer.RandomRule
```

步骤08 在 main 方法中添加如下代码：

```java
@SpringBootApplication
public class DemoApplication {

 public static void main(String[] args) throws Exception{
  SpringApplication.run(DemoApplication.class, args);
   //获取客户端
  RestClient client = (RestClient)ClientFactory.getNamedClient("my-client");
//调用 UserController 类的 getUserName 方法
  HttpRequest request = 
HttpRequest.newBuilder().uri("/user/getName").build();
        //循环调用
        for(int i = 0; i < 10; i++) {
            HttpResponse response = client.executeWithLoadBalancer(request);
            String result = response.getEntity(String.class);
            System.out.println(result);
        }
    }
}
```

步骤09 启动 spring-cloud-ribbon-consumer 项目，在控制台中将可以看到如下打印信息：

```
hello, ay-8081
hello, ay-8080
hello, ay-8081
```

```
hello, ay-8080
hello, ay-8081
hello, ay-8080
hello, ay-8081
hello, ay-8080
hello, ay-8081
hello, ay-8080
```

从打印信息中可以看出，服务通过轮询的方式调用，读者可在配置文件中配置其他的负载均衡算法来进行其他实验。

5.2 Ribbon 实战

5.2.1 Ribbon 自定义负载均衡策略

Ribbon 支持自定义负载均衡规则，本小节将结合 Nacos 讲解如何自定义负载均衡策略，具体步骤如下：

步骤 01 使用 Intellij IDEA 快速创建 Spring Boot 项目，项目名称为 spring-cloud-ribbon-custom，同时需要勾选 Spring Web 依赖和 Nacos Service Discovery 依赖，具体如图 5-11 所示。

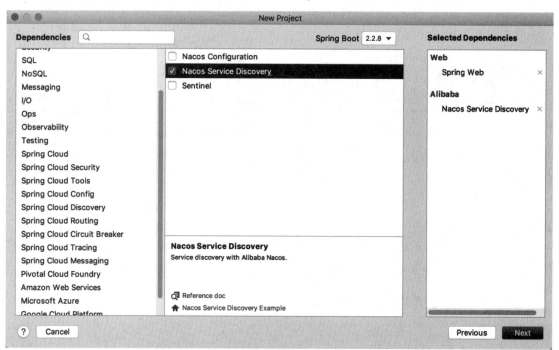

图 5-11　勾选 Spring Web 依赖和 Nacos Service Discovery 依赖

步骤 02 在 spring-cloud-ribbon-custom 项目中创建 RibbonConfiguration 配置类，该类为 Ribbon 配置类，不能被@ComponentScan 扫描到。

```java
/**
 * Ribbon 配置类,不能被@ComponentScan 扫描到
 * @author ay
 * @since 2020-07-14
 */
@Configuration
public class RibbonConfiguration {

    @Bean
    public IRule ribbonRule(){
        //ribbon 默认使用的是 zoneAvoidanceRule 规则,这里修改为自定义方式
        return new MyRule();
    }
}
```

步骤 03 在 spring-cloud-ribbon-custom 项目中创建 MyRule 类,用来自定义负载均衡规则。该负载均衡规则非常简单,就是始终返回第一个服务,具体代码如下:

```java
/**
 * 负载规则:始终返回第一个服务
 * @author ay
 * @since 2020-07-14
 */
public class MyRule extends AbstractLoadBalancerRule {

    private ILoadBalancer lb;
    public MyRule() {}

    @Override
    public Server choose(Object o) {
        ILoadBalancer lb = getLoadBalancer();
        // 获取所有的服务
        List<Server> servers = lb.getAllServers();
//        for (Server s : servers){
//            System.out.println("服务器的端口号: "+ s.getHostPort());
//        }
        return servers.get(0); // 始终返回第一个服务
    }
    @Override
    public void initWithNiwsConfig(IClientConfig iClientConfig) {
    }
}
```

步骤 04 自定义指定 Ribbon 客户端的配置,具体代码如下:

```java
/**
 * 使用 RibbonClient 为特定 name 的 Ribbon Client 自定义配置
 * 使用@RibbonClient 的 configuration 属性指定 Ribbon 的配置类
 * @author ay
 * @since 2020-07-14
 */
```

```java
@Configuration
@RibbonClient(name = "service-provider", configuration = 
RibbonConfiguration.class)
public class TestConfiguration {

}
```

步骤 05 在 spring-cloud-ribbon-custom 项目的 application.properties 配置文件中添加如下配置:

```
### 应用端口
server.port=7089
### 服务名称
spring.application.name=service-custom
### nacos 地址
spring.cloud.nacos.discovery.server-addr=192.168.38.190:8848
```

步骤 06 在启动类中添加@EnableDiscoveryClient 注解,具体代码如下:

```java
@SpringBootApplication
@EnableDiscoveryClient
public class DemoApplication {
    public static void main(String[] args) {
        SpringApplication.run(DemoApplication.class, args);
    }
}
```

步骤 07 创建 TestController 类,具体代码如下:

```java
@RestController
@RequestMapping("/test")
public class TestController {

    @Resource
    private LoadBalancerClient loadBalancerClient;

    @RequestMapping("/getUserName")
    public String getUserName(){
        for(int i=0; i< 20; i++){
            //获取 service-provider 服务
            ServiceInstance serviceInstance = 
                loadBalancerClient.choose("service-provider");
            //打印当前选择的是哪个节点
            System.out.println(serviceInstance.getServiceId()+
                serviceInstance.getHost()+ ": " +serviceInstance.getPort());
        }
        return "hello, ay";
    }
}
```

在上述步骤中,我们创建了 spring-cloud-ribbon-custom 项目(可以理解为服务消费者),并自定义了负载均衡规则 MyRule,服务启动后注册到 Nacos 中。接下来,我们创建服务生产者,具体步骤如下：

步骤01 使用 Intellij IDEA 快速创建 Spring Boot 项目，项目名称为 spring-cloud-Nacos-ribbon-provider，同时需要勾选 Spring Web 依赖和 Nacos Service Discovery 依赖。

步骤02 在启动类中添加@EnableDiscoveryClient 注解，将服务注册到 Nacos 中。在项目中创建 application-01.properties 和 application-02.properties 配置文件，并添加如下配置：

```
### application-01.properties 配置文件内容
### 服务端口
server.port=8070
### 服务名称
spring.application.name=service-provider
### nacos 地址
spring.cloud.nacos.discovery.server-addr=192.168.38.190:8848

### application-02.properties 配置文件内容
### 服务端口
server.port=8071
### 服务名称
spring.application.name=service-provider
### nacos 地址
spring.cloud.nacos.discovery.server-addr=192.168.38.190:8848
```

在上述配置中，定义了服务 service-provider，端口分别为 8070 和 8071，同时配置 Nacos 地址，将服务注册到 Nacos 中。

步骤03 创建 UserController 类，供服务消费者调用，具体代码如下：

```
@RestController
@RequestMapping("/user")
public class UserController {
    @RequestMapping("/getUserName")
    public String getUserName(){
        return "hello, ay";
    }
}
```

所有项目创建完成后，分别启动 spring-cloud-Nacos-ribbon-provider（服务名为 service-custom）和 spring-cloud-ribbon-custom（服务名为 service-provider）项目。在 Nacos 服务列表中，查看服务注册情况，如图 5-12 所示。

在浏览器中访问地址 "http://localhost:7089/test/getUserName"，在 spring-cloud-ribbon-custom 项目的控制台下打印如下信息：

```
service-provider192.168.0.103：8070
service-provider192.168.0.103：8070
service-provider192.168.0.103：8070
service-provider192.168.0.103：8070
service-provider192.168.0.103：8070
...
```

图 5-12　Nacos 服务注册信息

请求 /test/getUserName 路径时，service-custom 服务会调用 service-provider 服务的 getUserName 方法，service-provider 服务部署两个实例，service-custom 服务会通过自定义的负载均衡策略返回第一个服务进行调用，因此控制台打印的信息永远都是 8070 端口的服务信息。

5.2.2　Ribbon 饥饿加载

每个 Ribbon 客户端都有一个 Spring Cloud 维护的相应子应用程序上下文。此应用程序上下文在对客户端的第一个请求上延迟加载，严重的时候会引起调用超时。所以，我们可以通过指定 Ribbon 具体的客户端名称来开启饥饿加载，即在启动的时候便加载所有配置项的应用程序上下文。

```
ribbon:
  eager-load:
    ### 开启 Ribbon 的饥饿加载模式
    enabled: true
    ### 指定需要饥饿加载的服务名
    clients: client1, client2, client3
```

5.2.3　Ribbon 默认配置

通过 @RibbonClients 注解可以为所有的 Ribbon 客户端提供默认的配置，例如：

```
//配置类
@RibbonClients(defaultConfiguration = DefaultRibbonConfig.class)
public class RibbonClientDefaultConfigurationTestsConfig {
    public static class BazServiceList extends ConfigurationBasedServerList {
        public BazServiceList(IClientConfig config) {
            super.initWithNiwsConfig(config);
        }
    }
}

//作用于所有的客户端
```

```
@Configuration(proxyBeanMethods = false)
class DefaultRibbonConfig {

    @Bean
    public IRule ribbonRule() {
        return new BestAvailableRule();
    }

    @Bean
    public IPing ribbonPing() {
        return new PingUrl();
    }

    @Bean
    public ServerList<Server> ribbonServerList(IClientConfig config) {
        return new
RibbonClientDefaultConfigurationTestsConfig.BazServiceList(config);
    }

    @Bean
    public ServerListSubsetFilter serverListFilter() {
        ServerListSubsetFilter filter = new ServerListSubsetFilter();
        return filter;
    }

}
```

5.2.4 配置文件定义 Ribbon 客户端

从 1.2.0 版开始，Spring Cloud Netflix 通过将属性设置为与 Ribbon 文档兼容来支持自定义 Ribbon 客户端。这使我们可以在不同环境中更改行为，受支持的属性如下：

- \<clientName\>.ribbon.NFLoadBalancerClassName：应该实现 ILoadBalancer。
- \<clientName\>.ribbon.NFLoadBalancerRuleClassName：应该实现 IRule。
- \<clientName\>.ribbon.NFLoadBalancerPingClassName：应该实现 IPing。
- \<clientName\>.ribbon.NIWSServerListClassName：应该实现 ServerList。
- \<clientName\>.ribbon.NIWSServerListFilterClassName：应该实现 ServerListFilter。

这些属性中定义的类优先于使用@RibbonClient（configuration = MyRibbonConfig.class）定义的 bean 和 Spring Cloud Netflix 提供的默认值。

要为名为 user 的服务名称设置 IRule，可以设置以下属性：

```
users:
  ribbon:
    NIWSServerListClassName: com.netflix.loadbalancer.ConfigurationBasedServerList
    NFLoadBalancerRuleClassName: com.netflix.loadbalancer.WeightedResponseTimeRule
```

5.2.5 直接使用 Ribbon API

可以直接使用 LoadBalancerClient 对象进行负载均衡调用，如以下示例所示：

```
public class MyClass {
    @Autowired
    private LoadBalancerClient loadBalancer;

    public void doStuff() {
        ServiceInstance instance = loadBalancer.choose("stores");
        URI storesUri = URI.create(String.format("https://%s:%s", instance.getHost(), instance.getPort()));
        // ... do something with the URI
    }
}
```

5.2.6 Eureka/Nacos 整合 Ribbon

在 Spring Cloud 中，当 Ribbon 与 Eureka 配合使用时，Ribbon 可自动从 Eureka Server 获取服务提供者地址列表，并基于负载均衡算法请求其中一个服务提供者实例。图 5-13 展示了 Ribbon 与 Eureka 配合使用时的架构。

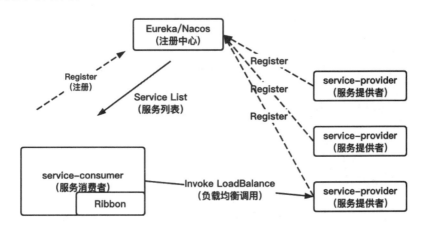

图 5-13　Eureka 与 Ribbon 负载均衡架构图

将 ribbon.eureka.enabled 属性设置为 false 会显式禁用 Ribbon 中对 Eureka 的使用，如以下示例所示。

application.yml

```
ribbon:
  eureka:
    enabled: false
```

Eureka 是一种抽象发现远程服务器的便捷方法，因此不必在客户端对它们的 URL 进行硬编码。

如果不想使用 Eureka，也可以使用 Ribbon 和 Feign。

在默认情况下，Ribbon 客户端会从 Eureka 注册中心获取服务列表来达到动态负载均衡的功能。有时可能需要绕过 Eureka 注册中心，直连某个服务，例如：

```
###ribbon客户端名称，可以随便写
user:
  ribbon:
    listOfServers: example.com,google.com
```

第 6 章

Spring Cloud OpenFeign 声明式调用

本章主要介绍 Spring Cloud OpenFeign 声明式调用，包括 OpenFeign 简介、快速创建第一个 Feign 程序、@FeignClient 注解详解、Feign @QueryMap 支持、Feign 请求响应压缩、Feign 日志配置（Java 方式、配置文件方式以及全局日志配置）、Feign 自定义错误、Feign 拦截器以及如何自定义 Feign 客户端等内容。

6.1 Spring Cloud Feign

6.1.1 Feign 简介

Spring Cloud Feign 是一个 HTTP 请求调用的轻量级框架，可以以 Java 接口注解的方式调用 HTTP 请求，而不用像 Java 中通过封装 HTTP 请求报文的方式直接调用。Feign 通过处理注解，将请求模板化，当实际调用的时候传入参数，根据参数再应用到请求上，进而转化成真正的请求，这种请求相对而言比较直观。

Feign 是 Netflix 开发的声明式、模板化的 HTTP 客户端，其灵感来自 Retrofit、JAXRS-2.0 以及 WebSocket。

Feign 被广泛应用在 Spring Cloud 的解决方案中，是学习基于 Spring Cloud 微服务架构不可或缺的重要组件。

项目使用 Feign，只需在 pom.xml 文件中添加如下依赖即可：

```xml
<dependency>
    <groupId>org.springframework.cloud</groupId>
    <artifactId>spring-cloud-starter-openfeign</artifactId>
</dependency>
```

6.1.2 第一个 Feign 程序

本小节介绍如何通过 Nacos+Feign 实现服务之间的调用。

1. 创建服务提供者

首先我们来创建服务提供者，具体步骤如下所示。

步骤 01 参考第 3 章的内容，启动并安装 Nacos 系统。

步骤 02 使用 Intellij IDEA 快速创建多模块项目，项目名称为 spring-boot-openfeign-provider，具体目录结构如图 6-1 所示。

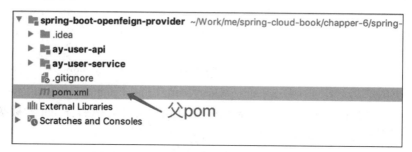

图 6-1　spring-boot-openfeign-provider 目录结构

项目 spring-boot-openfeign-provider 包含两大模块：ay-user-api 和 ay-user-service。

- ay-user-api：一个普通的 Maven 模块（jar 包），提供用户相关的 Feign 接口，这些接口提供相关的服务。
- ay-user-service：一个 Spring Boot 项目，模块内部实现 ay-user-api 相关的接口。

步骤 03 在父 pom.xml 文件中，添加 Spring Cloud 和 Spring Cloud Alibaba 依赖。

```xml
<?xml version="1.0" encoding="UTF-8"?>
<project xmlns="http://maven.apache.org/POM/4.0.0"
         xmlns:xsi="http://www.w3.org/2001/XMLSchema-instance"
         xsi:schemaLocation="http://maven.apache.org/POM/4.0.0 http://maven.apache.org/xsd/maven-4.0.0.xsd">
    <modelVersion>4.0.0</modelVersion>

    <parent>
        <groupId>org.springframework.boot</groupId>
        <artifactId>spring-boot-starter-parent</artifactId>
        <version>2.2.8.RELEASE</version>
        <relativePath/> <!-- lookup parent from repository -->
    </parent>
    <!-- 定义模块 -->
    <modules>
        <module>ay-user-service</module>
        <module>ay-user-api</module>
    </modules>
```

```xml
    <groupId>org.example</groupId>
    <artifactId>spring-boot-openfeign-provider</artifactId>
    <version>1.0-SNAPSHOT</version>
    <packaging>pom</packaging>
    <!-- 定义版本 -->
    <properties>
        <spring-cloud-alibaba.version>2.2.0.RELEASE
        </spring-cloud-alibaba.version>
        <spring-cloud.version>Hoxton.SR3</spring-cloud.version>
    </properties>
    <dependencyManagement>
        <dependencies>
            <!-- spring cloud 依赖管理 -->
            <dependency>
                <groupId>org.springframework.cloud</groupId>
                <artifactId>spring-cloud-dependencies</artifactId>
                <version>${spring-cloud.version}</version>
                <type>pom</type>
                <scope>import</scope>
            </dependency>
            <!-- spring cloud alibaba 依赖管理 -->
            <dependency>
                <groupId>com.alibaba.cloud</groupId>
                <artifactId>spring-cloud-alibaba-dependencies</artifactId>
                <version>${spring-cloud-alibaba.version}</version>
                <type>pom</type>
                <scope>import</scope>
            </dependency>
        </dependencies>
    </dependencyManagement>
</project>
```

步骤 04 在 ay-user-api 模块的 pom.xml 文件中添加 openfeign 依赖，具体代码如下：

```xml
<dependency>
    <groupId>org.springframework.cloud</groupId>
    <artifactId>spring-cloud-starter-openfeign</artifactId>
</dependency>
```

在 ay-user-api 模块中，开发 UserFeignApi 接口，具体代码如下：

```java
/**
 * 描述：用户接口
 * @author ay
 * @since 2020-07-24
 */
@FeignClient(name = "service-provider")
public interface UserFeignApi {

    @RequestMapping("/getUserName")
```

```
    String getUserName();
}
```

@FeignClient 注解的 name 属性用于指定 FeignClient 的名称，如果项目使用了 Ribbon，那么 name 属性会作为微服务的名称用于服务发现。

步骤 05 ay-user-service 是普通的 Spring Boot 项目，但是在创建的过程中需要使用 IDEA 勾选 Spring Web 选项，同时还需要依赖 ay-user-api 模块。pom.xml 部分配置如下。

```xml
<!-- Nacos 依赖 -->
<dependency>
    <groupId>com.alibaba.cloud</groupId>
    <artifactId>spring-cloud-starter-alibaba-nacos-discovery</artifactId>
</dependency>
<!-- Spring Web 依赖 -->
<dependency>
    <groupId>org.springframework.boot</groupId>
    <artifactId>spring-boot-starter-web</artifactId>
</dependency>
<!-- 依赖 ay-user-api 模块 -->
<dependency>
    <groupId>com.example</groupId>
    <artifactId>ay-user-api</artifactId>
    <version>0.0.1-SNAPSHOT</version>
</dependency>
```

在 ay-user-service 模块中，开发 UserFeignApiClient 类，具体代码如下：

```java
/**
 * 描述：UserFeignApiClient
 * @author ay
 * @since 2020-07-24
 */
@RestController
public class UserFeignApiClient implements UserFeignApi {

    @Override
    @RequestMapping("/getUserName")
    public String getUserName() {
        return "hello, ay";
    }
}
```

UserFeignApiClient 实现 UserFeignApi 接口，并实现 getUserName 方法，返回 "hello, ay" 字符串。

在 ay-user-service 模块中，在入口类中添加@EnableDiscoveryClient 和@EnableFeignClients 注解，具体代码如下：

```java
/**
 * 描述：入口类
 * @author ay
 * @since 2020-07-24
```

```
*/
@EnableDiscoveryClient
@SpringBootApplication
@EnableFeignClients
public class AyUserServiceApplication {

    public static void main(String[] args) {
        SpringApplication.run(AyUserServiceApplication.class, args);
    }
}
```

@EnableDiscoveryClient：可以开启服务注册发现功能。

@EnableFeignClients：用来启动 FeignClient，以支持 Feign。

步骤 06 在 ay-user-service 模块中，在 application.properties 配置文件中添加如下配置：

```
### 应用端口
server.port=8070
### 应用名称
spring.application.name=service-provider
### nacos 地址
spring.cloud.nacos.discovery.server-addr=localhost:8848
```

至此，服务提供者开发完成。

2. 创建服务消费者

接下来我们开发服务消费者，具体步骤如下所示。

步骤 01 使用 Intellij IDEA 开发工具创建 Spring Boot 项目，项目名为 spring-boot-openfeign-consumer。

步骤 02 在 pom.xml 文件中，添加 ay-user-api 依赖、Nacos 依赖，具体代码如下：

```xml
<dependencies>
    <!-- ay-user-api 依赖 -->
    <dependency>
        <groupId>com.example</groupId>
        <artifactId>ay-user-api</artifactId>
        <version>0.0.1-SNAPSHOT</version>
    </dependency>
    <!-- Spring Web 依赖 -->
    <dependency>
        <groupId>org.springframework.boot</groupId>
        <artifactId>spring-boot-starter-web</artifactId>
    </dependency>
    <!-- nacos 依赖 -->
    <dependency>
      <groupId>com.alibaba.cloud</groupId>
      <artifactId>spring-cloud-starter-alibaba-nacos-discovery</artifactId>
    </dependency>
</dependencies>
```

步骤 03 在 application.properties 配置文件中，添加如下依赖：

```properties
### 应用端口
server.port=8071
### 应用名称
spring.application.name=service-consumer
### nacos 地址
spring.cloud.nacos.discovery.server-addr=localhost:8848
```

步骤 04 在入口类中，添加@EnableDiscoveryClient 和@EnableFeignClients 注解，具体代码如下：

```java
/**
 * 描述：入口类
 * @author ay
 * @since 2020-07-24
 */
@EnableDiscoveryClient
@SpringBootApplication
@EnableFeignClients(basePackages = {"com.example.ayuserapi"})
public class DemoApplication {
    public static void main(String[] args) {
        SpringApplication.run(DemoApplication.class, args);
    }
}
```

@EnableFeignClients 注解默认会扫描注解所在包以及子包，如果需要扫描其他包下的 FeignClient，就需要单独使用属性指定。这里使用 basePackages 属性，指定扫描 ay-user-api 模块下 com.example.ayuserapi 包的 FeignClient。

步骤 05 开发 TestController 类，注入 UserFeignApi 接口，在 test 方法中调用 getUserName 方法，具体代码如下：

```java
/**
 * 描述：TestController 层
 * @author ay
 * @since 2020-07-24
 */
@RestController
public class TestController {

    @Resource
    private UserFeignApi userFeignApi;

    @RequestMapping("/test")
    public String test(){
        String userName = userFeignApi.getUserName();
        return userName;
    }
}
```

服务提供方 spring-boot-openfeign-provider 和服务消费方 spring-boot-openfeign-consumer 都开发

完成后分别启动它，在浏览器中访问 http://localhost:8071/test，打印出"hello, ay"字符串，说明 Feign 接口调用成功，具体流程如图 6-2 所示。

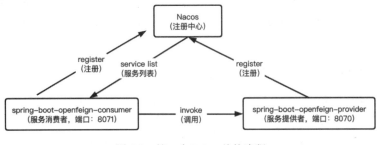

图 6-2　第一个 Feign 总体流程

6.2　FeignClient 详解与配置

6.2.1　@FeignClient 详解

本小节将详细讲解 6.1.2 小节提到的@FeignClient 注解，具体源码如下：

```
@Target(ElementType.TYPE)
@Retention(RetentionPolicy.RUNTIME)
@Documented
public @interface FeignClient {
    @AliasFor("name")
    String value() default "";
    //废弃
    @Deprecated
    String serviceId() default "";
    String contextId() default "";
    @AliasFor("value")
    String name() default "";
    String qualifier() default "";
    String url() default "";
    boolean decode404() default false;
    Class<?>[] configuration() default {};
    Class<?> fallback() default void.class;
    Class<?> fallbackFactory() default void.class;
    String path() default "";
    boolean primary() default true;
}
```

FeignClient 注解被@Target(ElementType.TYPE)修饰，表示 FeignClient 注解的作用目标在接口、类或者枚举上，其可以通过@Resource 注解注入。@FeignClient 注解属性主要包含如下内容：

1. name/value/serviceId

从源码可知，name 是 value 的别名，value 也是 name 的别名，两者的作用是一致的。name 用于指定 FeignClient 的名称，如果项目使用了 Ribbon，name 属性会作为微服务的名称，用于服务发现。serviceId 和 value 的作用一样，用于指定服务 ID，@Deprecated 说明已经废弃。

2. qualifier

该属性用来指定@Qualifier 注解的值，可以使用该值进行引用。例如：

```java
/**
 * 描述：TestFeignApi
 * @author ay
 * @since 2020-08-03
 */
@FeignClient(qualifier = "testFeignService", value = "test-service", path = "app/ay")
public interface TestFeignApi {

    @RequestMapping(value = "/test", method = RequestMethod.GET)
    String test();
}
```

调用方式如下：

```java
/**
 * @author ay
 * @since 2020-08-03
 */
@RestController
@RequestMapping("/openfeign/test")
public class TestEndpoints {
    @Autowired
    @Qualifier(value = "testFeignService")
    private TestFeignApi testFeignApi;
}
```

3. url

url 主要用于调试，可以手动指定@FeignClient 调用的地址。例如：

```java
/**
 * 描述：TestFeignApi
 * @author ay
 * @since 2020-08-03
 */
@FeignClient(value = "url-service", path = "app/ay", url = "http://localhost:8080")
public interface UrlFeignApi {

    @RequestMapping(value = "/test", method = RequestMethod.GET)
    String test();
}
```

名称和 URL 属性还支持占位符，例如：

```
@FeignClient(name = "${feign.name}", url = "${feign.url}")
public interface StoreClient {
    //...
}
```

4. path

path 用于定义当前 FeignClient 的统一前缀，例如：

```
/**
 * 描述：TestFeignApi
 * @author ay
 * @since 2020-08-03
 */
@FeignClient(value = "url-service", path = "app/ay")
public interface UrlFeignApi {

    @RequestMapping(value = "/test", method = RequestMethod.GET)
    String test();
}
```

可以通过 http://url-service/app/ay/test 请求 test()方法。

5. contextId

如果要创建多个具有相同名称或 URL 的 Feign 客户端，以便它们指向同一台服务器，但是每个客户端具有不同的自定义配置，则必须使用@FeignClient 的 contextId 属性，以避免这些配置的名称冲突，例如：

```
@FeignClient(contextId = "fooClient", name = "stores", configuration = FooConfiguration.class)
public interface FooClient {
    //...
}
```

和

```
@FeignClient(contextId = "barClient", name = "stores", configuration = BarConfiguration.class)
public interface BarClient {
    //...
}
```

6. fallback/fallbackFactory

- fallback：定义容错的处理类，当调用远程接口失败或超时时，就会调用对应接口的容错逻辑，fallback 指定的类必须实现@FeignClient 标记的接口，容错的相关内容会在接下来的章节中继续讲解。
- fallbackFactory：工厂类，用于生成 fallback 类，通过该属性可以实现每个接口通用的容错逻辑，减少重复的代码。

7. decode404

当发生 http 404 错误时,如果该字段为 true,就会调用 decoder 进行解码,否则抛出 FeignException。

8. Configuration

Feign 配置类可以自定义 Feign 的 Encoder、Decoder、LogLevel、Contract 等。Spring Cloud OpenFeign 默认为 Feign 提供以下对象(bean 类型 bean 名称:类名称):

- Decoder feignDecoder: ResponseEntityDecoder
- Encoder feignEncoder: SpringEncoder
- Logger feignLogger: Slf4jLogger
- Contract feignContract: SpringMvcContract
- Feign.Builder feignBuilder: HystrixFeign.Builder

如果 Ribbon 在类路径中且已启用,则 Client feignClient 是 LoadBalancerFeignClient,如果 Spring Cloud LoadBalancer 在类路径中,则使用 FeignBlockingLoadBalancerClient;如果 Ribbon 和 Spring Cloud LoadBalancer 都不在类路径中,则使用默认的 Feign Client。

spring-cloud-starter-openfeign 支持 spring-cloud-starter-netflix-ribbon 和 spring-cloud-starter-loadbalancer,但是由于它们是可选的依赖项,因此需要确保将要使用的依赖项添加到项目中。

通过将 feign.okhttp.enabled 或 feign.httpclient.enabled 分别设置为 true 并将它们放在类路径中来使用 OkHttpClient 和 ApacheHttpClient feign 客户端。

默认情况下,Spring Cloud OpenFeign 不会为 Feign 提供以下 bean 对象,但是仍然会从应用程序上下文中查找这些类型的 bean 以创建 Feign 客户端:

- Logger.Level。
- Retryer。
- ErrorDecoder。
- Request.Options。
- Collection<RequestInterceptor>。
- SetterFactory。
- QueryMapEncoder。

9. primary

将 Feign 与 Hystrix 一起使用时,ApplicationContext 中有多个相同类型的 bean。这将导致 @Autowired 无法正常工作,因为没有确切的一个 bean,也没有一个标记为主要的 bean。为了解决这个问题,Spring Cloud OpenFeign 将所有 Feign 实例标记为@Primary,因此 Spring Framework 将知道要注入哪个 bean。在某些情况下,这可能不是我们想要的。要关闭此行为,请将@FeignClient 的主要属性设置为 false。例如:

```
@FeignClient(name = "hello", primary = false)
public interface HelloClient {
    // methods here
}
```

以上是通过注解@FeignClient 的配置属性进行配置的，我们也可以使用 application.yml 配置文件进行配置，例如：

```yaml
feign:
  client:
    config:
      feignName:
        connectTimeout: 5000
        readTimeout: 5000
        loggerLevel: full
        errorDecoder: com.example.SimpleErrorDecoder
        retryer: com.example.SimpleRetryer
        requestInterceptors:
          - com.example.FooRequestInterceptor
          - com.example.BarRequestInterceptor
        decode404: false
        encoder: com.example.SimpleEncoder
        decoder: com.example.SimpleDecoder
        contract: com.example.SimpleContract
```

可以在@EnableFeignClients 属性 defaultConfiguration 中指定默认配置，不同之处在于此配置将适用于所有 Feign 客户端。

如果希望使用配置文件来配置所有@FeignClient，则可以使用默认 Feign 名称创建配置属性，例如：

```yaml
feign:
  client:
    config:
      default:
        connectTimeout: 5000
        readTimeout: 5000
        loggerLevel: basic
```

如果同时创建@Configuration bean 和配置文件，则配置文件将覆盖@Configuration 值。但是，如果要将优先级更改为@Configuration，就可以将 feign.client.default-to-properties 更改为 false。

还可以将 FeignClient 配置为不从父上下文继承。可以通过覆盖 FeignClientConfigurer 对象中的 InheritParentConfiguration()返回 false 做到这一点：

```java
@Configuration
public class CustomConfiguration{

@Bean
public FeignClientConfigurer feignClientConfigurer() {
        return new FeignClientConfigurer() {

            @Override
            public boolean inheritParentConfiguration() {
                return false;
            }
        };
```

```
        }
    }
```

6.2.2　Feign Hystrix 错误回退

如果 Hystrix 在类路径上并且 feign.hystrix.enabled = true，那么 Feign 将使用断路器包装所有方法，还可以返回 com.netflix.hystrix.HystrixCommand。

要以每个 Feign 客户端为基础禁用 Hystrix 支持，请创建一个具有"prototype"的范围。例如：

```
@Configuration
public class FooConfiguration {
    @Bean
    @Scope("prototype")
    public Feign.Builder feignBuilder() {
        return Feign.builder();
    }
}
```

Hystrix 支持回退的概念：当它们的电路断开或出现错误时执行的默认代码路径。要为给定的 @FeignClient 启用回退，就将 fallback 属性设置为实现回退的类名称。还需要将实现声明为 Spring bean，例如：

```
@FeignClient(name = "hello", fallback = HystrixClientFallback.class)
protected interface HystrixClient {
    @RequestMapping(method = RequestMethod.GET, value = "/hello")
    Hello iFailSometimes();
}

static class HystrixClientFallback implements HystrixClient {
    @Override
    public Hello iFailSometimes() {
        return new Hello("fallback");
    }
}
```

如果需要访问引起回退触发器的原因，则可以使用@FeignClient 中的 fallbackFactory 属性，例如：

```
@FeignClient(name = "hello", fallbackFactory =
HystrixClientFallbackFactory.class)
protected interface HystrixClient {
    @RequestMapping(method = RequestMethod.GET, value = "/hello")
    Hello iFailSometimes();
}

@Component
static class HystrixClientFallbackFactory implements
FallbackFactory<HystrixClient> {
```

```
        @Override
        public HystrixClient create(Throwable cause) {
            return new HystrixClient() {
                @Override
                public Hello iFailSometimes() {
                    return new Hello("fallback; reason was: " + cause.getMessage());
                }
            };
        }
    }
```

6.2.3 Feign @QueryMap 支持

OpenFeign 的@QueryMap 注解支持将 POJO 用作 GET 参数映射。不幸的是，默认的 OpenFeign QueryMap 注解与 Spring 不兼容，因为它缺少 value 属性。

Spring Cloud OpenFeign 提供等效的@SpringQueryMap 注解，该注解用于将 POJO 或 Map 参数注释为查询参数映射。

例如，用 Params 类定义参数 param1 和 param2：

```
public class Params {
    private String param1;
    private String param2;
    //省略 set、get 方法
}
```

紧接着，Feign 客户端通过使用@SpringQueryMap 注解来使用 Params 类：

```
@FeignClient(name = "service-provider")
public interface UserFeignApi {
    @GetMapping(path = "/demo")
    String demoEndpoint(@SpringQueryMap Params params);
}
```

调用方代码如下：

```
Params params = new Params();
params.setParam1("hello");
params.setParam2("ay");
String result = userFeignApi.demoEndpoint(params);
```

6.2.4 HATEOAS 支持

如果项目使用 org.springframework.boot:spring-boot-starter-hateoas starter 或 org.springframework.boot:spring-boot-starter-data-rest starter，则默认启用 Feign HATEOAS 支持。启用 HATEOAS 支持后，允许 Feign 客户端对 HATEOAS 表示模型进行序列化和反序列化，即 EntityModel、CollectionModel 和 PagedModel，具体实例如下：

```
@FeignClient("demo")
```

```
public interface DemoTemplate {

    @GetMapping(path = "/stores")
    CollectionModel<Store> getStores();
}
```

6.2.5　Spring @MatrixVariable 支持

Spring Cloud OpenFeign 为 Spring @MatrixVariable 注释提供支持。

如果将 Map 作为方法参数传递，则@MatrixVariable 路径段是通过将 Map 中的键/值对与=连接而创建的。

例如，请求地址"http://localhost:8080/objects/links/matrixVars;name=ay,sam"，对应的控制层代码如下所示：

```
@GetMapping("/objects/links/{matrixVars}")
Map<String, List<String>> getObjects(@MatrixVariable Map<String, List<String>> matrixVars);
```

注意，变量名称和路径段占位符都称为 matrixVars。

6.2.6　Feign 继承支持

Feign 支持接口继承方式快速生成客户端，例如：

1. UserService.java

```
public interface UserService {
    @RequestMapping(method = RequestMethod.GET, value ="/users/{id}")
    User getUser(@PathVariable("id") long id);
}
```

2. UserResource.java

```
@RestController
public class UserResource implements UserService {
}
```

3. UserClient.java

```
package project.user;

@FeignClient("users")
public interface UserClient extends UserService {

}
```

6.2.7 Feign CollectionFormat 支持

通过提供 @CollectionFormat 注解支持 feign.CollectionFormat。可以通过传递所需的 feign.CollectionFormat 作为注释值来对其进行注释。

在下面的示例中，使用 CSV 格式而不是默认的 EXPLODED 来处理该方法。

```
@FeignClient(name = "service-provider", configuration = LogConfiguration.class)
public interface UserFeignApi {
    @CollectionFormat(feign.CollectionFormat.CSV)
    @GetMapping(value = "/search/findByIdIn")
    String getByIds(@RequestParam("ids") Long[] ids);
}
```

其他服务调用该接口时，请求地址将是"http://service-provider/search/findByIdIn?ids=1,2,3"。当然还可以选择其他分隔符：

```
public enum CollectionFormat {
    CSV(","),
    SSV(" "),
    TSV("\t"),
    PIPES("|"),
    EXPLODED((String)null);
}
```

6.2.8 Feign 请求响应压缩

可以考虑为 Feign 请求启用请求或响应 GZIP 压缩。可以通过启用以下属性之一来做到这一点：

```
feign.compression.request.enabled=true
feign.compression.response.enabled=true
```

开启压缩可以有效节约网络资源，但是会增加 CPU 压力。

Feign 请求压缩提供的设置类似于 Web 服务器设置：

```
feign.compression.request.enabled=true
//配置压缩文档类型
feign.compression.request.mime-types=text/xml,application/xml,application/json
//配置最小压缩的文档大小
feign.compression.request.min-request-size=2048
```

这些属性使你可以选择压缩媒体类型和最小请求阈值长度。

对于 OkHttpClient 以外的 HTTP 客户端，可以启用默认的 gzip 解码器以 UTF-8 编码解码 gzip 响应：

```
feign.compression.response.enabled=true
feign.compression.response.useGzipDecoder=true
```

6.3　Feign 日志配置

Feign 可以非常灵活地处理日志，当 Feign Client 被创建时，就创建了 logger 日志记录策略。默认情况下，logger 的名称是创建 Feign Client 的服务接口类的全路径，即加了 @FeignClient 接口类的全路径。Feign 日志支持 Java 代码配置以及配置文件配置两种方式。

6.3.1　Java 代码方式

创建 LogConfiguration 配置类，具体代码如下：

```java
/**
 * 描述：日志配置
 * @author ay
 * @since 2020-08-03
 */
public class LogConfiguration {
    @Bean
    Logger.Level feignLoggerLevel(){
        return Logger.Level.FULL;
    }
}
```

Logger.Level 的值有以下几种选择：

- NONE：无记录（DEFAULT）。
- BASIC：只记录请求方法、URL 以及响应状态代码和执行时间。
- HEADERS：记录基本信息以及请求和响应标头。
- FULL：记录请求和响应的头文件、正文和元数据。

在 application.yml 配置文件中添加配置：

```yaml
### 日志级别
logging:
  level:
    com.example.ayuserapi.UserFeignApi: DEBUG
```

需要注意的是，Feign 的日志打印只会对 DEBUG 级别做出响应。Feign Client 代码如下：

```java
/**
 * 描述：用户接口
 * @author ay
 * @since 2020-07-24
 */
@FeignClient(name = "service-provider", configuration = LogConfiguration.class)
public interface UserFeignApi {
```

```
    @RequestMapping("/getUserName")
    String getUserName();
}
```

6.3.2 配置文件方式

配置文件方式更为简单，只需在 application.yml 中添加以下配置即可：

```yaml
logging:
  level:
    com.example.ayuserapi.UserFeignApi: DEBUG

feign:
  client:
    config:
      ### 要调用服务的名称
      service-provider:
        loggerLevel: full
```

6.3.3 全局日志配置

1. Java 代码方式

在启动类@EnableFeignClients 注解上配置 defaultConfiguration，具体代码如下：

```java
/**
 * 描述：入口类
 * @author ay
 * @since 2020-07-24
 */
@EnableDiscoveryClient
@SpringBootApplication
### 全局日志配置
@EnableFeignClients(defaultConfiguration = GlobalFeignConfiguration.class)
public class AyUserServiceApplication {

    public static void main(String[] args) {
        SpringApplication.run(AyUserServiceApplication.class, args);
    }
}
```

定义 GlobalFeignConfiguration 类，具体代码如下：

```java
/**
 * 全局日志配置
 * @author : ay
 * @since 2020-08-03
 **/
public class GlobalFeignConfiguration {
```

```
@Bean
Logger.Level feignLoggerLevel(){
    return Logger.Level.FULL;
}
```

在配置文件 application.yml 中需配置日志级别方能打印出 Feign 调用的日志信息：

```
### 全局日志配置
logging:
  level:
    com.example: DEBUG
```

2. 配置文件方式

只需在 application.yml 中添加如下配置：

```
### 全局日志配置
logging:
  level:
    com.example: DEBUG
feign:
  client:
    config:
      ### feign 全局日志级别
      default:
        loggerLevel: full
```

6.4 自定义处理

6.4.1 Feign 自定义错误

在 Spring Cloud 微服务中，Feign 接口之间相互调用，我们都希望 Feign 返回统一的报文格式，例如：

```
{
  "msg": "请求成功",
  "code": 200,
  "data": {}
}
```

当被调用的 Feign 接口抛出异常时，异常是被封装过的，无法获得有用的信息，所以需要自定义异常返回信息，具体步骤如下所示。

步骤 01 封装状态码：

```
/**
 * 描述：封装状态码
```

```
 * @author:ay
 * @since :2020-08-04
 */
public enum ResultCode {

    SUCCESS(0000, "成功"),
    FAIL(9999, "失败"),
    SYSTEM_EXCEPTION(4000, "系统异常");

    private Integer code;
    private String message;

    ResultCode(Integer code, String message) {
        this.code = code;
        this.message = message;
    }
    //省略 get 方法
}
```

步骤 02 封装 Result 对象:

```
/**
 * 描述：封装 result 对象
 * @author ay
 * @since 2020-08-04
 * @param <T>
 */
@Data
public class Result<T> implements Serializable {
    private Integer code;
    private String msg;
    private T data;
    public Result(){}
    public Result(Integer code, String msg, T data) {
        this.code = code;
        this.msg = msg;
        this.data = data;
    }
    public static <T> Result<T> success(String msg) {
        return new Result<T>(ResultCode.SUCCESS.getCode(), msg, null);
    }
    public static <T> Result<T> success(T model, String msg) {
        return new Result<>(ResultCode.SUCCESS.getCode(), msg, model);
    }
    public static <T> Result<T> fail(String msg, Integer code) {
        return new Result<>(code, msg, null);
    }
}
```

步骤 03 自定义异常 BaseException 类:

```java
/**
 * 描述：自定义异常
 * @author:ay
 * @since :2020-08-04
 */
@Data
public class BaseException extends RuntimeException {

    private Integer code;
    private String msg;

    public BaseException() {
        super();
    }

    public BaseException(ResultCode resultCode) {
        super(resultCode.getMessage());
        this.code = resultCode.getCode();
        this.msg = resultCode.getMessage();
    }

    public BaseException(Integer code, String message) {
        super(message);
        this.code = code;
        this.msg = message;
    }
}
```

步骤 04 自定义异常解码类 FeignErrorDecoder：

```java
/**
 * 自定义 Feign 异常处理
 * @author:ay
 * @since :2020-08-04
 */
@Slf4j
public class FeignErrorDecoder implements ErrorDecoder {

    @Override
    public Exception decode(String s, Response response) {
        BaseException baseException = null;
        try {
            String errorContent = Util.toString(response.body().asReader());
            Result result = JSONObject.parseObject(errorContent, Result.class);
            baseException = new BaseException(result.getCode(), result.getMsg());
        } catch (Exception ignore) {

        }
        return baseException;
    }
}
```

第 6 章 Spring Cloud OpenFeign 声明式调用

步骤 05 自定义全局异常处理类 GlobalExceptionHandler：

```java
/**
 * 描述：全局异常处理
 * @author:ay
 * @since :2020-08-04
 */
@Slf4j
@RestControllerAdvice
public class GlobalExceptionHandler {

    /**
     * 系统异常处理器
     * @param throwable
     * @param request
     * @return
     */
    @ExceptionHandler(Throwable.class)
    @ResponseStatus(HttpStatus.INTERNAL_SERVER_ERROR)
    public Result systemExceptionHandler(Throwable throwable,
HttpServletRequest request) {
        log.error("URL: {}, 系统异常", request.getRequestURI(), throwable);
        return Result.fail("系统异常", ResultCode.SYSTEM_EXCEPTION.getCode());
    }

    /**
     * 自定义异常处理器
     */
    @ExceptionHandler(BaseException.class)
    @ResponseStatus(HttpStatus.INTERNAL_SERVER_ERROR)
    public Result baseExceptionHandler(BaseException baseException,
HttpServletRequest request) {
        log.warn("URL: {}, 业务异常", request.getRequestURI());
        return Result.fail(baseException.getMsg(), baseException.getCode());
    }

}
```

步骤 06 定义配置类 FeignDecoderConfig：

```java
/**
 * 描述：feign 配置类
 * @author  ay
 * @since   2020-8-04
 */
@Configuration
public class FeignDecoderConfig {

    @Bean
```

```java
    public FeignErrorDecoder errorDecoder() {
        return new FeignErrorDecoder();
    }
}
```

6.4.2 Feign 拦截器

Feign 拦截器的作用主要是为了在微服务之间传递请求信息，比如头信息。创建 Feign 拦截器有以下几种方式：

（1）创建 FeignInterceptor 配置类，具体代码如下：

```java
/**
 * 描述：Feign 拦截器
 * @author ay
 * @since 2020-08-04
 */
public class FeignInterceptorConfig {
    @Bean
    public RequestInterceptor requestInterceptor() {
        return template -> {
            ServletRequestAttributes attributes = (ServletRequestAttributes) RequestContextHolder
                    .getRequestAttributes();
            HttpServletRequest request = attributes.getRequest();
            Enumeration<String> headerNames = request.getHeaderNames();
            if (headerNames != null) {
                while (headerNames.hasMoreElements()) {
                    String name = headerNames.nextElement();
                    String values = request.getHeader(name);
                    template.header(name, values);
                }
            }
        };
    }
}
```

在启动类上加@Import(FeignInterceptorConfig.class)。

（2）实现 RequestInterceptor 接口，具体代码如下：

```java
/**
 * 描述：Feign 拦截器
 * @author ay
 * @since 2020-08-04
 */
public class FeignRequestInterceptor implements RequestInterceptor {
    @Override
    public void apply(RequestTemplate requestTemplate) {
        ServletRequestAttributes attributes = (ServletRequestAttributes)
```

```
RequestContextHolder
                .getRequestAttributes();
        HttpServletRequest request = attributes.getRequest();
        Enumeration<String> headerNames = request.getHeaderNames();
        if (headerNames != null) {
            while (headerNames.hasMoreElements()) {
                String name = headerNames.nextElement();
                String values = request.getHeader(name);
                requestTemplate.header(name, values);
            }
        }
    }
}
```

（3）创建全局配置类 FeignInterceptorConfig，具体代码如下：

```
/**
 * 描述：Feign 拦截器配置类
 * @author ay
 * @since 2020-08-04
 */
@Configuration
public class FeignInterceptorConfig {
    @Bean
    public RequestInterceptor requestInterceptor() {
        return new FeignRequestInterceptor();
    }
}
```

6.4.3 自定义 Feign 客户端

在某些情况下，需要自定义 Feign 客户端，这时可以使用 Feign Builder API 来完成。下面给出一个示例，创建具有相同接口的两个 Feign Client，但为每个 Feign Client 配置一个单独的请求拦截器，具体代码如下：

```
@Import(FeignClientsConfiguration.class)
class FooController {

    private FooClient fooClient;

    private FooClient adminClient;

    @Autowired
    public FooController(Decoder decoder, Encoder encoder, Client client,
Contract contract) {
        this.fooClient = Feign.builder().client(client)
                .encoder(encoder)
                .decoder(decoder)
                .contract(contract)
//设置请求拦截器
```

```java
        .requestInterceptor(new BasicAuthRequestInterceptor("user", "user"))
                .target(FooClient.class, "https://PROD-SVC");

        this.adminClient = Feign.builder().client(client)
                .encoder(encoder)
                .decoder(decoder)
                .contract(contract)
//设置请求拦截器
        .requestInterceptor(new BasicAuthRequestInterceptor("admin", "admin"))
                .target(FooClient.class, "https://PROD-SVC");
    }
}
```

在上面的示例中，FeignClientsConfiguration.class 是 Spring Cloud OpenFeign 提供的默认配置，PROD-SVC 是客户将向其请求的服务名称。Feign Contract 对象定义接口上有效的注解和值。自动装配的 Contract bean 提供了对 SpringMVC 注解的支持。

第 7 章

熔断、限流、降级

本章主要介绍熔断、限流和降级的相关组件，包括 Hystrix 简介、Hystrix 请求缓存和请求合并、Spring Boot 应用配置 Hystrix 仪表盘、Turbine 集群监控、阿里 Sentinel 组件简介、常用的限流算法、Sentinel 与 Hystrix 的区别、Sentinel 如何进行限流和熔断降级等内容。

7.1 Spring Cloud Hystrix

7.1.1 Hystrix 简介

Netflix Hystrix 是 SOA/微服务架构中提供服务隔离、熔断、降级机制的工具/框架。Netflix Hystrix 是断路器的一种实现，用于高微服务架构的可用性，是防止服务出现雪崩的利器。

在分布式架构中，一个应用依赖多个服务是非常常见的。如果其中一个依赖由于延迟过高发生阻塞，调用该依赖服务的线程就会阻塞。如果相关业务的 QPS 较高，就可能产生大量阻塞，从而导致该应用/服务由于服务器资源被耗尽而拖垮。

另外，故障也会在应用之间传递，如果故障服务的上游依赖较多，就可能会引起服务的雪崩效应，跟数据瘫痪会引起依赖该数据库的应用瘫痪是一样的道理。

一个应用依赖多个外部服务并且一切都正常的情况如图 7-1 所示。如果其中一个依赖发生延迟，那么当前请求就会被阻塞，如图 7-2 所示。出现这种情况后，如果没有应对措施，后续的请求也会被持续阻塞，具体如图 7-3 所示。

每个请求都占用了系统的 CPU、内存、网络等资源。如果该应用的 QPS 较高，那么该应用所在的服务资源会被快速消耗完毕，直至应用死掉。如果这个出问题的依赖（Dependency I）不止一个应用或者受影响的应用上层也有更多的依赖，就会带来我们前面所提到的服务雪崩效应。为了应对以上问题，就需要有支持服务隔离、熔断等操作的工具。

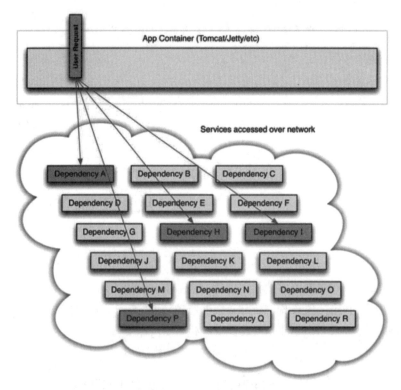

图 7-1　服务级联故障图 1

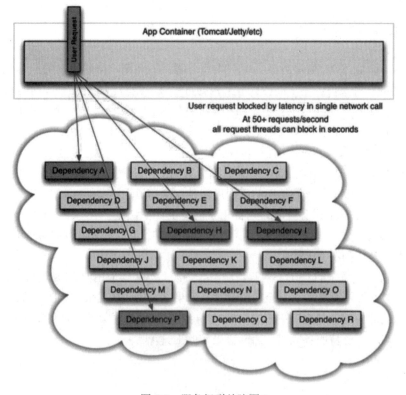

图 7-2　服务级联故障图 2

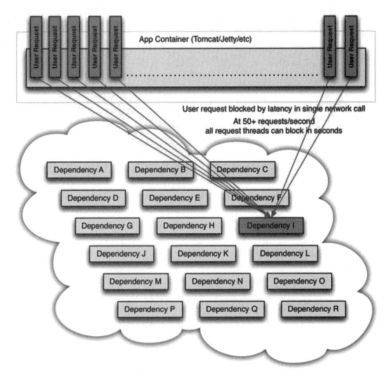

图 7-3　服务级联故障图 3

　　Hystrix 的中文名字是"豪猪"。豪猪周身长满了刺,能保护自己不受天敌的伤害,代表了一种防御机制,这与 Hystrix 本身的功能不谋而合,因此 Netflix 团队将该框架命名为 Hystrix,并使用对应的卡通形象作为 LOGO,如图 7-4 所示。

图 7-4　Hystrix 的 LOGO 图片

　　Hystrix 具有隔离(线程池隔离、信号量隔离)服务降级、熔断、限流、缓存等功能,基本上能覆盖到微服务中调用依赖服务会遇到的问题。

1. 隔离

　　线程池隔离:每个服务对应一个线程池,线程池满了将会进行降级。使用线程池存储当前的请求,线程池对请求做处理,设置任务返回处理超时时间,堆积的请求进入线程池队列。这种方式需要为每个依赖的服务申请线程池,有一定的资源消耗,好处是可以应对突发流量(流量洪峰来临时,处理不完可将数据存储到线程池队里慢慢处理)。

信号量隔离：基于 Tomcat 线程池来控制，当线程达到某个百分比时将拒绝访问走降级流程。信号量的资源隔离只是起到一个开关的作用，比如服务 A 的信号量大小为 10，就是说它同时只允许有 10 个 Tomcat 线程来访问服务 A，其他请求都会被拒绝，从而达到资源隔离和限流保护的作用。

2. 限流

限流就是信号量隔离（一般不会使用该模式）。

3. 熔断

触发快速失败，保证系统可用性。

4. 降级

使用回调方法返回托底数据。

Hystrix 被设计的目标是：对通过第三方客户端访问的依赖项（通常通过网络）的延迟和故障进行保护和控制；在复杂的分布式系统中阻止级联故障；快速失败，快速恢复；回退，尽可能优雅地降级；启用类似实时监控、警报和操作的控制。

7.1.2 Hystrix 初体验

了解 Hystrix 能为我们解决什么问题后，接下来学习如何使用 Hystrix 实现降级，具体步骤如下：

步骤 01 使用 Intellij IDEA 快速创建 Spring Boot 项目，项目名称为 spring-cloud-hystrix-demo，同时需要勾选 Spring Web 依赖和 Hystrix 依赖，如图 7-5 所示。

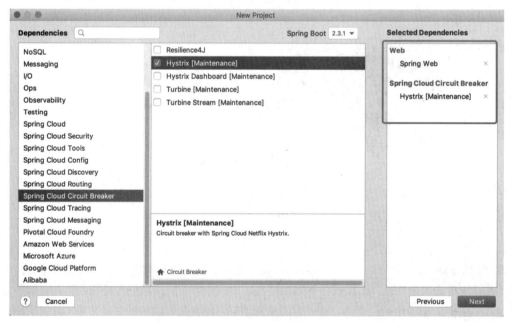

图 7-5　勾选 Spring Web 依赖和 Hystrix 依赖

步骤02 在项目的入口类中添加@EnableHystrix注解，具体代码如下：

```java
/**
 * 描述：入口类
 *
 * @author ay
 * @date 2019-03-25
 */
@SpringBootApplication
@EnableHystrix
public class DemoApplication {
    public static void main(String[] args) {
        SpringApplication.run(DemoApplication.class, args);
    }
}
```

在启动类上添加@EnableHystrix 来开启 Hystrix 的断路器功能。@EnableHystrix 继承了 @EnableCricuitBreaker，所以也可以在启动类上添加@EnableCircuitBreaker 注解。

步骤03 在项目的目录 src/main/java 下创建 HelloController 类，具体代码如下：

```java
/**
 * 描述
 *
 * @author ay
 * @date 2019-03-24
 */
@RestController
public class HelloController {
    @RequestMapping(value = "/hello")
    @HystrixCommand(fallbackMethod = "fallbackHello", commandProperties = {
        @HystrixProperty(name = "execution.isolation.thread.timeoutInMilliseconds",
        value = "1000")
    })
    public String hello() throws InterruptedException {
        //模拟访问超时
        Thread.sleep(3000);
        return "Hello Hystrix";
    }

    /**
     * 请求失败，调用方法
     */
    private String fallbackHello() {
        return "Request fails. It takes long time to response";
    }
}
```

在@HystrixCommand 中使用 execution.isolation.thread.timeoutInMilliseconds 设置超时时间，若业务执行时间超过这个时间，则执行 fallbackMethod 方法并返回客户端。在 hello 方法中，使用

Thread.sleep(3000)模拟访问超时。

@HystrixCommand 注解的常用参数如下：

- fallbackMethod：指定服务降级处理方法。
- ignoreExceptions：忽略某些异常，不发生服务降级。
- commandKey：命令名称，用于区分不同的命令。
- groupKey：分组名称，Hystrix 会根据不同的分组来统计命令的告警及仪表盘信息。
- threadPoolKey：线程池名称，用于划分线程池。

例如：

```
@RestController
public class HelloController {

    @GetMapping("/testHystrixCommand/{id}")
    @HystrixCommand(fallbackMethod = "fallbackMethod",
                commandKey = "testHystrixCommand",
                groupKey = "testHystrixCommandGroup",
                ignoreExceptions = {NullPointerException.class},
                threadPoolKey = "hystrixCommandThreadPool")
    public String testHystrixCommand(@PathVariable Long id){
        return "testHystrixCommand";
    }

    /**
     * 请求失败，调用方法
     */
    public String fallbackMethod(){
        //请求失败，自定义失败返回内容
        return "testHystrixCommand Request fails";
    }
}
```

testHystrixCommand 方法设置对于 NullPointerException 异常不进行异常降级。

步骤 04 运行入口类的 main 方法启动项目，在浏览器中输入请求路径"http://localhost:8080/hello"，或者在命令行窗口中输入 curl 命令 "curl http://localhost:8081/hello"，可看到浏览器或者命令行窗口返回"Request fails. It takes long time to response"。

7.1.3　Hystrix 请求缓存

Hystrix 请求缓存是指在一次请求中多次访问同一服务提供者的接口，请求将第一次服务提供者接口的结果进行缓存，缓存对象的生命周期在请求结束后自动清除。

Hystrix 请求缓存提供继承和注解的方式，注解方式比继承方式更方便使用。Hystrix 提供以下注解：

- @CacheResult：该注解用来标记请求命令返回的结果应该被缓存，它必须与

@HystrixCommand 注解结合使用。
- @CacheKey：该注解用来在请求命令的参数上进行标记，使其作为 cacheKey。如果没有使用此注解就会使用所有参数列表中的参数作为 cacheKey。
- @CacheRemove：该注解用来让请求命令的缓存失效，失效的缓存根据 commandKey 进行查找。

例如：

```java
/**
 * 描述：HystrixCacheController 控制层
 *
 * @author ay
 * @since 2020/9/13
 */
@RestController
public class HystrixCacheController {
    @Resource
    private UserService userService;

    @GetMapping("/testHystrixCache/{id}")
    public String testHystrixCache(@PathVariable Long id) {
        userService.getUserCache(id);
        userService.getUserCache(id);
        userService.getUserCache(id);
        return "操作成功";
    }
}
```

在 UserService 中添加具有缓存功能的 getUserCache 方法：

```java
/**
 * 描述：
 *
 * @author ay
 * @since 2020/9/13
 */
@Component
public class UserServiceImpl implements UserService {

    @CacheResult(cacheKeyMethod = "getCacheKey")
    @HystrixCommand(fallbackMethod = "getDefaultUser", commandKey = "getUserCache")
    public String getUserCache(Long id) {
        //打印日志
        Log.info("getUserCache id:{}", id);
        //数据库查询用户数据
        return "";
    }

    /**
```

```
 * 为缓存生成 key 的方法
 */
public String getCacheKey(Long id) {
    return String.valueOf(id);
}
```

在浏览器中输入访问地址"http://localhost:8080/testHystrixCache/1",虽然在控制层的 testHystrixCache 方法中调用三次 getUserCache 方法,但是控制台只会打印一次日志,后两次请求直接获取缓存数据。

在 UserService 中添加具有移除缓存功能的 removeCache 方法:

```
/**
 * 描述:HystrixCacheController 控制层
 *
 * @author ay
 * @since 2020/9/13
 */
@RestController
public class HystrixCacheController {
    @Resource
    private UserService userService;

    @GetMapping("/testHystrixCache/{id}")
    public String testRemoveHystrixCache(@PathVariable Long id) {
        userService.getUserCache(id);
        //清除缓存
        userService.removeCache(id);
        userService.getUserCache(id);
        return "操作成功";
    }
}


/**
 * 清除缓存
 * @param id
 * @return
 */
@CacheRemove(commandKey = "getUserCache", cacheKeyMethod = "getCacheKey")
@HystrixCommand
public String removeCache(Long id) {
    log.info("removeCache id:{}", id);
    return "";
}
```

注意,removeCache 方法必须指定 commandKey 才能清除指定缓存。

在缓存使用过程中,需要在每次使用缓存的请求前后对 HystrixRequestContext 进行初始化和关闭,否则会出现如下异常:

```
java.lang.IllegalStateException: Request caching is not available. Maybe you
need to initialize the HystrixRequestContext?
    at
com.netflix.hystrix.HystrixRequestCache.get(HystrixRequestCache.java:104)
~[hystrix-core-1.5.18.jar:1.5.18]
        at com.netflix.hystrix.AbstractCommand$7.call(AbstractCommand.java:478)
~[hystrix-core-1.5.18.jar:1.5.18]
        at com.netflix.hystrix.AbstractCommand$7.call(AbstractCommand.java:454)
~[hystrix-core-1.5.18.jar:1.5.18]
```

通过使用过滤器在每个请求前后初始化和关闭 HystrixRequestContext 来解决该问题：

```
/**
 * Created by macro on 2019/9/4.
 */
@Component
@WebFilter(urlPatterns = "/*",asyncSupported = true)
public class HystrixRequestContextFilter implements Filter {
    @Override
    public void doFilter(ServletRequest servletRequest, ServletResponse
servletResponse, FilterChain filterChain) throws IOException, ServletException {
        HystrixRequestContext context =
HystrixRequestContext.initializeContext();
        try {
            filterChain.doFilter(servletRequest, servletResponse);
        } finally {
            //关闭 HystrixRequestContext
            context.close();
        }
    }
}
```

7.1.4　Hystrix 请求合并

在分布式微服务架构中，服务消费者对服务提供者发起调用，线程池会申请与请求数量相同的线程数，具体如图 7-6 所示。

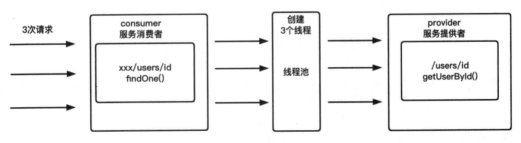

图 7-6　请求未合并调用

在高并发情况下，会导致服务提供者产生巨大的压力。Hystrix 请求合并就是将多个请求合并成一个请求去调用服务提供者（见图 7-7），从而降低服务提供者负载。

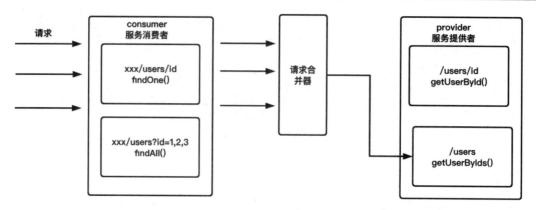

图 7-7 请求合并调用

在服务提供者提供了返回单个对象和多个对象的查询接口、单个对象的查询并发数很高并且服务提供者负载较高的时候，可以使用请求合并来降低服务提供者的负载。

Hystrix 通过请求合并器设置延迟时间，将时间内的多个 findOne 请求方法中的参数（id）取出来，拼成符合服务提供者的多个对象返回接口（getUsersByIds 方法）的参数，指定调用这个接口（getUsersByIds 方法），返回的对象 List 再通过 mapResponseToRequests 方法按照请求的次序将结果对象封装到 Response 返回结果。

Hystrix 合并请求也会带来相应弊端：

- Hystrix 在请求上设置延迟时间，在并发不高的接口上会降低响应速度。
- 如果返回 List 的方法并发比返回单个对象方法负载更高，有可能会提高服务提供者的负载。
- 实现请求合代码会比较复杂。

下面看一个具体实例：

```java
/**
 * 描述：请求合并
 *
 * @author ay
 * @since 2020/9/13
 */
@RestController
public class CollapserHystrixController {

    @Resource
    private UserService userService;

    @GetMapping("/testCollapser")
    public String testCollapser() throws Exception{
        //创建线程
        new Thread(){
            @Override
            public void run() {
                User user1 = userService.getUserById(1L);
            }
        }.start();
```

```java
        //创建线程
        new Thread(){
            @Override
            public void run() {
                User user1 = userService.getUserById(2L);
            }
        }.start();
        //创建线程
        new Thread(){
            @Override
            public void run() {
                User user1 = userService.getUserById(3L);
            }
        }.start();
        return "操作成功";
    }
}
```

在 CollapserHystrixController 类中,创建多个线程,同时调用 UserService 类的 getUserById 方法。getUserById 方法的具体代码如下:

```java
/**
 * 描述:
 *
 * @author ay
 * @since 2020/9/13
 */
@Component
public class UserServiceImpl implements UserService {

    Logger log = Logger.getLogger("");

    @HystrixCollapser(batchMethod = "getUserByIds",
            scope = com.netflix.hystrix.HystrixCollapser.Scope.GLOBAL,
            collapserProperties = {
        @HystrixProperty(name = "timerDelayInMilliseconds", value = "1000")
    })
    @Override
    public User getUserById(Long id) {
        //模拟查询数据库
        User user = new User();
        user.setId(id);
        log.info("getUserById,id:" + id);
        return user;
    }
    @HystrixCommand
    @Override
    public List<User> getUserByIds(List<Long> ids) {
        log.info("getUserByIds,ids:" + ids);
        List<User> users = new ArrayList<>();
        //模拟查询数据库
```

```
        for(Long id: ids){
            User user = new User();
            user.setId(id);
            users.add(user);
        }
        return users;
    }
}
```

@HystrixCollapser 注解的常用属性如下：

- batchMethod：用于设置请求合并的方法。
- collapserProperties：请求合并属性，用于控制实例属性有很多，例如 scope 属性，scope=Scope.GLOBAL 表示对所有线程请求中多次服务请求进行合并，scope=Scope.REQUEST 表示对一次请求的多次服务调用进行合并。
- timerDelayInMilliseconds：collapserProperties 中的属性，用于控制每隔多少时间合并一次请求。

使用@HystrixCollapser 实现请求合并，所有对 getUserById 方法的多次调用都会转化为对 getUserByIds 的单次调用。注意，getUserByIds 方法需要添加@HystrixCommand 注解。

在浏览器中多次快速访问地址 "http://localhost:8080/testCollapser"，控制台将输出如下信息：

```
2020-09-13 21:29:03.581  INFO 8651 ---: getUserByIds,ids:[3, 1, 2]
2020-09-13 21:29:06.582  INFO 8651 ---: getUserByIds,ids:[1, 3, 3, 1, 2, 2]
2020-09-13 21:29:07.582  INFO 8651 ---: getUserByIds,ids:[1, 3, 2]
2020-09-13 21:29:08.580  INFO 8651 ---: getUserByIds,ids:[2, 3, 1]
```

> **提 示**
>
> 使用 Hystrix 请求合并，需要开启上下文对象 HystrixRequestContext，其实就是使用 TheardLoacl 初始化一个上下文对象。

7.1.5 Hystrix 默认配置

将 spring.cloud.circuitbreaker.hystrix.enabled 设置为 false 可以禁用断路器 Hystrix 的自动配置。为所有的断路器提供一个默认配置，需要创建一个 Customizer 对象并传递一个 HystrixCircuitBreakerFactory 或者 ReactiveHystrixCircuitBreakerFactory。configureDefault 方法将提供一个默认配置，具体代码如下：

```
@Bean
public Customizer<HystrixCircuitBreakerFactory> defaultConfig() {
    return factory -> factory.configureDefault(id -> HystrixCommand.Setter
        .withGroupKey(HystrixCommandGroupKey.Factory.asKey(id))
        .andCommandPropertiesDefaults(HystrixCommandProperties.Setter()
        .withExecutionTimeoutInMilliseconds(4000)));
}
```

与提供默认配置类似，可以创建一个 Customize 对象，并将其传递给

HystrixCircuitBreakerFactory，具体代码如下：

```
@Bean
public Customizer<HystrixCircuitBreakerFactory> customizer() {
    return factory ->
    factory.configure(builder -> builder.commandProperties(
 HystrixCommandProperties.Setter().withExecutionTimeoutInMilliseconds(2000)),
"foo", "bar");
    }
```

7.1.6 Hystrix 配置详解

Hystrix 使用 Archaius 作为配置属性的默认实现，下面描述默认的 HystrixPropertiesStrategy 实现。

每个 Hystrix 参数都有 4 个地方可以配置，优先级从低到高如下，如果每个地方都配置相同的属性，则优先级高的值会覆盖优先级低的值：

- 全局默认值：如果未设置以下三个，则为默认设置，该属性是通过代码设置的。
- 动态全局默认属性：通过属性文件配置全局的值。
- 实例默认值：通过代码为实例定义的默认值。
- 动态实例属性：通过配置文件来为指定的实例进行属性配置，以覆盖前面的三个默认值。

Hystrix 配置如表 7-1 所示。

表 7-1　Hystrix 配置详解

	配置	描述
Command Properties（Execution）属性控制HystrixCommand.run()如何执行	execution.isolation.strategy	表示 HystrixCommand.run()执行时的隔离策略，有以下两种策略： ● THREAD：在单独的线程上执行，并发请求受线程池中的线程数限制 ● SEMAPHORE：在调用线程上执行，并发请求量受信号量计数限制 在默认情况下，推荐 HystrixCommands 使用 thread 隔离策略，HystrixObservableCommand 使用 semaphore 隔离策略。只有在高并发（单个实例每秒达到几百个调用）的调用时才需要修改 HystrixCommands 的隔离策略为 semaphore。semaphore 隔离策略通常只用于非网络调用
	execution.isolation.thread.timeoutInMilliseconds	设置调用者执行的超时时间（单位毫秒），默认值为 1000
	execution.timeout.enabled	表示是否开启超时设置，默认值为 true
	execution.isolation.thread.interruptOnTimeout	表示设置是否在执行超时时中断 HystrixCommand.run() 的执行，默认值为 true

(续表)

	配置	描述
Command Properties（Fallback）控制 HystrixCommand.getFallback() 如何执行。这些属性对隔离策略 THREAD 和 SEMAPHORE 都起作用	execution.isolation.thread.interruptOnCancel	表示设置是否在取消任务执行时中断 HystrixCommand.run() 的执行，默认值为 false
	execution.isolation.semaphore.maxConcurrentRequests	当 HystrixCommand.run() 使用 SEMAPHORE 的隔离策略时，设置最大的并发量，默认值为 10
	fallback.isolation.semaphore.maxConcurrentRequests	设置从调用线程允许 HystrixCommand.getFallback() 方法允许的最大并发请求数。如果达到最大的并发量，则接下来的请求会被拒绝并且抛出异常，默认值为 10
	fallback.enabled	是否开启 fallback 功能，默认值为 true
Command Properties（Circuit Breaker）控制断路器的行为	circuitBreaker.enabled	是否开启断路器功能，默认值为 true
	circuitBreaker.requestVolumeThreshold	设置滚动窗口中将使断路器跳闸的最小请求数量。如果此属性值为 20，在窗口时间内（如 10s 内）只收到 19 个请求且都失败了，则断路器也不会开启，默认值为 20
	circuitBreaker.sleepWindowInMilliseconds	断路器跳闸后，在此值的时间内 Hystrix 会拒绝新的请求，只有过了这个时间断路器才会打开闸门，默认值为 5000
	circuitBreaker.errorThresholdPercentage	设置失败百分比的阈值。如果失败比率超过这个值，则断路器跳闸并且进入 fallback 逻辑，默认值为 50
	circuitBreaker.forceOpen	如果设置为 true，则强制使断路器跳闸，会拒绝所有的请求。此值会覆盖 circuitBreaker.forceClosed 的值，默认值为 false
	circuitBreaker.forceClosed	如果设置为 true，则强制使断路器进行关闭状态，此时会允许执行所有请求，无论失败的次数是否达到 circuitBreaker.errorThresholdPercentage 值，默认值为 false
Metrics HystrixCommand 和 HystrixObservableCommand 执行信息相关的配置属性	metrics.rollingStats.timeInMilliseconds	设置统计滚动窗口的时间长度。如果此值为 10s，将窗口分成 10 个桶，每个桶表示 1s 时间，默认值为 10000
	metrics.rollingStats.numBuckets	设置统计滚动窗口的桶数量，默认值为 10
	metrics.rollingPercentile.enabled	设置执行延迟是否被跟踪，并且被计算在失败百分比中。如果设置为 false，则所有的统计数据返回-1，默认值为 true
	metrics.rollingPercentile.timeInMilliseconds	此属性设置统计滚动百分比窗口的持续时间，默认值为 60000
	metrics.rollingPercentile.numBuckets	设置统计滚动百分比窗口的桶数量，默认值为 6

(续表)

	配置	描述
	metrics.rollingPercentile.bucketSize	此属性设置每个桶保存的执行时间的最大值。如果桶数量是 100，统计窗口为 10s，10s 里有 500 次执行，那么只有最后 100 次执行会被统计到 bucket 里去，默认值为 100
	metrics.healthSnapshot.intervalInMilliseconds	采样时间间隔，默认值为 500
Request Context 属性控制 HystrixCommand 使用到的 Hystrix 的上下文	requestCache.enabled	是否开启请求缓存功能，默认值为 true
	requestLog.enabled	表示是否开启日志，打印执行 HystrixCommand 的情况和事件，默认值为 true
Collapser Properties 设置请求合并的属性	maxRequestsInBatch	设置同时批量执行的请求的最大数量，默认值为 Integer.MAX_VALUE
	timerDelayInMilliseconds	批量执行创建多久之后再触发真正的请求，默认值为 10
	requestCache.enabled	是否对 HystrixCollapser.execute() 和 HystrixCollapser.queue()开启请求缓存，默认值为 true
Thread Pool Properties 设置 Hystrix Commands 的线程池行为，大部分情况线程数量是 10。线程池数量的计算公式如下：最高峰时每秒的请求数量 × 99%命令执行时间 + 喘息空间，设置线程池数量的主要原则是保持线程池越小越好，因为它是减轻负载并防止资源在延迟发生时被阻塞的主要工具	coreSize	设置线程池 core 的大小，默认值为 10
	maximumSize	设置最大线程池的大小，只有设置 allowMaximumSizeToDivergeFromCoreSize 时此值才起作用，默认值为 10
	maxQueueSize	设置最大的 BlockingQueue 队列的值：如果设置为-1，则使用 SynchronousQueue 队列；如果设置为正数，则使用 LinkedBlockingQueue 队列；默认值为-1
	queueSizeRejectionThreshold	maxQueueSize 值不能被动态修改，通过设置此值可以实现动态修改等待队列长度，即等待队列的数量大于 queueSizeRejectionThreshold 时（但是没有达到 maxQueueSize 值）开始拒绝后续的请求进入队列。如果设置为-1，则属性不起作用，默认值为 5
	keepAliveTimeMinutes	设置线程多久没有服务后需要释放（maximumSize-coreSize）个线程，默认值为 1
	allowMaximumSizeToDivergeFromCoreSize	设置 allowMaximumSizeToDivergeFromCoreSize 值为 true 时 maximumSize 才有作用，默认值为 false
	metrics.rollingStats.timeInMilliseconds	设置滚动窗口的时间，默认值为 10000
	metrics.rollingStats.numBuckets	设置滚动静态窗口分成桶的数量，默认值为 10，建议每个桶的时间长度大于 100ms

更多 Hystrix 的配置属性优先级和详解可参考官方文档：https://github.com/Netflix/Hystrix/wiki/Configuration。

7.2 Hystrix 工作流程

通过 Hystrix 向服务依赖项请求时发生的情况如图 7-8 所示。

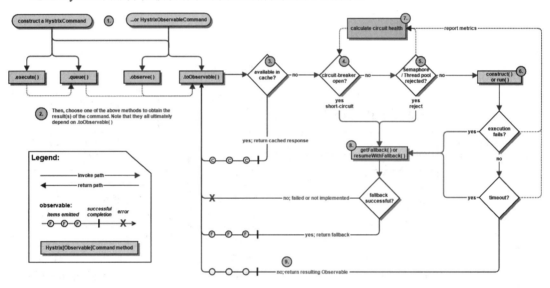

图 7-8　Hystrix 工作流程

1. 包装请求，构造一个 HystrixCommand 或 HystrixObservableCommand 对象

每次调用都会创建一个 HystrixCommand 或 HystrixObservableCommand 对象，以表示对依赖项的请求，向构造函数传递发出请求时所需的任何参数。

如果期望依赖项返回简单响应，则构造一个 HystrixCommand 对象，例如：

```
HystrixCommand command = new HystrixCommand(arg1, arg2);
```

如果期望依赖项返回一个发出响应的 Observable，则构造一个 HystrixObservableCommand 对象，例如：

```
HystrixObservableCommand command = new HystrixObservableCommand(arg1, arg2);
```

2. 执行命令

执行 execute 或 queue 做同步异步调用：

- execute()：阻塞型方法，返回单个结果（或者抛出异常）。
- queue()：异步方法，返回一个 Future 对象，可以从中取出单个结果（或者抛出异常）。
- observe() 和 toObservable()：都返回对应的 Observable 对象，代表（多个）操作结果。注意，observe 方法在调用的时候就开始执行对应的指令，而 toObservable 方法相当于是 watch 方法的 lazy 版本，当我们去订阅的时候对应的指令才会被执行并产生结果。

```
K               value   = command.execute();
Future<K>       fValue  = command.queue();
Observable<K>   ohValue = command.observe();        //hot observable
Observable<K>   ocValue = command.toObservable();   //cold observable
```

执行同步调用 execute 方法，会调用 queue().get()方法，queue()又会调用 toObservable().toBlocking().toFuture()。所以，所有的方法调用都依赖 Observable 的方法调用，只是取决于是需要同步还是异步调用。

3. 缓存处理

当请求来到后，会判断请求是否启用了缓存（默认是启用的），再判断当前请求是否携带了缓存 Key。如果命中缓存就直接返回；否则进入剩下的逻辑。

4. 判断断路器是否打开（熔断）

在结果没有命中缓存的时候，Hystrix 在执行命令前检查断路器是否为打开状态：

（1）如果断路器是打开的，Hystrix 不会执行命令，而是转接到 Fallback 处理逻辑（对应下面的第 8 步）。

（2）如果断路器是关闭的，那么 Hystrix 跳到第 5 步，检查是否有可用资源来执行命令。

5. 线程池/队列/信号量是否已满

如果与该命令关联的线程池和队列（或信号量，如果未在线程中运行）已满，则 Hystrix 将不执行该命令，而是转接到 Fallback 处理逻辑（对应下面的第 8 步）。

6. HystrixObservableCommand.construct() or HystrixCommand.run()

调用 HystrixCommand 的 run 方法，如果调用超时，当前处理线程将会抛出一个 TimeoutException。在这种情况下，Hystrix 会转接到 Fallback 处理逻辑，即第 8 步。如果命令没有抛出异常并返回结果，那么 Hystrix 在记录一些日志采集监控报告之后将结果返回。

7. 计算电路器的健康度

计算熔断器状态，所有的运行状态（成功、失败、拒绝、超时）上报给熔断器，断路器统计数据来决定熔断器的状态。

8. Fallback 处理

降级处理逻辑，根据上方的步骤可以得出以下 4 种情况会进入降级处理：①熔断器打开；②线程池/信号量跑满；③调用超时；④调用失败。

9. 返回成功的响应

如果 Hystrix 命令成功执行，就将以 Observable 的形式将一个或多个响应返回给调用方。

根据在第 2 步中调用命令的方式，此 Observable 可能会在返回之前进行转换，如图 7-9 所示。

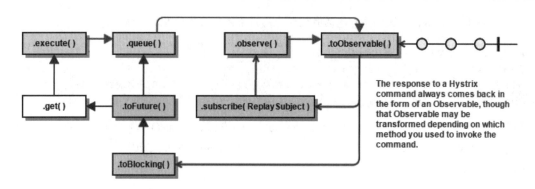

图 7-9　四种调用方式之间的依赖关系

这四种调用关系之间的依赖关系具体如下：

- toObservable()：返回最原始的 Observable 对象，必须订阅它才能真正开始执行命令的流程。
- observe()：在 toObservable 产生的原始 Observable 之后立即订阅它，让命令能够马上开始异步执行，并返回一个 Observable 对象，当调用它的 subscribe 时，将重新产生结果并通知给订阅者。
- queue()：将 Observable 转换为 BlockingObservable，并调用它的 toFuture 方法返回异步的 Future 对象。
- execute()：在 queue() 产生异步结果 Future 对象之后，通过调用 get 方法阻塞并等待结果返回。

7.3　Hystrix 监控

  Hystrix Dashboard 是 Spring Cloud 的仪表盘组件，可以查看 Hystrix 实例的执行情况，支持查看单个实例和查看集群实例，但是需要结合 spring-boot-actuator 模块一起使用。

  Hystrix Dashboard 主要用来实时监控 Hystrix 的各项指标信息。Hystrix Dashboard 可以有效地反映出每个 Hystrix 实例的运行情况，帮助我们快速发现系统中的问题，从而采取对应措施。

  Hystrix 的主要优点之一是它收集的有关每个 HystrixCommand 的一组指标。Hystrix 仪表板以有效的方式显示每个断路器的运行状况，具体如图 7-10 所示。

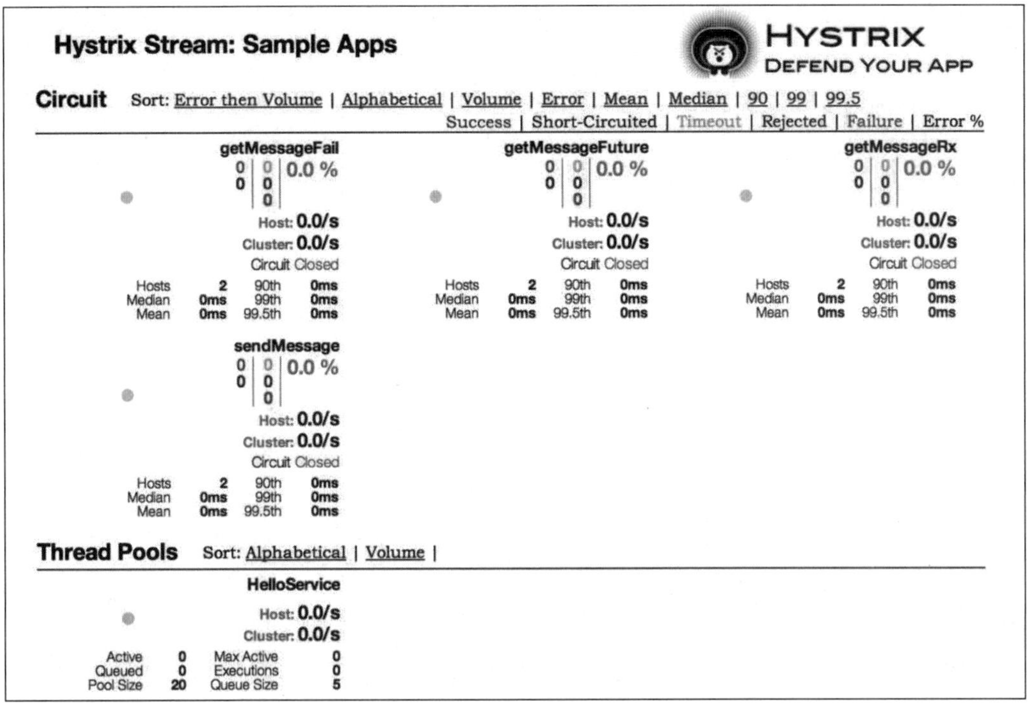

图 7-10　断路器的运行状况

7.3.1　Spring Boot 应用配置 Hystrix 仪表板

Hystrix 仪表板为监视仪表板上的一组指标提供了好处。它以非常简单的方式显示每个断路器的运行状况。在本小节中，我们将学习如何在 Spring Boot 项目中使用它。

步骤01 创建 Spring Boot 项目，项目名称为 spring-boot-hystrix-dashboard，包括以下依赖项：

```xml
<dependencies>
    <dependency>
        <groupId>org.springframework.boot</groupId>
        <artifactId>spring-boot-starter-web</artifactId>
    </dependency>
    <dependency>
        <groupId>org.springframework.cloud</groupId>
        <artifactId>spring-cloud-starter-netflix-hystrix</artifactId>
    </dependency>
    <dependency>
        <groupId>org.springframework.cloud</groupId>
        <artifactId>spring-cloud-starter-netflix-hystrix-dashboard</artifactId>
    </dependency>
    <dependency>
        <groupId>org.springframework.boot</groupId>
        <artifactId>spring-boot-starter-actuator</artifactId>
    </dependency>
```

```xml
    <dependency>
        <groupId>org.springframework.boot</groupId>
        <artifactId>spring-boot-starter-test</artifactId>
        <scope>test</scope>
    </dependency>
</dependencies>
```

步骤 02 启动类的代码如下：

```java
/**
 * 启动类
 * @author ay
 * @since 2020-12-14
 */
@SpringBootApplication
@EnableHystrix
@EnableHystrixDashboard
public class AyUserServiceApplication {
    public static void main(String[] args) {
        SpringApplication.run(AyUserServiceApplication.class, args);
    }
}
```

- @EnableHystrix：专门用于在类路径上使用 Hystrix 实现断路器模式。
- @EnableHystrixDashboard：将提供 Hystrix 流的仪表板视图。

步骤 03 定义 REST 控制器，具体代码如下：

```java
@RestController
public class UserController {

    @RequestMapping(value = "/hello")
    @HystrixCommand(fallbackMethod = "planb", commandProperties = {
            @HystrixProperty(name =
"execution.isolation.thread.timeoutInMilliseconds", value = "1000")
    })
    public String hello() throws InterruptedException {
        Thread.sleep(2000);
        return "Hello World";
    }
    private String planb() {
        return "Sorry our Systems are busy! try again later.";
    }
}
```

Hystrix 提供了一个注解@HystrixCommand，我们可以在服务层使用它来添加断路器模式的功能。如果使用@HystrixCommand 注解的方法失败，则将执行配置的回调方法。在上述情况下，如果经过 1000 毫秒的时间，则将执行方法 planb。

步骤 04 在 application.properties 中配置 Hystrix 端点公开：

```
management.endpoint.health.enabled=true
management.endpoints.jmx.exposure.include=*
management.endpoints.web.exposure.include=*
management.endpoints.web.base-path=/actuator
management.endpoints.web.cors.allowed-origins=true
management.endpoint.health.show-details=always
```

步骤 05 构建并启动应用程序，让我们向端点/hello 发出一些请求，然后检查执行器流是否已收集度量。执行器流可从 http://localhost:8070/actuator/hystrix.stream 获得，如图 7-11 所示。

```
$ curl http://localhost:8080/actuator/hystrix.stream

ping:

data: {"type":"HystrixCommand","name":"hello","group":"SampleController","curre

data: {"type":"HystrixThreadPool","name":"SampleController","currentTime":15657
```

图 7-11　/actuator/hystrix.stream 获取执行器流

现在我们有了流，并且记录了一些请求，下面进入 Hystrix 仪表盘。Hystrix 仪表盘可以通过 http://localhost:8080/hystrix 找到，如图 7-12 所示。

图 7-12　Hystrix 仪表盘

①Hystrix 仪表盘共支持三种不同的监控方式：

- 默认集群监控：通过 URL（https://turbine-hostname:port/turbine.stream）开启，实现对默认集群的监控。
- 指定的集群监控：通过 URL（https://turbine-hostname:port/turbine.stream?cluster=[clusterName]）开启，实现对 clusterName 集群的监控。
- 单体应用的监控：通过 URL（https://hystrix-app:port/actuator/hystrix.stream）开启，实现对具

体某个服务实例的监控。

②Delay:默认是 2000ms,主要用来控制服务器上轮询监控信息的延迟时间,通过该配置可以降低客户端的网络和 CPU 消耗。

③Title:对应图 7-11 头部标题 Hystrix Stream 之后的内容,默认使用具体监控实例的 URL。

复制 URL 地址 http://localhost:8070/actuator/hystrix.stream 到图 7-12 中,单击 Monitor Stream 按钮,跳转到断路器监控面板,如图 7-13 所示。

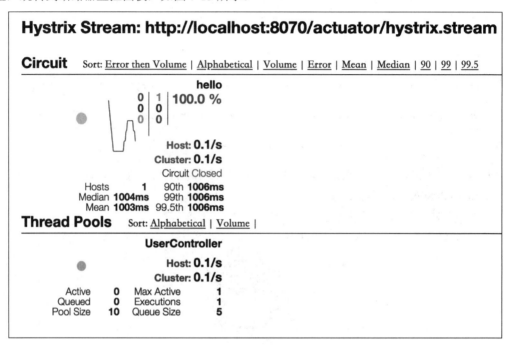

图 7-13 Hystrix 监控面板

从 Hystrix 监控面板可以看出,监控信息有两个重要的图形信息:实心圆和曲线。

- 实心圆:实心圆大小会根据实例请求流量发生变化,流量越大,实心圆越大。同时,实心圆的颜色代表实例的健康程度,监控度从绿色、黄色、橙色到红色递减。
- 曲线:记录 2 分钟内流量的相对变化,通过观察它来判断流量的上升和下降。

Hystrix 监控面板还有几个重要参数,具体如下:

- Circuit Closed:表示断路器状态。
- 100.0%:表示最近 10s 的错误比例。
- Host:0.1/s:表示请求频率。
- Hosts 1、Median 1004ms、Mean 1003ms:表示集群下的主机报告。

7.3.2 Turbine 集群监控

从系统的整体运行状况来看,查看单个实例的 Hystrix 数据不是很有用。Turbine 是一种将所有相关的/hystrix.stream 端点聚合到组合的/turbine.stream 中的应用程序,可在 Hystrix 仪表板中使用。
集群监控的具体步骤如下所示。

步骤01 参考 3.6.3 小节的内容,快速创建 Eureka 服务,充当服务注册中心,项目名称为 spring-boot-turbine-eureka-service,端口为 8001。启动 Eureka 服务并访问 Eureka 首页地址（http://localhost:8001/）,查看 Eureka 服务是否启动成功。spring-boot-turbine-eureka-service 项目的 application.yml 配置文件内容如下:

```yaml
# 指定运行端口
server:
  port: 8001

# 指定服务名称
spring:
  application:
    name: eureka-server

# 指定主机地址
eureka:
  instance:
    hostname: localhost
  client:
    # 指定是否从注册中心获取服务(注册中心不需要开启)
    fetch-registry: false
    # 指定是否将服务注册到注册中心(注册中心不需要开启)
    register-with-eureka: false
```

步骤02 参考 7.3.1 小节的内容,创建 spring-boot-turbine-user-serivce 服务,端口为 9090,服务名称为 user-service,该服务已实现 Hystrix 熔断回调机制,并通过 Actuator 公开了/hystrix.stream 端点。可访问 http://localhost:9090/actuator/hystrix.stream 进行验证。

步骤03 创建 turbine 服务,项目名称为 spring-boot-hystrix-turbine-dashboard,端口为 8101。在项目的 pom.xml 配置文件中添加如下依赖:

```xml
<dependency>
    <groupId>org.springframework.cloud</groupId>
    <artifactId>spring-cloud-starter-netflix-hystrix-dashboard</artifactId>
</dependency>
<dependency>
    <groupId>org.springframework.cloud</groupId>
    <artifactId>spring-cloud-starter-netflix-eureka-client</artifactId>
</dependency>
<dependency>
    <groupId>org.springframework.cloud</groupId>
    <artifactId>spring-cloud-starter-netflix-turbine</artifactId>
</dependency>
```

spring-cloud-starter-netflix-turbine 包含使 Turbine 服务器运行所需的所有依赖项。
要启用 Turbine,只需使用@EnableTurbine 注解主类即可:

```java
@SpringBootApplication
@EnableDiscoveryClient
@EnableTurbine
@EnableHystrixDashboard
public class TurbineDashboardApplication {
    public static void main(String[] args) {
        SpringApplication.run(TurbineDashboardApplication.class, args);
    }
}
```

application.yml 配置文件的内容如下：

```yaml
# 指定运行端口
server:
  port: 8101

# 指定服务名称
spring:
  application:
    name: turbine-dashboard

eureka:
  client:
    # 注册到 Eureka 的注册中心
    register-with-eureka: true
    # 获取注册实例列表
    fetch-registry: true
    service-url:
      # 配置注册中心地址
      defaultZone: http://localhost:8001/eureka
hystrix:
  dashboard:
    proxyStreamAllowList=*:

turbine:
  combineHostPort: true
  appConfig: user-service
  #使用单引号和双引号进行转义
  clusterNameExpression: "'default'"

#另一种写法
#turbine:
#  app-config: USER-SERVICE
#  combine-host-port: true
#  cluster-name-expression: new String("default")
```

- turbine.app-config：指定需要收集监控信息的服务名。
- turbine.cluster-name-expression：指定集群名称为 default，如果服务数量非常多，可以启动多个 turbine 服务来构建不同的聚合集群，同时该参数可以在 Hystrix 仪表盘中用来定位不同的聚合集群，在 Hystrix Stream 的 URL 中通过 cluster 参数来指定即可。

- turbine.cluster-name-expression：设置为 true，可以让同一个主机上的服务通过主机名和端口号的组合来进行区分。

步骤 04 启动 spring-boot-turbine-eureka-service、spring-boot-hystrix-turbine-dashboard 以及 spring-boot-turbine-user-serivce 服务，在浏览器中访问 http://localhost:9090/turbine.stream?cluster=default 来查看汇总流，还可以在 Hystrix 仪表板上使用此 URL 生成一个很好的聚合视图，具体如图 7-14 所示。

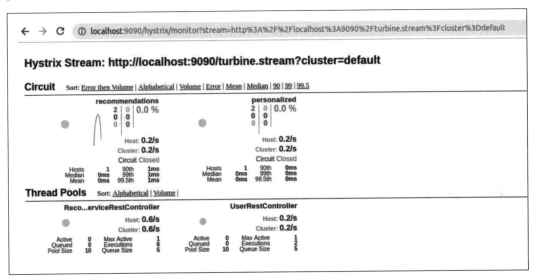

图 7-14　Hystrix 仪表板聚合视图

在某些情况下，其他应用程序了解在 Turbine 中配置了哪些 custers 可能会很有用。可以使用 /clusters 端点，该端点将返回所有已配置集群的 JSON 数组。例如：

```
[
  {
    "name": "RACES",
    "link": "http://localhost:8383/turbine.stream?cluster=RACES"
  },
  {
    "name": "WEB",
    "link": "http://localhost:8383/turbine.stream?cluster=WEB"
  }
]
```

可以将 turbo.endpoints.clusters.enabled 设置为 false 来禁用此端点。

7.4 Sentinel

7.4.1 Sentinel 简介

Sentinel 是阿里中间件团队开源的，面向分布式服务架构的高可用流量防护组件，主要以流量为切入点，从限流、流量整形、熔断降级、系统负载保护、热点防护等多个维度来帮助开发者保障微服务的稳定性。

Sentinel 具有以下特征：

- 丰富的应用场景：Sentinel 承接了阿里巴巴近十年的双十一大促流量的核心场景，例如秒杀（突发流量控制在系统容量可以承受的范围）、消息削峰填谷、集群流量控制、实时熔断下游不可用应用等。
- 完备的实时监控：Sentinel 同时提供实时的监控功能。我们可以在控制台中看到接入应用的单台机器秒级数据，甚至 500 台以下规模的集群的汇总运行情况。
- 广泛的开源生态：Sentinel 提供开箱即用的与其他开源框架/库的整合模块，例如与 Spring Cloud、Dubbo、gRPC 的整合。我们只需要引入相应的依赖并进行简单的配置即可快速接入 Sentinel。
- 完善的 SPI 扩展点：Sentinel 提供简单易用、完善的 SPI 扩展接口。我们可以通过实现扩展接口来快速定制逻辑，例如定制规则管理、适配动态数据源等。

7.4.2 限流算法

限流的方式有很多，常用的有计数器、漏桶和令牌桶等。

1. 计数器

采用计数器是一种比较简单的限流算法，一般我们会限制一秒钟能够通过的请求数。比如限流 QPS 为 100，算法的实现思路就是从第一个请求进来开始计时，在接下来的 1 秒内每来一个请求就把计数加 1，如果累加的数字达到了 100，后续的请求就会被全部拒绝。等到 1 秒结束后，把计数恢复成 0，重新开始计数。如果在单位时间 1 秒内的前 10 毫秒处理了 100 个请求，那么后面的 990 毫秒会请求拒绝所有的请求，我们把这种现象称为 "突刺现象"。

2. 漏桶算法

漏桶算法的思路很简单，一个固定容量的漏桶按照常量固定速率流出水滴。如果桶是空的，就不需要流出水滴。我们可以按照任意速率流入水滴到漏桶。如果流入的水滴超出了桶的容量，流入的水滴就会溢出（被丢弃），而漏桶容量是不变的。漏桶算法的大致原理如图 7-15 所示。

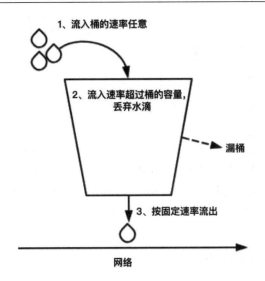

图 7-15　漏桶算法简单原理

漏桶算法提供了一种机制，通过它可以让突发流量被整形，以便为网络提供稳定的流量。

3．令牌桶算法

令牌桶算法是比较常见的限流算法之一，可以使用它进行接口限流，其大致原理如图 7-16 所示。

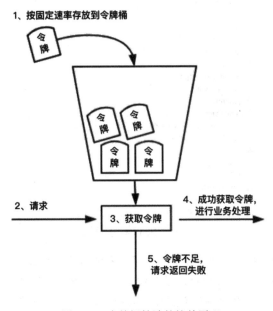

图 7-16　令牌桶算法的简单原理

令牌按固定的速率被放入令牌桶中，例如 tokens/s。桶中最多存放 b 个令牌（Token），当桶装满时，新添加的令牌会被丢弃或拒绝。当请求到达时，将从桶中删除 1 个令牌。令牌桶中的令牌不仅可以被移除，还可以往里添加，所以为了保证接口随时有数据通过，必须不停地往桶里加令牌。由此可见，往桶里加令牌的速度决定了数据通过接口的速度。我们通过控制往令牌桶里加令牌的速度来控制接口的流量。

4．漏桶算法和令牌桶算法的区别

漏桶算法和令牌桶算法的主要区别在于：

- 漏桶算法是按照常量固定速率流出请求的，流入请求速率任意，当流入的请求数累积到漏桶容量时，新流入的请求被拒绝。
- 令牌桶算法是按照固定速率往桶中添加令牌的，请求是否被处理需要看桶中的令牌是否足够，当令牌数减为零时，拒绝新的请求。
- 令牌桶算法允许突发请求，只要有令牌就可以处理，允许一定程度的突发流量。
- 漏桶算法限制的是常量流出速率，从而使突发流入速率平滑。

7.4.3 Sentinel 项目结构

从 GitHub 地址中下载 Sentinel 源码，具体如 7-17 所示。

图 7-17 Sentinel 项目结构

- sentinel-core：核心模块，包含限流、降级、系统保护等功能。
- sentinel-dashboard：控制台模块，可以对 sentinel 客户端实现可视化的管理。
- sentinel-extension：扩展模块，主要对 DataSource 进行部分扩展实现。
- sentinel-demo：样例模块，包含如何使用 sentinel 进行限流、降级等案例。
- sentinel-transport：传输模块，提供了基本的监控服务端和客户端的 API 接口，以及一些基于不同库的实现。
- sentinel-adapter：适配器模块，实现了对一些常见框架的适配。
- sentinel-benchmark：基准测试模块，对核心代码的精确性提供基准测试。

7.4.4　Sentinel 与 Hystrix 的区别

前面提到 Hystrix 也包含熔断和降级功能，那么阿里巴巴的 Sentinel 和 Netflix Hystrix 之间有何区别呢？具体内容如表 7-2 所示。

表 7-2　Sentinel 与 Netflix Hystrix 的区别

	Sentinel	Netflix Hystrix
隔离策略	信号量隔离	线程池隔离/信号量隔离
熔断降级策略	基于响应时间或失败比率	基于失败比率
实时指标实现	滑动窗口	滑动窗口（基于 RxJava）
规则配置	支持多种数据源	支持多种数据源
扩展性	多个扩展点	插件的形式
基于注解的支持	支持	支持
限流	基于 QPS，支持基于调用关系的限流	有限的支持
流量整形	支持慢启动、匀速器模式	不支持
系统负载保护	支持	不支持
控制台	开箱即用，可配置规则、查看秒级监控、机器发现	不完善
常见框架的适配	Servlet、Spring Cloud、Dubbo、gRPC 等	Servlet、Spring Cloud Netflix

7.4.5　Sentinel 控制台

Sentinel 提供一个轻量级的开源控制台，提供机器发现以及健康情况管理、监控（单机和集群）、规则管理和推送的功能。另外，鉴权在生产环境中必不可少。

Sentinel 控制台包含如下功能：

- 查看机器列表以及健康情况：收集 Sentinel 客户端发送的心跳包，用于判断机器是否在线。
- 监控（单机和集群聚合）：通过 Sentinel 客户端暴露的监控 API，定期拉取并且聚合应用监控信息，最终可以实现秒级的实时监控。
- 规则管理和推送：统一管理推送规则。
- 鉴权：在生产环境中，鉴权非常重要。这里每个开发者需要根据自己的实际情况进行定制。

启动 Sentinel 控制台也非常简单，具体步骤如下：

步骤 01　从 Git 地址 https://github.com/alibaba/Sentinel 下载 Sentinel 最新源码，或者从 releases 界面 https://github.com/alibaba/Sentinel/releases 下载 Sentinel 安装包。

步骤 02　通过源码方式启动 Sentinel 控制台，只需执行 sentinel-dashboard 模块的 main 方法，具体如图 7-18 所示。

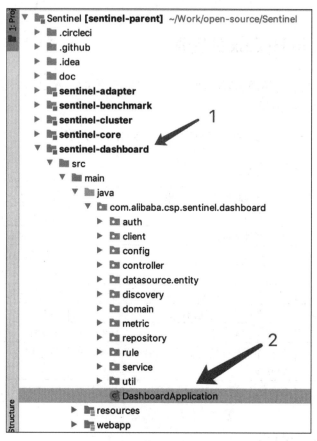

图 7-18　以源码方式启动 Sentinel

通过安装包启动 Sentinel 控制台，只需在安装包所在目录执行如下命令：

```
java -Dserver.port=8089 -Dcsp.sentinel.dashboard.server=localhost:8080 -Dproject.name=sentinel-dashboard -jar sentinel-dashboard.jar
```

其中，-Dserver.port=8080 用于指定 Sentinel 控制台端口为 8080。

步骤 03　启动成功后，在浏览器中输入访问地址 http://localhost:8089/，进入登录界面，用户名和密码都是 sentinel，登录成功后进入欢迎界面，如图 7-19 所示。

图 7-19　Sentinel 欢迎界面

步骤 04 至此,Sentinel 控制台安装完成。

7.4.6 客户端接入控制台

Sentinel 控制台安装完成之后,接下来创建 Spring Boot 项目客户端,接入 Sentinel。

步骤 01 使用 Intellij IDEA 创建 Spring Boot 项目,项目名称为 spring-boot-sentinel-01。在 Intellij IDEA 创建向导中,需要勾选 Spring Web 和 Sentinel 依赖,具体如图 7-20 所示。

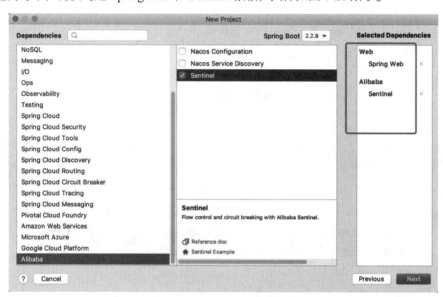

图 7-20　勾选 Sentinel 依赖

勾选 Sentinel 依赖,Intellij IDEA 将在项目中引入如下依赖包:

```xml
<dependency>
    <groupId>com.alibaba.cloud</groupId>
    <artifactId>spring-cloud-starter-alibaba-sentinel</artifactId>
</dependency>
```

步骤 02 开发控制层类 TestController,具体代码如下:

```java
/**
 * @author ay
 * @since 2020-08-07
 */
@RestController
public class TestController {

    @GetMapping(value = "/hello/{name}")
    @SentinelResource(value = "sayHello")
    public String apiHello(@PathVariable String name) {
        return "hello," + name;
    }
```

}

@SentinelResource 是 Sentinel 提供的限流相关的注解，后面章节会详细讲解。

步骤 03 在 application.yml 配置文件中，添加如下配置：

```yaml
### 服务端口
server:
  port: 18090
### 应用名称
spring:
  application:
    name: sentinel
  cloud:
    sentinel:
      transport:
        ### 指定应用与Sentinel控制台交互的端口
        port: 8719
        ### sentinel后台地址
        dashboard: localhost:8089
      eager: true
```

步骤 04 运行 main 方法，启动项目，在浏览器中输入访问地址"http://localhost:18090/hello/ay"，重新刷新 Sentinel 控制台，便可以看到 Sentinel 提供的相关功能菜单，具体如图 7-21 所示。

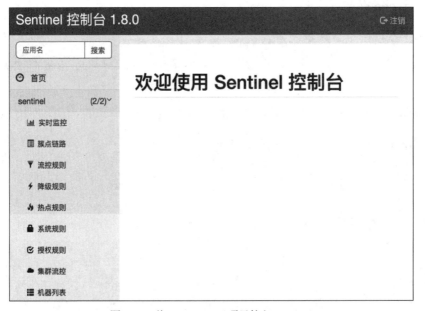

图 7-21　将 Spring Boot 项目接入 Sentinel

7.4.7　Sentinel 微服务限流

Sentinel 流量控制（flow control）的原理是监控应用流量的 QPS 或并发线程数等指标，当达到指定的阈值时对流量进行控制，以避免被瞬时的流量高峰冲垮，从而保障应用的高可用性。下面演

示如何通过 Sentinel 对微服务接口进行限流,具体步骤如下:

步骤 01 使用 Intellij IDEA 创建 Spring Boot 项目,项目名称为 spring-boot-sentinel-02。在 Intellij IDEA 创建向导中,勾选 Spring Web 和 Sentinel 依赖。

步骤 02 开发 RateLimitController 控制层类,具体代码如下:

```java
/**
 * 描述:限流功能
 * @author ay
 * @since 2020-08-07
 */
@RestController
@RequestMapping("/rateLimit")
public class RateLimitController {
    /**
     * 按资源名称限流,需要指定限流处理逻辑
     */
    @GetMapping("/byResource")
    @SentinelResource(value = "byResource",blockHandler = "handleException")
    public CommonResult byResource() {
        return new CommonResult("按资源名称限流", 200);
    }
    public CommonResult handleException(BlockException exception){
        return new CommonResult(exception.getClass().getCanonicalName(),200);
    }
}
```

@SentinelResource:value 属性指定资源名称,blockHandler 属性指定处理 BlockException 异常的函数名称。函数要求必须是 public,返回类型与原方法必须一致,函数参数类型需要和原方法相匹配,并在最后加 BlockException 类型的参数。函数默认和原方法在同一个类中。若希望使用其他类的函数,可配置 blockHandlerClass,并指定 blockHandlerClass 里面的方法。

CommonResult 类的具体代码如下:

```java
/**
 * @author ay
 * @since 2020-08-07
 */
public class CommonResult<T> {
    private T data;
    private String message;
    private Integer code;
    public CommonResult(){}
    public CommonResult(T data, String message, Integer code) {
        this.data = data;
        this.message = message;
        this.code = code;
    }
    public CommonResult(String message, Integer code) {
        this(null, message, code);
```

```
    }
    public CommonResult(T data) {
        this(data, "操作成功", 200);
    }
}
```

步骤 03 在 application.yml 配置文件中,添加如下配置:

```
### 服务端口
server:
  port: 18091
### 应用名称
spring:
  application:
    name: sentinel
  cloud:
    sentinel:
      transport:
        ### 指定应用与Sentinel控制台交互的端口
        port: 8719
        ### sentinel后台地址
        dashboard: localhost:8089
      eager: true
```

步骤 04 在 Sentinel 控制台的流控规则中,新增流控规则,如图 7-22 所示。

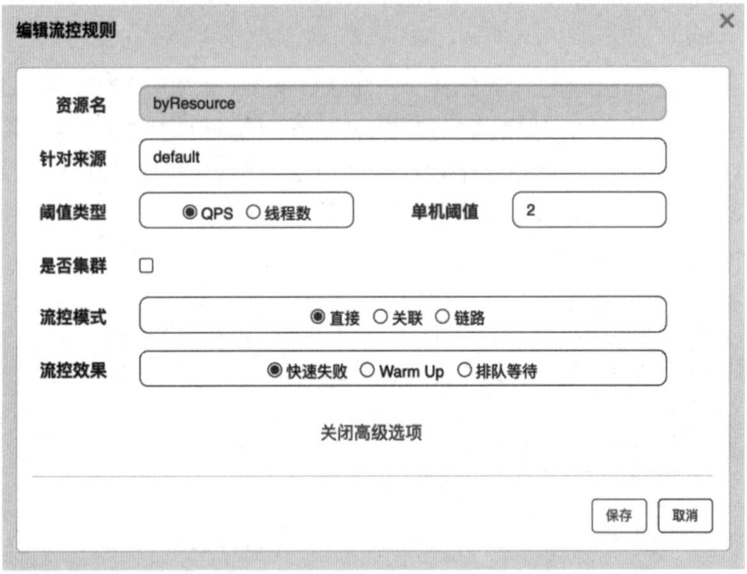

图 7-22 Sentinel 流控规则设置

"资源名"需要和@SentinelResource 注解的 value 属性值保持一致,"阈值类型"默认为 QPS,"单机阈值"设为 2,"流控效果"设为"快速失败"。

Sentinel 限流策略有两种统计类型:一种是统计并发线程数,另一种是统计 QPS。

当 QPS 超过某个阈值的时候,采取措施进行流量控制。流量控制的效果包括直接拒绝、Warm Up

和匀速排队。

- 直接拒绝：默认的流量控制方式，当 QPS 超过任意规则的阈值后，新的请求就会被立即拒绝，拒绝方式为抛出 FlowException。
- Warm Up：预热/冷启动方式。在系统长期处于低水位的情况下，当流量突然增加时，直接把系统拉升到高水位可能瞬间把系统压垮。通过冷启动让通过的流量缓慢增加，在一定时间内逐渐增加到阈值上限，给冷系统一个预热的时间，避免冷系统被压垮。
- 匀速排队：严格控制请求通过的间隔时间，即让请求以均匀的速度通过，对应的是漏桶算法。该方式的作用如图 7-23 所示。

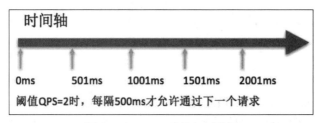

图 7-23　Sentinel 流控规则设置

并发线程数控制：并发数控制用于保护业务线程池不被慢调用耗尽。例如，当应用所依赖的下游应用由于某种原因导致服务不稳定、响应延迟增加，对于调用者来说意味着吞吐量下降和更多的线程数占用，极端情况下甚至会导致线程池耗尽。为应对太多线程占用的情况，业内有使用隔离的方案，比如通过不同业务逻辑使用不同线程池来隔离业务自身之间的资源争抢（线程池隔离）。这种隔离方案虽然隔离性比较好，但是代价是线程数目太多，线程上下文切换的 overhead 比较大，特别是对低延时的调用有比较大的影响。Sentinel 并发控制不负责创建和管理线程池，而是简单统计当前请求上下文的线程数目（正在执行的调用数目）。如果超出阈值，新的请求就会被立即拒绝，效果类似于信号量隔离。

步骤 05　运行 main 方法启动项目，在浏览器中输入访问地址"http://localhost:18091/rateLimit/byResource"，应用返回如下信息：

```
{"data":null,"message":"按资源名称限流","code":200}
```

快速刷新浏览器页面，保证 1s 内大于 2 次请求，超过设置的阈值时，应用抛出 FlowException 异常，返回如下信息：

```
{"data":null,"message":"com.alibaba.csp.sentinel.slots.block.flow.FlowException","code":200}
```

更多内容可参考 Sentinel 官网（https://github.com/alibaba/Sentinel/wiki）。

第 8 章

Spring Cloud Bus 消息总线

本章主要介绍 Spring Cloud Bus 消息总线、Spring 事件机制、Spring Cloud Bus 原理、如何使用 Kafka 实现消息总线、Kafka 介绍与安装、Spring Cloud Stream 简介和核心概念讲解、Stream 应用编程模型/Binder 抽象、Stream 快速入门和 Stream 原理。

8.1 Kafka 实现消息总线

8.1.1 Kafka 概述

Kafka 是一款开源、轻量级、可分区、具有复制备份（Replicated）、基于 ZooKeeper 协调管理的分布式流平台的功能强大的消息系统。与传统的消息系统相比，Kafka 能够很好地处理活跃的流数据，使得数据在各个子系统中高性能、低延迟地不停流转。据 Kafka 官方网站介绍，Kafka 的定位就是一个分布式流处理平台。在官方看来，作为一个流式处理平台，必须具备以下 3 个关键特性。

- 能够允许发布和订阅流数据。从这个角度来讲，平台更像一个消息队列（MQ）或者企业级消息系统。
- 存储流数据时提供相应的容错机制。
- 当流数据到达时能够被及时处理。

Kafka 能够很好地满足以上 3 个特性，很好地建立实时流式数据通道，由该通道可靠地获取系统或应用程序的数据，也可以通过 Kafka 方便地构建实时流数据应用来转换或对流式数据进行响应处理。特别是在 0.10 版本之后，Kafka 推出了 Kafka Streams，这让 Kafka 对流数据处理变得更加方便。

Kafka 消息系统基本的体系结构如图 8-1 所示。

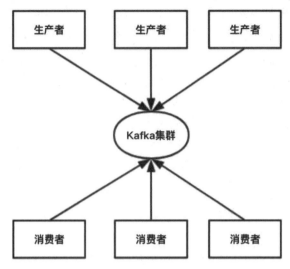

图 8-1　Kafka 消息系统基本的体系结构

在对 Kafka 基本体系结构有了一定了解之后，我们对 Kafka 的基本概念进行详细阐述。

1. 主题

Kafka 将一组消息抽象归纳为一个主题（Topic），一个主题就是对消息的一个分类。生产者将消息发送到特定主题，消费者订阅主题或主题的某些分区进行消费。

2. 消息

消息是 Kafka 通信的基本单位，由一个固定长度的消息头和一个可变长度的消息体构成。在老版本中，每一条消息称为 Message；在由 Java 重新实现的客户端中，每一条消息称为 Record。

3. 分区和副本

Kafka 将一组消息归纳为一个主题，而每个主题又被分成一个或多个分区（Partition）。每个分区由一系列有序、不可变的消息组成，是一个有序队列。每个分区在物理上对应为一个文件夹，分区的命名规则为主题名称后接 "-" 连接符，之后接分区编号，分区编号从 0 开始，编号最大值为分区的总数减 1。每个分区又有一至多个副本（Replica），分区的副本分布在集群的不同代理上，以提高可用性。从存储角度上分析，分区的每个副本在逻辑上抽象为一个日志（Log）对象，即分区的副本与日志对象是一一对应的。每个主题对应的分区数可以在 Kafka 启动时所加载的配置文件中配置，也可以在创建主题时指定。当然，客户端还可以在主题创建后修改主题的分区数。分区使得 Kafka 在并发处理上变得更加容易。从理论上来说，分区数越多，吞吐量越高，但这要根据集群的实际环境及业务场景而定。同时，分区也是 Kafka 保证消息被顺序消费以及对消息进行负载均衡的基础。Kafka 只能保证一个分区之内消息的有序性，并不能保证跨分区消息的有序性。每条消息被追加到相应的分区中，是按顺序写入磁盘，因此效率高。同时与传统消息系统不同的是，Kafka 并不会立即删除已被消费的消息，由于磁盘的限制，消息不会一直被存储（事实上这是没有必要的），因此 Kafka 提供两种删除老数据的策略：

- 基于消息已存储的时间长度。
- 基于分区的大小。

4. Leader 副本和 Follower 副本

由于 Kafka 副本的存在，因此需要保证一个分区的多个副本之间数据的一致性。Kafka 会选择该分区的一个副本作为 Leader 副本、其他副本作为 Follower 副本，只有 Leader 副本才负责处理客户端的读/写请求，Follower 副本从 Leader 副本同步数据。如果没有 Leader 副本，就需要所有的副本同时负责读/写请求处理，同时还得保证这些副本之间数据的一致性。假设有 n 个副本，就需要有 $n \times n$ 条通路来同步数据，这样数据的一致性和有序性就很难保证。引入 Leader 副本后，客户端只需要与 Leader 副本进行交互，这样数据的一致性及顺序性就有了保证。Follower 副本从 Leader 副本同步消息，对于 n 个副本只需 $n-1$ 条通路即可，这样就使得系统更加简单而且高效。Follower 副本与 Leader 副本的角色并不是固定不变的，如果 Leader 副本失效，那么将通过相应的选举算法从其他 Follower 副本中选出新的 Leader 副本。

5. 偏移量

任何发布到分区的消息都会被直接追加到日志文件（分区目录下以 ".log" 为文件名后缀的数据文件）的尾部，而每条消息在日志文件中的位置都会对应一个按序递增的偏移量。偏移量是一个分区下严格有序的逻辑值，它并不表示消息在磁盘上的物理位置。由于 Kafka 几乎不允许对消息进行随机读写，因此 Kafka 并没有提供额外索引机制存储偏移量，也就是说并不会给偏移量再提供索引。消费者可以通过控制消息偏移量来对消息进行消费，如消费者可以指定消费的起始偏移量。为了保证消息被顺序消费，消费者已消费的消息对应的偏移量也需要保存。需要说明的是，消费者对消息偏移量的操作并不会影响消息本身的偏移量。旧版消费者将消费偏移量保存到 ZooKeeper 中，而新版消费者将消费偏移量保存到 Kafka 内部的一个主题中。当然，消费者也可以自己在外部系统保存消费偏移量，而无须保存到 Kafka 中。

6. 日志段

一个日志被划分为多个日志段（Log Segment）。日志段是 Kafka 日志对象分片的最小单位。与日志对象一样，日志段也是一个逻辑概念，一个日志段对应磁盘上一个具体日志文件和两个索引文件。日志文件是以 ".log" 为文件名后缀的数据文件，用于保存消息实际数据。两个索引文件分别以 ".index" 和 ".timeindex" 作为文件名后缀，分别表示消息偏移量索引文件和消息时间戳索引文件。

7. 代理

在 Kafka 基本体系结构中，提到了 Kafka 集群。Kafka 集群是由一个或多个 Kafka 实例构成的，我们将每一个 Kafka 实例称为代理（Broker），通常也称代理为 Kafka 服务器（Kafka Server）。在生产环境中，Kafka 集群一般包括一台或多台服务器，我们可以在一台服务器上配置一个或多个代理。每一个代理都有唯一的标识 ID，这个 ID 是一个非负整数。在一个 Kafka 集群中，每增加一个代理就需要为这个代理配置一个与该集群中其他代理不同的 ID，ID 值可以选择任意非负整数，只要保证它在整个 Kafka 集群中唯一，这个 ID 就是代理的名字，也就是在启动代理时配置的 broker.id 对应的值。

8. 生产者

生产者（Producer）负责将消息发送给代理，也就是向 Kafka 代理发送消息的客户端。

9. 消费者和消费组

消费者（Consumer）以拉取方式获取数据，是消费的客户端。在 Kafka 中，每一个消费者都属于一个特定的消费组（Consumer Group），我们可以为每个消费者指定一个消费组，以 groupId 代表消费组名称，通过 group.id 配置项设置。如果不指定消费组，该消费者就属于默认消费组 test-consumer-group。同时，每个消费者也有一个全局唯一的 ID，通过配置项 client.id 指定，如果客户端没有指定消费者的 ID，Kafka 就会自动为该消费者生成一个全局唯一的 ID，格式为 ${groupId}-${hostName}-${timestamp}-${UUID 前 8 位字符}。同一个主题的一条消息只能被同一个消费组下的某一个消费者消费，但不同消费组的消费者可同时消费该消息。消费组是 Kafka 用来实现对一个主题消息进行广播和单播的手段，实现消息广播只需指定各消费者均属于不同的消费组，消息单播则只需让各消费者属于同一个消费组。

10. ISR

Kafka 在 ZooKeeper 中动态维护了一个 ISR（In-Sync Replica），即保存同步的副本列表。该列表中保存的是与 Leader 副本保持消息同步的所有副本对应的代理节点 ID。若一个 Follower 副本宕机或落后太多，则该 Follower 副本节点将从 ISR 列表中移除。

8.1.2 Kafka 安装

学习 Spring Cloud Bus 消息总线之前，需要先学会如何安装 Kafka，具体步骤如下：

步骤 01 由于 Kafka 是用 Scala 语言开发的，运行在 JVM 上，因此在安装 Kafka 之前需要先安装 JDK。到官网（http://apache.fayea.com/kafka/2.1.0/kafka_2.12-2.1.0.tgz）下载 Kafka 安装包到本地，然后执行 tar 命令解压安装包到指定的目录，具体命令如下：

```
### 解压安装包
tar zxvf kafka_2.12-2.1.0.tgz
```

步骤 02 从官方地址（https://zookeeper.apache.org/releases.html）下载 ZooKeeper 安装包并解压。这里编者使用的 ZooKeeper 版本是 zookeeper-3.4.12。打开解压后的安装包，可以看到如图 8-2 所示的目录结构。

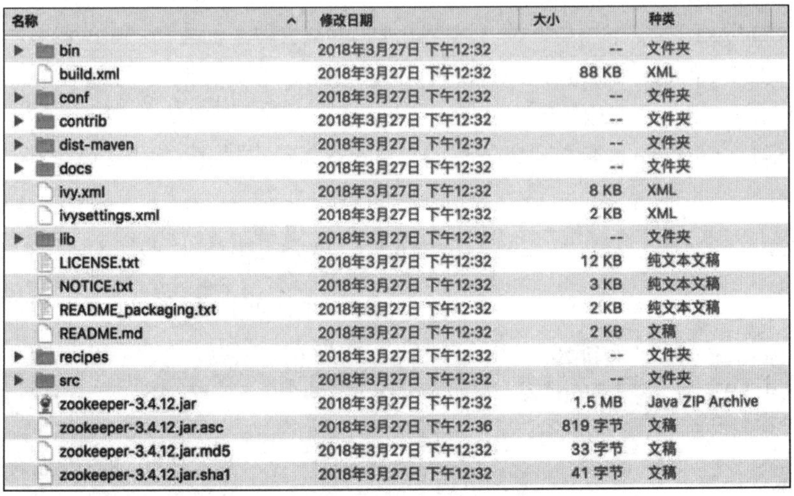

图 8-2 ZooKeeper 安装包目录结构

ZooKeeper 在/conf 目录下提供了默认的配置文件 zoo_sample.cfg，需要稍微调整才能够运行 ZooKeeper。复制 zoo_sample.cfg 配置文件并重新命名为 zoo.cfg，具体 Shell 命令如下：

```
cp zoo_sample.cfg zoo.cfg
```

在 ZooKeeper 安装目录 zookeeper-3.4.12/bin 下启动 ZooKeeper 服务器，具体命令如下：

```
sh zkServer.sh start
ZooKeeper JMX enabled by default
Using config: /Users/ay/Downloads/soft/zookeeper-3.4.12/bin/../conf/zoo.cfg
-n Starting zookeeper ...
### STARTED 代表 zk 服务器已启动
STARTED
```

步骤 03 修改配置。修改 kafka_2.11-2.1.0/config 目录下的 server.properties 文件，为了便于后续集群环境搭建的配置，需要保证同一个集群下的 broker.id 唯一，因此这里手动配置 broker.id，直接保持与 ZooKeeper 的 myid 值一致，同时配置日志存储路径。server.properties 修改的配置如下：

```
### 指定代理的 id
broker.id=1
### 开启注释
listeners = PLAINTEXT://localhost:9092
```

步骤 04 在 kafka_2.11-2.1.0 目录下执行命令，启动 Kafka 服务器，具体命令如下：

```
### 启动 Kafka
bin/kafka-server-start.sh config/server.properties &

### 命令格式
bin/kafka-server-start.sh [-daemon] server.properties [--override property=value]*
```

这个命令后面可以有多个参数，[-daemon]是可选参数，该参数可以让当前命令以后台服务方式执行，第二个参数必须是 Kafka 的配置文件。后面还可以有多个--override 开头的参数，其中的 property 可以是 Broker Configs 中提供的所有参数。这些额外的参数会覆盖配置文件中的设置。

在 kafka_2.11-2.1.0/logs 目录下的 server.log 会看到 Kafka Server 启动日志。在 Kafka 启动日志中会记录 Kafka Server 启动时加载的配置信息。

步骤 05 通过 ZooKeeper 客户端登录 ZooKeeper 查看目录结构，执行以下命令：

```
### 登录 ZooKeeper
sh zkCli.sh -server localhost:2181
```

在 Kafka 启动之前，ZooKeeper 中只有一个 zookeeper 目录节点，Kafka 启动后目录节点如下：

```
### 查看 zookeeper 目录节点
[zk: localhost:2181(CONNECTED) 0] ls /
[cluster, controller_epoch, controller, brokers, zookeeper, admin,
isr_change_notification, consumers, log_dir_event_notification,
latest_producer_id_block, config]
```

执行以下命令，查看当前已启动的 Kafka 代理节点：

```
[zk: localhost:2181(CONNECTED) 1] ls /brokers/ids
[1]
```

输出信息显示当前只有一个 Kafka 代理节点，当前代理的 brokerId 为 1。至此，Kafka 安装配置介绍完毕。

8.1.3　Docker 安装 ZooKeeper 和 Kafka

可使用如下 Docker 命令下载运行 ZooKeeper 和 Kafka：

```
docker run -p 2181:2181 -p 9092:9092 --env ADVERTISED_HOST=`docker-machine ip
\`docker-machine active\`` --env ADVERTISED_PORT=9092 spotify/kafka
```

spotify/kafka 镜像已经整合 ZooKeeper 和 Kafka。

8.2　Stream 简介

　　Spring Cloud Stream 是用于构建消息驱动的微服务应用程序的框架。Spring Cloud Stream 基于 Spring Boot 来创建独立的生产级 Spring 应用程序，并使用 Spring Integration 提供与消息代理的连接。它提供了来自多家供应商的中间件的合理配置，并介绍了发布–订阅、消费组和分区的概念。

8.2.1　核心概念

　　Spring Cloud Stream 提供了许多抽象和原语，简化了消息驱动的微服务应用程序的编写，主要包括以下内容：

- Spring Cloud Stream 的应用程序模型。
- Binder 抽象。
- 持久的发布–订阅支持。
- 消费者组支持。
- 分区支持。
- 可插拔的 Binder API。

要了解编程模型，应该熟悉以下核心概念：

- Destination Binders：目标绑定器，目标指的是 Kafka 或者 RabbitMQ，绑定器就是封装了目标中间件的包。如果操作的是 Kafka，就使用 Kafka Binder；如果操作的是 RabbitMQ，就使用 RabbitMQ Binder。
- Bindings：外部消息传递系统和应用程序之间的桥梁，提供消息的"生产者"和"消费者"（由目标绑定器创建）。
- Message：一种规范化的数据结构，生产者和消费者基于这个数据结构通过外部消息系统与目标绑定器和其他应用程序通信。

8.2.2 Stream 应用编程模型

应用程序通过 inputs 或者 outputs 与 Spring Cloud Stream 中的 binder 交互，通过配置来 binding，Spring Cloud Stream 的 binder 负责与消息中间件交互，如图 8-3 所示。

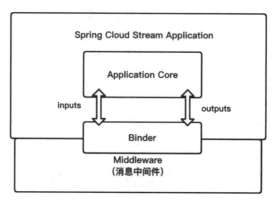

图 8-3　Stream 应用编程模型

Binder 层负责和 MQ 中间件的通信，应用程序 Application Core 通过 inputs 接收 Binder 包装后的 Message，相当于是消费者 Consumer；通过 outputs 投递 Message 给 Binder，然后由 Binder 转换后投递给 MQ 中间件，相当于是生产者 Producer。

Channel 描述的是消息从应用程序和 Binder 之间的流通通道，也就是图 8-1 中的 input 和 output。

8.2.3 Binder 抽象

Spring Cloud Stream 为 Kafka 和 RabbitMQ 提供了 Binder 实现。该框架还包括一个测试 Binder，用于对 Spring Cloud Stream 应用进行集成测试。

Binder 抽象也是框架的扩展点之一，这意味着你可以在 Spring Cloud Stream 之上实现自己的 Binder。

Spring Cloud Stream 使用 Spring Boot 进行配置，而 Binder 抽象使 Spring Cloud Stream 应用程序可以灵活地连接到中间件。例如，部署者可以在运行时动态选择外部目标（例如：Kafka topics 或 RabbitMQ exchanges）与消息处理器的输入和输出之间的映射。可以通过外部配置属性以及 Spring Boot 支持的任何形式（包括应用程序参数、环境变量和 application.yml 或 application.properties 文件）提供此类配置。

通过定义绑定器 Binder 作为中间层，实现了应用程序与消息中间件细节之间的隔离。在没有 Binder 概念绑定器的情况下，Spring Boot 应用要直接与消息中间件进行信息交互。各消息中间件在实现细节方面有差异性，当消息中间件有变动时，都需要修改应用程序逻辑。使用绑定器 Binder 作为中间层，向应用程序暴露统一的 Channel 通道，应用程序不需要再考虑各种不同的消息中间件实现。当需要升级消息中间件或是更换其他消息中间件产品时，只需要更换对应的 Binder 绑定器而不需要修改任何 Spring Boot 的应用逻辑。

Spring Cloud Stream 自动检测并使用在类路径上找到的 Binder。你可以用相同代码使用不同类型的中间件。因此，在构建时可以包含一个不同的 Binder。对于更复杂的情形，你还可以在应用程序中打包多个 Binder，并在运行时选择 Binder（甚至选择对不同的绑定使用不同的 Binder）。

8.2.4 发布-订阅

Spring Cloud Stream 支持的是共享 topics 的 publish-subscribe 模型，并没有采用 point-to-point 的 queues 模型，因为 pub-sub 模型在微服务中更具有普适性，而且 pub-sub 模型也能通过只有一个消费者来变相支持 p2p 模型。

8.2.5 消费组

在分布式微服务架构中，每个服务都不会以单节点的方式运行在生产环境。当同一个服务启动多个实例的时候，这些实例会绑定到同一个消息通道的目标主题（Topic）上。

默认情况下，当生产者发出一条消息到绑定通道上时，这条消息会产生多个副本被每个消费者实例接收和处理，但是在有些业务场景之下我们希望生产者产生的消息只被其中一个实例消费，这时我们需要为这些消费者设置消费组来实现这样的功能。实现的方式非常简单，只需要在服务消费者端给 spring.cloud.stream.bindings.<bindingName>.group 属性指定组名即可。

默认情况下，当未指定组时，Spring Cloud Stream 会给应用程序分配一个匿名且独立的单成员消费者组。

8.2.6 分区支持

Spring Cloud Stream 支持在给定应用程序的多个实例之间对数据进行分区。在分区方案中，每个 topic 均可以被构建为多个分区。一个或者多个生产者应用实例给多个消费者应用实例发送消息并确保相同特征的数据被同一消费者实例处理。

Spring Cloud Stream 为分区提供了通用的抽象实现，用来在消息中间件的上层实现分区处理，所以它对于消息中间件自身是否实现了消息分区并不关心，这使得 Spring Cloud Stream 为不具备分区功能的消息中间件也增加了分区功能扩展（如具备分区特性的 Kafka 或者不带分区特性的 rabbitmq）。

8.2.7 健康指标

Spring Cloud Stream 为 binders 提供了健康指标。它以 binders 名称注册，可以通过设置 management.health.binders.enabled 属性来启用或禁用。

为了查看健康指标，需要在项目中添加 Actuator 依赖项：

```xml
<dependency>
    <groupId>org.springframework.boot</groupId>
    <artifactId>spring-boot-starter-actuator</artifactId>
```

```
</dependency>
```

如果应用程序未明确设置 management.health.binders.enabled，则将 management.health.defaults.enabled 匹配为 true 并启用 binders 运行状况指示器。如果要完全禁用运行状况指示器，就必须将 management.health.binders.enabled 设置为 false。

通常情况下，可以使用/actuator/health 端点获取健康指标，但只会收到顶级应用程序状态。为了接收完整的详细信息，需要在应用程序中添加 management.endpoint.health.show-details=ALWAYS 配置。

如果要完全禁用所有可用的健康状况指标，而提供自己的健康状况指标，那么可以将属性 management.health.binders.enabled 设置为 false，然后在应用程序中实现 HealthIndicator 类来自定义健康指标。

另外，必须通过设置以下属性来启用 Binders 执行器端点：management.endpoints.web.exposure.include=bindings。在浏览器中访问 http://\<host\>:\<port\>/actuator/bindings 即可获取所有的 Binders 信息。

还可以通过 POST 请求或者 curl 命令来停止、开始、暂停和继续 Binders，例如：

```
curl -d '{"state":"STOPPED"}' -H "Content-Type: application/json" -X POST http://<host>:<port>/actuator/bindings/myBindingName
curl -d '{"state":"STARTED"}' -H "Content-Type: application/json" -X POST http://<host>:<port>/actuator/bindings/myBindingName
curl -d '{"state":"PAUSED"}' -H "Content-Type: application/json" -X POST http://<host>:<port>/actuator/bindings/myBindingName
curl -d '{"state":"RESUMED"}' -H "Content-Type: application/json" -X POST http://<host>:<port>/actuator/bindings/myBindingName
```

8.3 Spring Cloud Stream 实战

8.3.1 Stream 快速入门

步骤01 确保已经在电脑上安装 Kafka 并成功启动。

步骤02 使用 Intellij IDEA 快速创建 Spring Boot 项目，项目名称分别为 spring-cloud-stream-output、spring-cloud-stream-input。

步骤03 在 spring-cloud-stream-output、spring-cloud-stream-input 项目的 pom.xml 文件中，添加如下依赖：

```xml
<?xml version="1.0" encoding="UTF-8"?>
<project xmlns="http://maven.apache.org/POM/4.0.0"
xmlns:xsi="http://www.w3.org/2001/XMLSchema-instance"
         xsi:schemaLocation="http://maven.apache.org/POM/4.0.0
https://maven.apache.org/xsd/maven-4.0.0.xsd">
    <modelVersion>4.0.0</modelVersion>
    <parent>
        <groupId>org.springframework.boot</groupId>
```

```xml
        <artifactId>spring-boot-starter-parent</artifactId>
        <version>2.2.6.RELEASE</version>
        <relativePath/> <!-- lookup parent from repository -->
    </parent>
    <groupId>com.example</groupId>
    <artifactId>demo</artifactId>
    <version>0.0.1-SNAPSHOT</version>
    <name>demo</name>
    <description>Demo project for Spring Boot</description>

    <properties>
        <java.version>1.8</java.version>
        <spring-cloud.version>Hoxton.SR3</spring-cloud.version>
    </properties>

    <dependencies>
        <dependency>
            <groupId>org.springframework.boot</groupId>
            <artifactId>spring-boot-starter-web</artifactId>
        </dependency>
        <dependency>
            <groupId>org.apache.kafka</groupId>
            <artifactId>kafka-streams</artifactId>
        </dependency>
        <dependency>
            <groupId>org.springframework.cloud</groupId>
            <artifactId>spring-cloud-stream-binder-kafka</artifactId>
        </dependency>
        <dependency>
            <groupId>org.springframework.boot</groupId>
            <artifactId>spring-boot-starter-test</artifactId>
            <scope>test</scope>
            <exclusions>
                <exclusion>
                    <groupId>org.junit.vintage</groupId>
                    <artifactId>junit-vintage-engine</artifactId>
                </exclusion>
            </exclusions>
        </dependency>
        <dependency>
            <groupId>org.springframework.kafka</groupId>
            <artifactId>spring-kafka-test</artifactId>
            <scope>test</scope>
        </dependency>
    </dependencies>

    <dependencyManagement>
        <dependencies>
            <dependency>
                <groupId>org.springframework.cloud</groupId>
```

```xml
            <artifactId>spring-cloud-dependencies</artifactId>
            <version>${spring-cloud.version}</version>
            <type>pom</type>
            <scope>import</scope>
        </dependency>
    </dependencies>
</dependencyManagement>

<build>
    <plugins>
        <plugin>
            <groupId>org.springframework.boot</groupId>
            <artifactId>spring-boot-maven-plugin</artifactId>
        </plugin>
    </plugins>
</build>

</project>
```

步骤 04 在 spring-cloud-stream-output 项目中，开发 MQSource 类，具体代码如下：

```java
/**
 * @author ay
 * @since 2020-04-26
 */
public interface MQSource {

    String OUTPUT = "ay-topic";

    /**
     * 指定 TOPIC
     *
     * @return
     */
    @Output(MQSource.OUTPUT)
    MessageChannel output();
}
```

在 Spring Cloud Stream 中，可以在接口中通过@Output、@Input 注解定义消息通道，通过使用这两个接口中的成员变量来定义输入和输出通道的名称。定义输出通道需要返回 MessageChannel 接口对象，它定义了向消息通道发送消息的方法。默认情况下，通道的名称就是注解的方法的名称。也能够自己定义通道名称，只需要给@Input 和@Output 注解传入 String 类型参数通道名称就可以了。

同理，在 spring-cloud-stream-input 项目中开发 MqSink 类，具体代码如下：

```java
/**
 * @author ay
 */
public interface MqSink {
    String INPUT = "ay-topic";
    /**
```

```
 * 消费者接口
 */
@Input(MqSink.INPUT)
SubscribableChannel input();
}
```

定义输入通道需要返回 SubscribaleChannel 接口对象，这个接口继承自 MessageChannel 接口。它定义了维护消息通道订阅者的方法。

除了使用自定义接口外，我们可以使用 Spring Cloud Stream 提供的默认实现（Sink、Source 以及 Processor 接口），具体源码如下：

```
//Sink 接口
public interface Sink {
    String INPUT = "input";

    @Input("input")
    SubscribableChannel input();
}
//Source 接口
public interface Source {
    String OUTPUT = "output";

    @Output("output")
    MessageChannel output();
}
//Processor 接口
public interface Processor extends Source, Sink {
}
```

从上面的源码可知，Sink 和 Source 使用@Input 和@Output 注解定义了输入通道和输出通道，而 Processor 通过继承 Source 和 Sink 的方式同时定义了一个输入通道和输出通道。Sink 和 Source 中指定的通道名称分别为 input 和 output。

步骤 05 在 spring-cloud-stream-output 项目中开发 SendMsgProducer 类，具体代码如下：

```
/**
 * @author ay
 * @date 2020-04-26
 */
@EnableBinding(MQSource.class)
public class SendMsgProducer {

    @Resource
    private MQSource mQSource;

    public void sendMessage(String msg) {
        try {
            mQSource.output().send(MessageBuilder.withPayload(msg).build());
        } catch (Exception e) {
            e.printStackTrace();
```

```
        }
    }
}
```

只需要使用@EnableBinding 就能创建和绑定通道（channel）。其中，绑定通道是指将通道和 Binder 进行绑定。Middleware 不止一种，比如 Kafka、RabbitMQ 等。不同的 Middleware 有不同的 Binder 实现，通道与 Middleware 连接需要经过 Binder，所以通道要与明确的 Binder 绑定。

如果类路径下只有一种 Binder，那么 Spring Cloud Stream 会找到并绑定它，不需要进行配置。如果有多个就需要明确配置。

使用@EnableBinding 注解后，Spring Cloud Stream 就会自动帮我们实现接口。可以通过 Spring 支持的任何一种方式获取接口的实现，例如自动注入、getBean 等方式。

查看@EnableBinding 源码：

```
@Target({ElementType.TYPE, ElementType.ANNOTATION_TYPE})
@Retention(RetentionPolicy.RUNTIME)
@Documented
@Inherited
@Configuration
@Import({BindingBeansRegistrar.class,
BinderFactoryAutoConfiguration.class})
@EnableIntegration
public @interface EnableBinding {
    Class<?>[] value() default {};
}
```

从源码中可以看出，@EnableBinding 注解包含@Configuration 注解，使用它注解的类会成为 Spring 的基本配置类。@EnableBinding 注解接收一个参数，参数类型是 class。@EnableBinding(MQSource.class) 整段代码代表创建 MQSource 定义的通道，并将通道和 Binder 绑定。

接着看如下代码：

```
mQSource.output().send(MessageBuilder.withPayload(msg).build());
```

首先调用 output()方法获取输出通道对象，接着调用 send 方法发送数据。send 方法接收一个 Message 对象，这个对象不能直接新建，需要使用 MessageBuilder 获取。

同理，在 spring-cloud-stream-input 项目中开发 MqSinkReceiver 类，具体代码如下：

```
/**
 * @author ay
 * @date 2020/03/17
 */
@EnableBinding(value = {MqSink.class})
public class MqSinkReceiver {

    @StreamListener(MqSink.INPUT)
    public void messageListen(String json) {
        //处理请求的类，对消息进行处理
        System.out.println(json);
    }
```

@StreamListener 接收的参数是要处理的通道名，所注解的方法就是处理从通道获取数据的方法。方法的参数就是获取到的数据。

步骤 06 在 spring-cloud-stream-output 项目中开发 HelloController 类，具体代码如下：

```java
/**
 * @author ay
 * @since 2020-04-26
 */
@RestController
@RequestMapping("test")
public class HelloController {

    @Resource
    private SendMsgProducer sendMsgProducer;

    @RequestMapping("/send")
    public void send() {
        sendMsgProducer.sendMessage("hello ay!");
    }
}
```

所有代码开发完成后，分别启动 spring-cloud-stream-output 生产者和 spring-cloud-stream-input 消费者项目，在浏览器中输入请求地址"http://localhost:18091/test/send"，生产者会向 ay-topic 发出一条消息"hello ay!"，消费者订阅该 topic，获取消息并打印消息到控制台。以上实例是最简单的发布-订阅模式。

8.3.2 生产者的另一种实现

前面章节生产者通过直接注入定义的生产接口获取 MessageChannel 实例，然后发送消息，其实也可以直接注入 MessageChannel 实例来完成消息的发送，具体步骤如下：

步骤 01 在生产者项目的 MQSource 类中，继续添加如下通道定义：

```java
/**
 * @author ay
 * @since 2020-04-26
 */
public interface MQSource {

    //省略部分代码
    String OUTPUT_2 = "ay-topic-2";

    @Output(MQSource.OUTPUT_2)
    MessageChannel output2();
}
```

步骤 02 在 SendMsgProducer 类中使用 @Resource 注解直接注入 MessageChannel,具体代码如下:

```java
/**
 * @author ay
 * @date 2020-04-26
 */
@EnableBinding({MQSource.class})
public class SendMsgProducer {

    @Resource(name = MQSource.OUTPUT_2)
    private MessageChannel messageChannel;

    public void sendMessage2(String msg){
        try{
            messageChannel.send(MessageBuilder.withPayload(msg).build());
        }catch (Exception e){
            e.printStackTrace();
        }
    }
}
```

> **注 意**
>
> @Resource 的 name 与 MQSource 接口定义中 @Output 的名称应保持一致。

步骤 03 在消费者项目的 MqSink 类中添加如下代码:

```java
/**
 * @author ay
 */
public interface MqSink {

    //省略部分代码
    String INPUT_2 = "ay-topic-2";

    @Input(MqSink.INPUT_2)
    SubscribableChannel input2();

}
```

步骤 04 在消费者项目的 MqSinkReceiver 类中添加通道监听,具体代码如下:

```java
/**
 * @author ay
 * @date 2020/03/17
 */
@EnableBinding(value = {MqSink.class})
public class MqSinkReceiver {

    //省略部分代码
```

```
    @StreamListener(MqSink.INPUT_2)
    public void messageListen2(String json) {
        //处理请求的类,对消息进行处理
        System.out.println(json);
    }
}
```

步骤 05 在 HelloController 中添加 send2 方法,具体代码如下:

```
/**
 * @author ay
 * @since 2020-04-26
 */
@RestController
@RequestMapping("test")
public class HelloController {

    @Resource
    private SendMsgProducer sendMsgProducer;

    //省略部分代码

    @RequestMapping("/send2")
    public void send2() {
        sendMsgProducer.sendMessage2("hello ay 2!");
    }
}
```

最后,在浏览器中输入访问地址"http://localhost:18091/test/send2",控制台会打印"hello ay 2!"字符串。

8.3.3 生产和消费消息

除了调用通道的 sned 方法发布消息外,还可以使用 Spring Intergration 的方式生产数据。Spring Cloud Stream 是 Spring Integration 和 Spring Boot 的整合,支持 Integration 原生应用,具体实例如下所示。

定义接收通道:

```
public interface MQSource {
    //省略部分代码

    String OUTPUT_3 = "ay-topic-3";

    @Output(MQSource.OUTPUT_3)
    MessageChannel output3();
}
```

定义生产者,采用轮询方式发送消息,消息体为 Date 数据:

```
/**
```

```
 * 描述
 *
 * @author ay
 * @since 2020/10/3
 */
@EnableBinding(Source.class)
public class IntegrationSource {

    @Bean
    @InboundChannelAdapter(value = MQSource.OUTPUT_3, poller =
@Poller(fixedDelay = "10", maxMessagesPerPoll = "1"))
    public MessageSource<String> timerMessageSource() {
        System.out.println(123);
        return () -> new GenericMessage<>(new
SimpleDateFormat("yyyy-MM-dd").format(new Date()));
    }
}
```

fixedDelay 表示多少毫秒发送 1 次，maxMessagesPerPoll 属性表示一次发送几条消息。消费者通过@ServiceActivator 注解实现：

```
@ServiceActivator(inputChannel = MqSink.INPUT_3)
public void messageListen3(String json){
    System.out.println(json);
}
```

启动生产者和消费者服务，控制台便会打印日期信息。想对接收的 Date 类型数据进行格式化时，可以通过@Transformer 注解实现：

```
@Transformer(inputChannel = MqSink.INPUT_3, outputChannel = MqSink.INPUT_3)
public Object transform(Date message){
    return new SimpleDateFormat("yyyy-MM-dd hh:mm:ss").format(message);
}
```

再次启动消费者服务，日期被成功格式化。在 Integration 原生应用中，生产和消费间的数据转换需要一定的代码成本，接下来演示如何通过 Spring Cloud Stream 完成数据转换。

定义接收通道：

```
public interface MqSink {
    //省略代码...
    String INPUT_4 = "ay-topic-4";

    @Input(MqSink.INPUT_4)
    SubscribableChannel input4();
}
```

定义实体对象 AyUser，作为发送消息的实体：

```
public class AyUser {
    //省略 set、get 以及构造方法
    //主键
    private Integer id;
```

```
    //姓名
    private String name;
}
```

发送消息的方法:

```
@EnableBinding({MQSource.class})
public class SendMsgProducer {

    @Resource(name = MQSource.OUTPUT_4)
    private MessageChannel messageChannel_4;

    public void sendMessage4(){
        try{
            AyUser user = new AyUser(1, "ay");
            messageChannel_4.send(MessageBuilder.withPayload(user).build());
        }catch (Exception e){
            e.printStackTrace();
        }
    }
}
```

控制层代码:

```
@RestController
@RequestMapping("test")
public class HelloController {

    @RequestMapping("/send4")
    public void send4() {
        sendMsgProducer.sendMessage4();
    }
}
```

消费者代码:

```
@ServiceActivator(inputChannel = MqSink.INPUT_4)
public void messageListen4(@Payload AyUser user, @Headers Map headers){
    System.out.println(user.toString());
    System.out.println(headers.toString());
}
```

@Payload 注解作用于接收的消息体,@Headers 注解获取消息头信息,两者为一个 Map 键值对;@Header 和@Headers 的区别是一个获取单个属性(需要指明哪个属性),一个获取全部属性。

启动生产者和消费者服务,在浏览器中输入访问地址 "http://localhost:18091/test/send4",控制台打印如下信息:

```
com.example.demo.model.AyUser@42e1ba0b
    {kafka_offset=4, scst_nativeHeadersPresent=true,
kafka_consumer=org.apache.kafka.clients.consumer.KafkaConsumer@3e21665d,
deliveryAttempt=1, kafka_timestampType=CREATE_TIME,
kafka_receivedMessageKey=null, kafka_receivedPartitionId=0,
```

```
contentType=application/json, kafka_receivedTopic=ay-topic-4,
kafka_receivedTimestamp=1602162189794,
kafka_groupId=anonymous.80f3e6b6-9059-4b04-bcce-38dc6bb9324a}
    //省略部分消息
```

从上面的消息可以看出，AyUser 对象转换成功，其中还包含消息的其他信息，例如 offset、contentType=application/json 指明消息类型等。使用 Spring Cloud Stream 无须做任何处理即可完成对象转换。

8.4　Bus 简介

8.4.1　Bus 消息总线

Spring Cloud Bus 是 Spring Cloud 体系内的消息总线，支持 RabbitMQ 和 Kafka 两种消息中间件。所谓消息总线，简单理解就是一个消息中心，众多微服务实例都可以连接到总线上，实例可以往消息中心发送或接收信息（通过监听）。例如：实例 A 发送一条消息到总线上，总线上的实例 B 可以接收到信息（实例 B 订阅了实例 A），消息总线充当一个中间者的角色，使得实例 A 和实例 B 解耦，如图 8-4 所示。

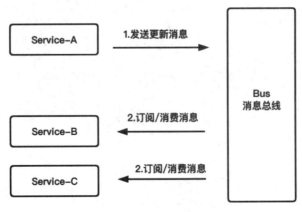

图 8-4　消息总线应用

8.4.2　Spring 事件机制

Spring Cloud Bus 在 Stream 基础之上再次进行抽象封装，使得我们可以在不用理解消息发送、监听等概念的基础上使用消息来完成业务逻辑的处理。Spring Cloud Bus 内部实现原理是事件机制。Spring 框架有一个事件机制，当使用 Spring 的事件机制时需要了解其中的接口或者抽象：

- ApplicationEventPublisher：接口类，用来发布事件。
- ApplicationEvent：抽象类，用来定义事件，里面只有一个构造函数和一个长整型的 timestamp。
- ApplicationListener：接口类，用来实现事件的监听。

Spring 应用的上下文 ApplicationContext 默认实现了 ApplicationEventPublisher 接口。因此，在发布事件时可以直接使用 ApplicationContext.publishEvent() 方法来发送。

下面是基于 Spring 事件发送的实例，具体步骤如下：

步骤01 创建 Spring Boot 项目（参考 1.3.1 节内容），项目名称为 spring-event-demo，在项目中引入 lombok 依赖，简化代码。

步骤02 定义事件 UserEvent，具体代码如下：

```java
@Data
public class UserEvent extends ApplicationEvent {

    private String action;
    private User user;

    public UserEvent(User user) {
        super(user);
    }
}
```

@Data 为 lombok 提供注解，帮助我们生成 set、get 方法。

用户实体类代码如下：

```java
@Data
@AllArgsConstructor
public class User {
    private Integer id;
    private String name;
}
```

步骤03 发布事件类 PublishEvent，具体代码如下：

```java
@Component
public class PublishEvent {

    private static Logger logger = LoggerFactory.getLogger(UserEvent.class);

    @Resource
    private ApplicationContextHolder applicationContextHolder;

    public void fire(UserEvent userEvent) {
        ApplicationContext applicationContext = applicationContextHolder.getApplicationContext();
        if(null != applicationContext) {
            logger.info("==> 发布事件:{} <==", userEvent);
            applicationContext.publishEvent(userEvent);
        }else {
            logger.warn("==> 无法获取 ApplicationContext 的实例. <==");
        }
    }
}
```

ApplicationContextHolder 用来获取上下文对象 ApplicationContext，并调用 ApplicationContext 的 publishEvent 方法发布事件。

```java
@Component
public class ApplicationContextHolder implements ApplicationContextAware {

    private ApplicationContext applicationContext;
    @Override
    public void setApplicationContext(ApplicationContext applicationContext) throws BeansException {
            this.applicationContext = applicationContext;
    }

    public ApplicationContext getApplicationContext() {
        return applicationContext;
    }
}
```

一个类实现了 ApplicationContextAware 接口，就可以方便地获得 ApplicationContext 中的所有 Bean。对于 Spring 框架的基础知识，不做过多介绍。

步骤 04 用户事件监听只需要实现 ApplicationListener 接口并进行相应处理即可，具体代码如下：

```java
@Component
public class UserEventListener implements ApplicationListener<UserEvent> {

    private static final Logger logger =
LoggerFactory.getLogger(UserEventListener.class);

    @Override
    public void onApplicationEvent(UserEvent userEvent) {
        logger.debug("收到用户事件{}", userEvent);
        //todo 具体业务代码
    }
}
```

步骤 05 定义 PushController 类，具体代码如下：

```java
@RestController
public class PushController {

    @Resource
    private PublishEvent publishEvent;

    @GetMapping("/push")
    public void push(){
        User user = new User(1, "ay");
        publishEvent.fire(new UserEvent(user));
    }
}
```

在浏览器中访问 http://localhost:8080/push，验证代码，最后控制台会打印如下信息：

```
    --- [nio-8080-exec-1] com.ay.spring.UserEvent         : ==> 发布事件:User(id=1,
name=ay) <==
    --- [nio-8080-exec-1] com.ay.spring.UserEventListener  : 收到用户事件User(id=1,
name=ay)
```

8.4.3 Spring Cloud Bus 实战

Spring Cloud Bus 可以将事件机制和 Stream 结合在一起，具体机制如下：

（1）在需要发布或者监听事件的应用中增加@RemoteApplicationEventScan 注解，通过该注解就可以启动 Stream 中消息通道的绑定。

（2）对于事件发布，需要继承 ApplicationEvent 的扩展类 RemoteApplicationEvent。通过 ApplicationContext.publishEvent()发布事件时，Spring Cloud Bus 会对所要发布的事件进行包装，形成消息，通过默认的 springCloudBus 消息通道发送到消息中间件。

（3）对于事件监听者，则不需要进行任何变更，仍旧按照上面的方式就可以实现消息的监听。需要注意的是，在消费的微服务工程中必须定义第 2 步所定义的事件，并且需要保障全类名一致。

通过 Bus 就可以像编写单体架构应用一样进行开发，而不需要关心什么消息中间件、主题、消息、通道等一系列概念。

接下来，我们通过 Spring Cloud Bus 来实现简单的消息通知，具体步骤如下：

步骤01 创建 Maven 项目，项目名称为 spring-cloud-bus-kafka。该项目包含以下 3 个模块：

①eureka-server：注册中心服务，之前章节都是使用 Nacos 作为注册中心，本节使用 eureka 演示如何创建 eureka 注册中心（可参考 3.6.3 小节的内容）。

②user-service：用户服务，主要用来发布消息。

③user-consumer：消费用户服务发布的消息。

步骤02 安装并启动 ZooKeeper 和 Kafka。

步骤03 创建 user-service 模块，在 pom.xml 文件中添加如下依赖：

```xml
<dependency>
        <groupId>org.springframework.cloud</groupId>
        <artifactId>spring-cloud-starter-bus-kafka</artifactId>
</dependency>
<!-- eureka -->
<dependency>
        <groupId>org.springframework.cloud</groupId>
        <artifactId>spring-cloud-starter-netflix-eureka-client</artifactId>
</dependency>
<dependency>
        <groupId>org.springframework.cloud</groupId>
        <artifactId>spring-cloud-starter-netflix-eureka-server</artifactId>
</dependency>
```

定义用户事件类 UserEvent，实现 RemoteApplicationEvent：

```
@Data
```

```java
public class UserEvent extends RemoteApplicationEvent {

    private static Logger logger = LoggerFactory.getLogger(UserEvent.class);

    public static final String ET_UPDATE="USER_UPDATE";
    public static final String ET_DELETE="USER_DELETE";

    private String action;

    private String uniqueKey;

    public UserEvent(Object source, String originService, String destinationService, String action, String id) {
        super(source, originService, destinationService);
        logger.info("==> User Event {},{},{} <==", source, originService, destinationService);
        this.action = action;
        this.uniqueKey = id;
    }
}
```

- originService：对于事件发布者来说 originService 就是自己。
- destinationService：将事件发布到哪些微服务实例。destinationService 配置的格式为 {serviceId}:{appContextId}，在配置时 serviceId 和 appContextId 可以使用通配符，比如 userservice:** 会将事件发布给 userservice 微服务。

开发 UserService 类实现事件发布，具体代码如下：

```java
@Service
public class UserService {

    public static final Logger logger = LoggerFactory.getLogger(UserService.class);

    private List<User> users;

    @Resource
    private ApplicationContextHolder holder;

    public UserService() {
        List<User> users = new ArrayList<>();
        users.add(new User(1, "ay"));
        users.add(new User(2, "al"));
        this.users = users;
    }

    public List<User> findAll() {
        return this.users;
    }
```

```java
    public User save(User userDTO) {
        for (User user : this.users) {
            if (user.getId() == userDTO.getId()) {
                user.setName(userDTO.getName());
                break;
            }
        }
        this.users.add(userDTO);
        //这是关键,保存用户之后就发布更新事件
        this.fireEvent(UserEvent.ET_UPDATE, userDTO);
        return userDTO;
    }

    private void fireEvent(String eventAction, User user) {
        logger.info("==> context:{} <==",holder.getApplicationContext().getId());
        //源对象
        UserEvent userEvent = new UserEvent(user,
                //上下文 ID
                holder.getApplicationContext().getId(),
                //将消息发往所有服务
                "*:**",
                //事件
                eventAction,
                //源对象的唯一标识符
                String.valueOf(user.getId()));
        RemoteApplicationEventPublisher.publishEvent(userEvent,holder);
    }
}
```

消息发送类 RemoteApplicationEventPublisher,具体代码如下:

```java
/**
 * 事件发布
 * @author ay
 * @since 2020-1103
 */
public class RemoteApplicationEventPublisher {
    public static final Logger logger =
LoggerFactory.getLogger(RemoteApplicationEventPublisher.class);

    public static void publishEvent(RemoteApplicationEvent event,
ApplicationContextHolder ach) {
        ApplicationContext cxt = ach.getApplicationContext();
        if(null != cxt) {
            cxt.publishEvent(event);
            logger.info("已经发布事件:{}", event);
        }else {
            logger.warn("无法获取到应用上下文实例,不能发布事件。");
        }
    }
}
```

User 类和 ApplicationContextHolder 类的代码和 8.4.2 小节的代码一样，这里不再赘述。

在 application.properties 文件中添加如下配置：

```
server.port=8081
spring.application.name=user-service
### eureka 地址
eureka.client.service-url.defaultZone=http://localhost:8888/eureka
### kafka broker 地址
spring.cloud.stream.kafka.binder.brokers=localhost
### kafka broker 端口
spring.cloud.stream.kafka.binder.defaultBrokerPort=9092
### zookeeper 地址
spring.cloud.stream.kafka.binder.zk-nodes=localhost
```

最后，在启动类中添加@RemoteApplicationEventScan 注解：

```
/**
 * 描述：启动类
 * @author ay
 * @since 2020-11-03
 */
@SpringBootApplication
@EnableDiscoveryClient
@RemoteApplicationEventScan(basePackages = "com.ay")
public class UserServiceApplication {
    public static void main(String[] args) {
        SpringApplication.run(UserServiceApplication.class, args);
    }
}
```

步骤 04 创建 user-consumer 模块，将 UserService 类复制到该模块下，实现事件监听类 UserEventListener，具体代码如下：

```
@Component
public class UserEventListener implements ApplicationListener<UserEvent> {

    private static final Logger logger =
LoggerFactory.getLogger(UserEventListener.class);

    @PostConstruct
    public void doSomething() {
        logger.info("==> do something <==");
    }
    @Override
    public void onApplicationEvent(UserEvent userEvent) {
        logger.debug("==> 收到用户事件{} <==", userEvent);
        //todo 具体业务代码
    }
}
```

在 application.properties 文件中添加相同的配置，只是换个端口和服务名称。最后，在启动类中开启远程消息扫描，具体代码如下：

```
@SpringBootApplication
@EnableDiscoveryClient
@EnableFeignClients
@RemoteApplicationEventScan(basePackages = "com.ay")
public class UserConsumerApplication {

    public static void main(String[] args) {
        SpringApplication.run(UserConsumerApplication.class, args);
    }
}
```

步骤 05 依次启动 eureka-server、user-service 和 user-consumer 服务，查看 user-service 控制台启动日志：

```
    2020-11-07 16:05:51.818  INFO 5694 --- [           main] 
o.s.i.endpoint.EventDrivenConsumer       : started bean 
'_org.springframework.integration.errorLogger'
    2020-11-07 16:05:51.819  INFO 5694 --- [           main] 
o.s.c.s.binder.DefaultBinderFactory      : Creating binder: kafka
    2020-11-07 16:05:51.956  INFO 5694 --- [           main] 
o.s.c.s.binder.DefaultBinderFactory      : Caching the binder: kafka
    2020-11-07 16:05:51.956  INFO 5694 --- [           main] 
o.s.c.s.binder.DefaultBinderFactory      : Retrieving cached binder: kafka
    2020-11-07 16:05:52.014  INFO 5694 --- [           main] 
o.s.c.s.b.k.p.KafkaTopicProvisioner      : Using kafka topic for outbound: 
springCloudBus
    2020-11-07 16:05:52.016  INFO 5694 --- [           main] 
o.a.k.clients.admin.AdminClientConfig    : AdminClientConfig values:
```

可以看到 user-service 服务连接到 kafka，并创建名为 springCloudBus 的 topic。

8.4.4 Spring Cloud Bus 原理

Spring Cloud Bus 整合 Spring 的事件处理机制和消息中间件，其简单原理如图 8-5 所示。

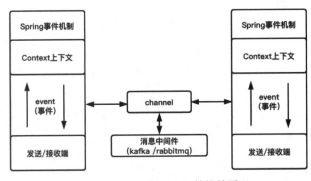

图 8-5 Spring Cloud Bus 的简单原理

完整流程如下所示：

（1）发送端构造事件 Event，将其发布到 Context 上下文中，然后将事件转化为 Json 格式的消息，发送到 Channel 中。

（2）接收端从 Channel 中获取到消息并转化为事件，然后将事件推送到 Context 上下文中。

（3）接收端收到事件后调用服务进行处理。

在整个流程中，只有发送/接收端从 Context 上下文中取事件和发送事件是需要自行开发的，其他部分都由框架封装完成。

8.4.5 Spring Cloud Bus 端点

Spring Cloud Bus 提供了两个端点——/actuator/bus-refresh 和/actuator/bus-env，分别对应于 Spring Cloud Commons 中/actuator/refresh 和/actuator/env 的各个执行器端点。

/actuator/bus-refresh 端点清除 RefreshScope 缓存并重新绑定@ConfigurationProperties。要公开 /actuator/bus-refresh 端点，需要在应用程序中添加以下配置：

```
management.endpoints.web.exposure.include=bus-refresh
```

相对于/actuator/env，/actuator/bus-env 使用键值对更新每个实例的 Environment，默认不暴露，需配置 management.endpoints.web.exposure.include=bus-env 来开放接口访问。

/actuator/bus-env 端点 POST 请求参数如下：

```
{
    "name": "key1",
    "value": "value1"
}
```

8.4.6 Bus 事件追踪

Bus 事件（RemoteApplicationEvent 的子类）可以通过设置 spring.cloud.bus.trace.enabled=true 进行跟踪。如果这样做，那么 Spring Boot TraceRepository（如果存在）将显示每个服务实例发送的所有事件和所有的 ack。以下示例来自/ trace 端点：

```
{
  "timestamp": "2015-11-26T10:24:44.411+0000",
  "info": {
    "signal": "spring.cloud.bus.ack",
    "type": "RefreshRemoteApplicationEvent",
    "id": "c4d374b7-58ea-4928-a312-31984def293b",
    "origin": "stores:8081",
    "destination": "*:**"
  }
},
{
  "timestamp": "2015-11-26T10:24:41.864+0000",
```

```
    "info": {
      "signal": "spring.cloud.bus.sent",
      "type": "RefreshRemoteApplicationEvent",
      "id": "c4d374b7-58ea-4928-a312-31984def293b",
      "origin": "customers:9000",
      "destination": "*:**"
    }
  },
  {
    "timestamp": "2015-11-26T10:24:41.862+0000",
    "info": {
      "signal": "spring.cloud.bus.ack",
      "type": "RefreshRemoteApplicationEvent",
      "id": "c4d374b7-58ea-4928-a312-31984def293b",
      "origin": "customers:9000",
      "destination": "*:**"
    }
  }
}
```

以上信息表明 RefreshRemoteApplicationEvent 已从 customers:9000 发送，广播到所有服务，并被 customers:9000 和 stores:8081 收到（确认）。

Bus 可以携带任何类型为 RemoteApplicationEvent 的事件，传输格式是 JSON，并且反序列化时需要知道使用的是哪些类型。要注册一个新的类型，需要把它放在 org.springframework.cloud.bus.event 的子包中。要自定义事件名称，可以在自定义类上使用 @JsonTypeName 或者使用默认策略（使用类的简单名称）。

如果不能或不想为自定义事件使用 org.springframework.cloud.bus.event 的子包，就必须使用 @RemoteApplicationEventScan 指定要扫描的包（包括子包），扫描类型为 RemoteApplicationEvent 的事件。

例如，有一个名为 MyEvent 的自定义事件：

```
package com.acme;

public class MyEvent extends RemoteApplicationEvent {
    ...
}
```

可以通过以下方式注册该事件：

```
package com.acme;

@Configuration
@RemoteApplicationEventScan
public class BusConfiguration {
    ...
}
```

不指定值时，将注册使用@RemoteApplicationEventScan 类的包。在此示例中 com.acme 通过使用 BusConfiguration 包进行注册。

还可以使用@RemoteApplicationEventScan 上的 value、basePackages 或 basePackageClasses 属

性来显式指定要扫描的软件包，如以下示例所示：

```
package com.acme;

@Configuration
//@RemoteApplicationEventScan({"com.acme", "foo.bar"})
//@RemoteApplicationEventScan(basePackages = {"com.acme", "foo.bar", "fizz.buzz"})
@RemoteApplicationEventScan(basePackageClasses = BusConfiguration.class)
public class BusConfiguration {
    ...
}
```

@RemoteApplicationEventScan 的所有前述示例都是等效的，因为 com.acme 软件包是通过在 @RemoteApplicationEventScan 上显式指定软件包来注册的。

第 9 章

Spring Cloud Alibaba Seata 分布式事务

本章主要介绍 Spring Cloud Alibaba Seata 分布式事务组件，包括 Seata 简介、Seata 部署、Seata 原理与设计以及如何通过 Seata 解决分布式事务问题。

9.1 Seata 基础知识

9.1.1 Seata 简介

在传统的单体应用中，业务操作使用同一条连接操作不同的数据表，一旦出现异常就可以整体回滚，如图 9-1 所示。

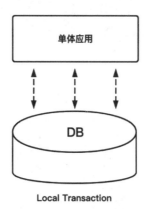

图 9-1　单体应用本地事务

随着公司的快速发展、业务需求的变化，单体应用被拆分成微服务应用，原来的单体应用被拆分成多个独立的微服务，分别使用独立的数据源，业务操作需要调用三个服务来完成。此时每个服务内部的数据一致性由本地事务来保证，但是全局的数据一致性问题无法保证。

在微服务架构中，一次业务请求需要操作多个数据源或需要进行远程调用，就会产生分布式事务问题，如图 9-2 所示。

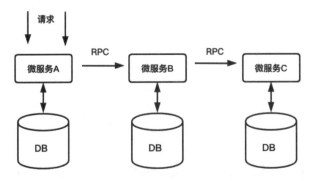

图 9-2　分布式架构微服务调用

Seata 是一款开源的分布式事务解决方案，致力于提供高性能和简单易用的分布式事务服务。Seata 将为用户提供 AT、TCC、SAGA 和 XA 事务模式，为用户打造一站式的分布式解决方案。

9.1.2　Seata 部署

Seata 分 TC、TM 和 RM 三个角色，TC（Server 端）为单独服务端部署，TM 和 RM（Client 端）由业务系统集成。

Seata 的安装步骤如下所示。

步骤 01 通过 https://github.com/seata/seata/releases 地址下载 Seata 安装包，本文使用的 Seata 版本是 seata-server-1.3.0。

步骤 02 解压 seata-server-1.3.0 安装包，可以看到如下目录结构：

```
### seata 目录存在 4 个文件或文件夹
> ls
LICENSE bin conf    lib
```

步骤 03 修改 conf/registry.conf 文件：

```
### 进入 conf 目录
>cd conf
### registry.conf 是注册配置文件
>vi registry.conf
registry {
  # file 、nacos 、eureka、redis、zk、consul、etcd3、sofa
  ### 使用 nacos 作为注册中心
  type = "nacos"
  ### 配置 nacos 信息
  nacos {
    ### 应用名称
    application = "seata-server"
    ### nacos 服务地址
    serverAddr = "127.0.0.1:8848"
    group = "SEATA_GROUP"
```

```
      namespace = ""
      cluster = "default"
      username = ""
      password = ""
    }
    file {
      name = "file.conf"
    }
  }

  config {
    # file、nacos 、apollo、zk、consul、etcd3
    type = "file"

    nacos {
      serverAddr = "127.0.0.1:8848"
      namespace = ""
      group = "SEATA_GROUP"
      username = ""
      password = ""
    }
    file {
      name = "file.conf"
    }
  }
```

步骤04 Seata Server 目前存在三种存储模式（file、db、redis），其中 file 模式无须改动，直接启动即可。

不同的存储模式，性能也存在差异：

- file 模式为单机模式，全局事务会话信息内存中读写并持久化本地文件 root.data，性能较高。
- db 模式为高可用模式，全局事务会话信息通过 db 共享，相应性能会差。
- redis 模式 Seata-Server 1.3 及以上版本支持，性能较高，存在事务信息丢失风险，需要提前配置适合当前场景的 redis 持久化配置。

步骤05 进入 bin 目录启动 Seata，具体如下所示：

```
nohup sh seata-server.sh -p 8091 -h localhost -m file &> seata.log &
```

- -p：指定启动 seata server 的端口号。
- -h：指定 seata server 所绑定的主机。
- -m：指定全局事务会话信息存储模式。

步骤06 查看启动日志：

```
> tail -1000f seata.log
  2020-09-01 08:39:42.297 INFO --- [           main]
io.seata.config.FileConfiguration        : The configuration file used is
registry.conf
```

```
2020-09-01 08:39:42.355  INFO --- [           main] 
io.seata.config.FileConfiguration        : The configuration file used is file.conf
2020-09-01 08:39:42.668  INFO --- [           main] 
i.s.core.rpc.netty.NettyServerBootstrap  : Server started, listen port: 8091
```

当看到"Server started, listen port: 8091"信息时,表示 Seata 已启动。

9.1.3 Seata 原理与设计

一个分布式事务理解成一个包含了若干分支事务的全局事务(见图 9-3)。全局事务的职责是协调其下管辖的分支事务达成一致,要么一起成功提交,要么一起失败回滚。通常分支事务本身就是一个满足 ACID 的本地事务。

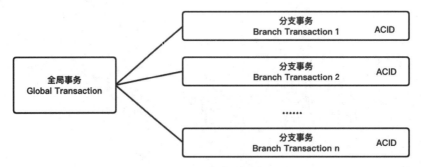

图 9-3 分布式架构微服务调用

Seata 包含以下组件:

- Transaction Coordinator(TC):事务协调器,维护全局事务的运行状态,负责协调并驱动全局事务的提交或回滚。
- Transaction Manager(TM):事务管理者,控制全局事务的边界,负责开启一个全局事务,并最终发起全局提交或全局回滚的决议。
- Resource Manager(RM):资源管理器,控制分支事务,负责分支注册、状态汇报,并接收事务协调器的指令,驱动分支(本地)事务的提交和回滚。

典型的分布式事务过程(见图 9-4)如下:

(1) TM 向 TC 申请开启一个全局事务,全局事务创建成功并生成一个全局唯一的 XID。
(2) XID 在微服务调用链路的上下文中传播。
(3) RM 向 TC 注册分支事务,将其纳入 XID 对应全局事务的管辖。
(4) TM 向 TC 发起针对 XID 的全局提交或回滚决议。
(5) TC 调度 XID 下管辖的全部分支事务完成提交或回滚请求。

第 9 章 Spring Cloud Alibaba Seata 分布式事务

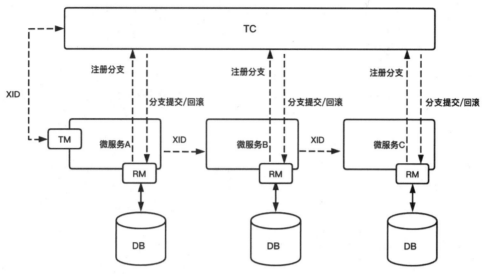

图 9-4 典型分布式事务过程

9.2 Seata 使用

本节以经典的用户下单为例,演示如何通过 Seata 实现分布式事务控制。总体流程分为三个重要步骤:

(1)在订单服务中创建订单。
(2)通过远程调用库存服务,扣减库存。
(3)通过远程调用用户服务,扣减用户账户金额。

任何一个步骤出现异常,在没有分布式事务控制下都会出现脏数据。

9.2.1 数据库准备

创建三个数据库:

- seata-user:该数据库用于存储用户信息。
- seata-order:该数据库用于存储订单信息。
- seata-stock:该数据库用于存储库存信息。

seata-user 数据库创建 account 表,具体 SQL 如下:

```
CREATE TABLE `account` (
  `id` bigint(11) NOT NULL AUTO_INCREMENT COMMENT 'id',
  `user_id` bigint(11) DEFAULT NULL COMMENT '用户id',
  `total` decimal(10,0) DEFAULT NULL COMMENT '总额度',
  `used` decimal(10,0) DEFAULT NULL COMMENT '已用余额',
```

```
  `residue` decimal(10,0) DEFAULT '0' COMMENT '剩余可用额度',
  PRIMARY KEY (`id`)
) ENGINE=InnoDB AUTO_INCREMENT=2 DEFAULT CHARSET=utf8;
```

account 表的数据初始化，具体 SQL 如下：

```
INSERT INTO `seat-account`.`account` (`id`, `user_id`, `total`, `used`, `residue`) VALUES ('1', '1', '300', '0', '300');
```

seata-order 数据库创建 order 表，具体 SQL 如下：

```
CREATE TABLE `order` (
  `id` bigint(11) NOT NULL AUTO_INCREMENT,
  `user_id` bigint(11) DEFAULT NULL COMMENT '用户id',
  `product_id` bigint(11) DEFAULT NULL COMMENT '产品id',
  `count` int(11) DEFAULT NULL COMMENT '数量',
  `money` decimal(11,0) DEFAULT NULL COMMENT '金额',
  `status` int(1) DEFAULT NULL COMMENT '订单状态：0：创建中；1：完成',
  PRIMARY KEY (`id`)
) ENGINE=InnoDB AUTO_INCREMENT=2 DEFAULT CHARSET=utf8;
```

seata-stock 数据库创建 storage 表，具体 SQL 如下：

```
CREATE TABLE `storage` (
  `id` bigint(11) NOT NULL AUTO_INCREMENT,
  `product_id` bigint(11) DEFAULT NULL COMMENT '产品id',
  `total` int(11) DEFAULT NULL COMMENT '总库存',
  `used` int(11) DEFAULT NULL COMMENT '已用库存',
  `residue` int(11) DEFAULT NULL COMMENT '剩余库存',
  PRIMARY KEY (`id`)
) ENGINE=InnoDB AUTO_INCREMENT=2 DEFAULT CHARSET=utf8;
-- 初始化数据
INSERT INTO `seat-storage`.`storage` (`id`, `product_id`, `total`, `used`, `residue`) VALUES ('1', '1', '100', '0', '100');
```

9.2.2 创建微服务

创建 seata-user-service、seata-order-service 以及 seata-stock-service 微服务，具体步骤如下：

步骤01 使用 Intellij IDEA 工具，快速创建 Spring Boot 项目，项目名称分别为 seata-user-service、seata-order-service 以及 seata-stock-service 服务。

步骤02 修改 application.yml 文件，自定义事务组的名称：

```yaml
server:
  port: 9000
spring:
  application:
    name: seata-user-service
  cloud:
    alibaba:
      seata:
```

```yaml
      #自定义事务组名称，需要与seata-server中的对应
      tx-service-group: fsp_tx_group
  #nacos注册中心地址
  nacos:
    discovery:
      server-addr: localhost:8848
  #数据库连接信息
  datasource:
    driver-class-name: com.mysql.jdbc.Driver
    password: 123456
    url: jdbc:mysql://localhost:3306/seata-user
    username: root
logging:
  level:
    io:
      seata: info
mybatis:
  mapperLocations: classpath:mapper/*.xml
```

seata-user-service、seata-order-service 以及 seata-stock-service 的 application.yml 配置基本一样，服务端口分别修改为 9000、9001、9002，服务名称为项目名称。数据库连接信息修改为对应数据库名称。

步骤 03 在项目的 pom.xml 文件中添加如下依赖：

```xml
<?xml version="1.0" encoding="UTF-8"?>
<project xmlns="http://maven.apache.org/POM/4.0.0"
xmlns:xsi="http://www.w3.org/2001/XMLSchema-instance"
         xsi:schemaLocation="http://maven.apache.org/POM/4.0.0
https://maven.apache.org/xsd/maven-4.0.0.xsd">
    <modelVersion>4.0.0</modelVersion>
    <parent>
        <groupId>org.springframework.boot</groupId>
        <artifactId>spring-boot-starter-parent</artifactId>
        <version>2.2.1.RELEASE</version>
        <relativePath/> <!-- lookup parent from repository -->
    </parent>
    <groupId>com.macro.cloud</groupId>
    <artifactId>seata-account-service</artifactId>
    <version>0.0.1-SNAPSHOT</version>
    <name>seata-account-service</name>
    <description>Demo project for Spring Boot</description>

    <properties>
        <java.version>1.8</java.version>
        <spring-cloud.version>Hoxton.SR5</spring-cloud.version>
        <mysql-connector-java.version>5.1.37</mysql-connector-java.version>
        <mybatis-spring-boot-starter.version>2.0.0</mybatis-spring-boot-starter.version>
        <druid-spring-boot-starter.version>1.1.10</druid-spring-boot-starter.version>
        <lombok.version>1.18.8</lombok.version>
```

```xml
            <seata.version>1.3.0</seata.version>
        </properties>

        <dependencies>
            <!--nacos-->
            <dependency>
                <groupId>com.alibaba.cloud</groupId>
                <artifactId>spring-cloud-starter-alibaba-nacos-discovery</artifactId>
            </dependency>
            <!--seata-->
            <dependency>
                <groupId>com.alibaba.cloud</groupId>
                <artifactId>spring-cloud-starter-alibaba-seata</artifactId>
                <exclusions>
                    <exclusion>
                        <artifactId>seata-all</artifactId>
                        <groupId>io.seata</groupId>
                    </exclusion>
                </exclusions>
            </dependency>
            <dependency>
                <groupId>io.seata</groupId>
                <artifactId>seata-all</artifactId>
                <version>${seata.version}</version>
            </dependency>
            <!--feign-->
            <dependency>
                <groupId>org.springframework.cloud</groupId>
                <artifactId>spring-cloud-starter-openfeign</artifactId>
            </dependency>
            <dependency>
                <groupId>org.springframework.boot</groupId>
                <artifactId>spring-boot-starter-web</artifactId>
            </dependency>
            <dependency>
                <groupId>org.springframework.boot</groupId>
                <artifactId>spring-boot-starter-test</artifactId>
                <scope>test</scope>
            </dependency>
            <dependency>
                <groupId>org.mybatis.spring.boot</groupId>
                <artifactId>mybatis-spring-boot-starter</artifactId>
                <version>${mybatis-spring-boot-starter.version}</version>
            </dependency>
            <dependency>
                <groupId>mysql</groupId>
                <artifactId>mysql-connector-java</artifactId>
                <version>${mysql-connector-java.version}</version>
            </dependency>
```

```xml
    <dependency>
        <groupId>com.alibaba</groupId>
        <artifactId>druid-spring-boot-starter</artifactId>
        <version>${druid-spring-boot-starter.version}</version>
    </dependency>
    <dependency>
        <groupId>org.projectlombok</groupId>
        <artifactId>lombok</artifactId>
        <version>${lombok.version}</version>
    </dependency>
</dependencies>

<dependencyManagement>
    <dependencies>
        <dependency>
            <groupId>org.springframework.cloud</groupId>
            <artifactId>spring-cloud-dependencies</artifactId>
            <version>${spring-cloud.version}</version>
            <type>pom</type>
            <scope>import</scope>
        </dependency>
        <dependency>
            <groupId>com.alibaba.cloud</groupId>
            <artifactId>spring-cloud-alibaba-dependencies</artifactId>
            <version>2.2.0.RELEASE</version>
            <type>pom</type>
            <scope>import</scope>
        </dependency>
    </dependencies>
</dependencyManagement>

<build>
    <plugins>
        <plugin>
            <groupId>org.springframework.boot</groupId>
            <artifactId>spring-boot-maven-plugin</artifactId>
        </plugin>
    </plugins>
</build>
</project>
```

三个微服务都需要引入 Nacos、数据库 MySQL 以及驱动、Seata、MyBatis、Feign 等相关的依赖包。微服务启动时会注册到 Nacos 注册中心，数据库 CRUD 操作通过 MyBatis 完成，微服务之间通过 Feign 接口调用。这里需要注意的是，Spring Boot、Spring Cloud 和 Spring Cloud Alibaba 之间的版本号需要对应，具体可以参考第 2 章的内容。

步骤 04 在 resources 目录下添加并修改 registry.conf 配置文件，使用 Nacos 作为注册中心，使用 file.conf 作为配置中心。

```
registry {
  # file 、nacos 、eureka、redis、zk
  type = "nacos"

  nacos {
    serverAddr = "localhost:8848"
    namespace = ""
    cluster = "default"
  }
  #省略部分代码
}

config {
  # file、nacos 、apollo、zk
  type = "file"

  file {
    name = "file.conf"
  }
  #省略部分代码
}
```

步骤 05 file.conf 配置文件代码如下所示：

```
#省略部分代码
service {
  #重要配置
  vgroupMapping.fsp_tx_group = "default"
  #重要配置
  default.grouplist = "127.0.0.1:8091"
  #degrade current not support
  enableDegrade = false
  #disable
  disable = false
  #unit ms,s,m,h,d represents milliseconds, seconds, minutes, hours, days, default permanent
  max.commit.retry.timeout = "-1"
  max.rollback.retry.timeout = "-1"
  disableGlobalTransaction = false
}
#省略部分代码
```

事务分组是 seata 的资源逻辑，类似于服务实例。在 file.conf 中的 **fsp_tx_group** 就是一个事务分组，一个 seata-server 可以管理多个事务分组。微服务中配置了事务分组，通过用户配置的配置中心查询 service.vgroupMapping .[事务分组配置项]，取得配置项的值就是 TC 集群的名称，拿到集群名称程序通过一定的前后缀+集群名称去构造服务名，各配置中心的服务名实现不同。拿到服务名去相应的注册中心去拉取相应服务名的服务列表，获得后端真实的 TC 服务列表。

这里多了一层获取事务分组到映射集群的配置。这样设计后，事务分组可以作为资源的逻辑隔离单位，出现某集群故障时可以快速 failover，只切换对应分组，可以把故障缩减到服务级别，但前提是要有足够的 server 集群。

对于上面的 file.conf 配置，seata 具体处理流程为：

①读取配置：通过 FileConfiguration 本地加载 file.conf 的配置参数。

②获取事务分组：spring 配置，springboot 可配置在 yml、properties 中，服务启动时加载配置，对应的值"fsp_tx_group"即为一个事务分组名，若不配置，默认获取属性 spring.application.name 的值+"-fescar-service-group"。

③查找 TC 集群名：将事务分组名"fsp_tx_group"拼接成"service.vgroupMapping.fsp_tx_group"来查找 TC 集群名，clusterName 为"default"。

④查询 TC 服务：拼接"service."+clusterName+".grouplist"，找到真实 TC 服务地址 127.0.0.1:8091。

步骤 06 在启动类中取消数据源的自动创建，具体代码如下：

```java
@SpringBootApplication(exclude = DataSourceAutoConfiguration.class)
@EnableDiscoveryClient
@EnableFeignClients
public class SeataAccountServiceApplication {
    public static void main(String[] args) {
        SpringApplication.run(SeataAccountServiceApplication.class, args);
    }
}
```

步骤 07 创建配置使用 Seata 对数据源进行代理，具体代码如下：

```java
@Configuration
public class DataSourceProxyConfig {

    @Value("${mybatis.mapperLocations}")
    private String mapperLocations;

    @Bean
    @ConfigurationProperties(prefix = "spring.datasource")
    public DataSource druidDataSource(){
        return new DruidDataSource();
    }

    @Bean
    public DataSourceProxy dataSourceProxy(DataSource dataSource) {
        return new DataSourceProxy(dataSource);
    }

    @Bean
    public SqlSessionFactory sqlSessionFactoryBean(DataSourceProxy dataSourceProxy) throws Exception {
        SqlSessionFactoryBean sqlSessionFactoryBean = new SqlSessionFactoryBean();
        sqlSessionFactoryBean.setDataSource(dataSourceProxy);
        sqlSessionFactoryBean.setMapperLocations(new PathMatchingResourcePatternResolver()
```

```
                .getResources(mapperLocations));
        sqlSessionFactoryBean.setTransactionFactory(new
SpringManagedTransactionFactory());
        return sqlSessionFactoryBean.getObject();
    }
}
```

步骤 08 使用@GlobalTransactional 注解开启分布式事务,具体代码如下:

```
/**
 * 订单业务实现类
 * @author ay
 * @since 2020-09-27
 */
@Service
public class OrderServiceImpl implements OrderService {

    @Resource
    private OrderDao orderDao;
    @Resource
    private StorageService storageService;
    @Resource
    private AccountService accountService;

    /**
     * 总体流程为:创建订单->调用库存服务扣减库存->调用账户服务扣减账户余额->修改订单状态
     */
    @Override
    @GlobalTransactional(name = "fsp-create-order",rollbackFor = Exception.class)
    public void create(Order order) {
        System.out.println("--->>>下单开始");
        //本应用创建订单
        orderDao.create(order);

        //远程调用库存服务扣减库存
        System.out.println("------>order-service 中扣减库存开始");
        storageService.decrease(order.getProductId(),order.getCount());
        System.out.println("------>order-service 中扣减库存结束");

        //远程调用账户服务扣减余额
        System.out.println("------>order-service 中扣减余额开始");
        accountService.decrease(order.getUserId(),order.getMoney());
        System.out.println("------>order-service 中扣减余额结束");

        //修改订单状态为已完成
        System.out.println("------>order-service 中修改订单状态开始");
        orderDao.update(order.getUserId(),0);
        System.out.println("------>order-service 中修改订单状态结束");

        System.out.println("------>下单结束");
```

```
    }
}
```

只需要使用一个@GlobalTransactional 注解在业务方法上即可。create()方法总体为：创建订单→调用库存服务扣减库存→调用账户服务扣减账户余额→修改订单状态。

步骤09 这里只展示部分代码，没办法把所有代码全部写到书里，完整的代码地址在本书"前言"中查找。所有的微服务开发配置完成后，启动 seata-user-service、seata-order-service、seata-stock-service 以及 seata-service。

步骤10 在浏览器中访问 http://localhost:9001/order/create?userId=1&productId=1&count=10&money=100，用户正常下单，购买 productId=1 的商品，数量为 10 个，花了 100 元，数据库中的数据变化如图 9-5、图 9-6 和图 9-7 所示。

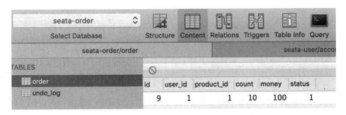

图 9-5 seata-order 表数据

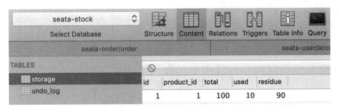

图 9-6 seata-stock 表数据

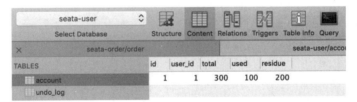

图 9-7 seata-user 表数据

步骤11 修改 seata-user-service 服务的 decrease()方法，在方法内部使用 Thread.sleep(30*1000)模拟超时，具体代码如下：

```
/**
 * 账户业务实现类
 * @author ay
 * @since 2020/09/27
 */
@Service
public class AccountServiceImpl implements AccountService {

    @Autowired
```

```java
    private AccountDao accountDao;

    /**
     * 扣减账户余额
     */
    @Override
    public void decrease(Long userId, BigDecimal money) {
        LOGGER.info("------>account-service中扣减账户余额开始");
        // 模拟超时异常，全局事务回滚
        try {
            Thread.sleep(50*1000);
        } catch (InterruptedException e) {
            e.printStackTrace();
        }
        accountDao.decrease(userId,money);
        LOGGER.info("------>account-service中扣减账户余额结束");
    }
}
```

在浏览器中访问 http://localhost:9001/order/create?userId=1&productId=1&count=10&money=100，此时发现下单后数据库的数据并没有任何改变，因为 seata-user-service 服务的 decrease() 方法出现超时异常，Seata 进行全局事务回滚。

将 @GlobalTransactional 注解注释掉，仍然访问 http://localhost:9001/order/create?userId=1&productId=1&count=10&money=100，由于 seata-user-service 的超时会导致当库存和账户金额扣减后订单状态并没有设置为已经完成，而且由于远程调用的重试机制，账户余额还会被多次扣减。

第 10 章

Spring Cloud Sleuth 服务链路追踪

本章主要介绍 Spring Cloud Sleuth 服务链路追踪，包括 Sleuth 和 Zipkin 简介、Zipkin 安装与快速启动、Spring Cloud Sleuth 整合 Zipkin、Spring Cloud Sleuth 整合 ELK、Sleuth 原理浅析等内容。

10.1 Spring Cloud Sleuth 简介

随着业务的发展，系统规模变得越来越大，微服务拆分越来越细，各微服务间的调用关系也越来越复杂。客户端请求在后端系统中会经过多个不同的微服务调用来协同产生最后的请求结果。几乎每一个请求都会形成一个复杂的分布式服务调用链路，在每条链路中任何一个依赖服务出现延迟超时或者错误都有可能引起整个请求最后的失败，具体如图 10-1 所示。

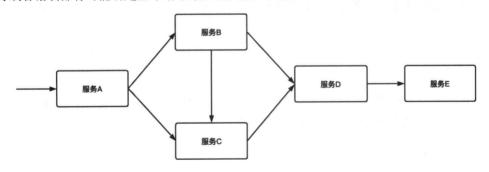

图 10-1 微服务复杂调用关系

这时需要一个能够监控微服务整个调用链的工具，跟踪一个用户请求的全过程（包括数据采集、数据传输、数据存储、数据分析、数据可视化），捕获这些跟踪数据，构建微服务整个调用链的视图。Spring Cloud Sleuth 就是这样一个工具。

服务追踪系统的实现主要包括三个部分：

- 埋点数据收集：负责在服务端进行埋点，以收集服务调用的上下文数据。

- 实时数据处理：负责将收集到的链路信息按照 TraceId 和 spanId 进行串联和存储。
- 数据链路展示：把处理后的服务调用数据按照调用链的形式展示出来。

下面我们再来看一下 Sleuth 的核心概念。

（1）Trace：一组 Span 的集合，表示一条调用链路。例如，服务 A 调用服务 B，再调用服务 C，A→B→C 链路就是一条 Trace，每个服务（例如 B）就是一个 Span。如果在服务 B 中再加入两个线程，分别调用了 D、E，那么 D、E 就是 B 的子 Span。

（2）TraceId：全局跟踪 ID，用来标记一次完整服务调用，所以和一次服务调用相关的 Span 中的 TraceId 都是相同的。Zipkin 将具有相同 TraceId 的 Span 组装成跟踪树来直观地将调用链路图展现在我们面前。

（3）Span：基本工作单元，通过 64 位 ID 唯一标识。Trace 以另一个 64 位 ID 表示，Span 还包含其他数据信息，比如摘要、时间戳事件、关键值注释（tags）、Span 的 ID 以及进度 ID（通常是 IP 地址）。

（4）Id：Span 的 ID，只要做到一个 TraceId 下唯一就可以。

（5）ParentId：父 Span 的 ID，调用有层级关系，所以 Span 作为调用节点的存储结构，也有层级关系。跟踪链是采用跟踪树的形式来展现的，树的根节点就是调用的顶点。从开发者的角度来说，顶级 Span 是从接入了 Zipkin 的应用中最先接触到服务调用的应用中采集的。所以，顶级 Span 没有 ParentId 字段。

（6）Annotation：基本标注列表，用来及时记录一个事件的存在，一个标注可以理解成 Span 生命周期中重要时刻的数据快照，比如一个标注中一般包含发生时刻（timestamp）、事件类型（value）、端点（endpoint）等信息。事件类型包括以下几种：

- cs（Client Sent）：客户端发起一个请求，这个 Annotion 注解描述 Span 的开始。
- sr（Server Received）：服务端获得请求并准备开始处理它，sr 减去 cs 即网络延迟时间。
- ss（Server Sent）：表明请求处理的完成（请求返回客户端），ss 减去 sr 即服务端需要的处理请求时间。
- cr（Client Received）：表明 Span 的结束，客户端成功接收到服务端的回复，cr 减去 cs 即客户端从服务端获取回复的所有时间。

Spring Cloud Sleuth 将 Trace 和 Span ID 添加到 Slf4J MDC，因此可以在日志聚合器中从给定的 Trace 或者 Span ID 提取所有日志，如以下示例日志所示：

```
  2016-02-02 15:30:57.902  INFO [bar,6bfd228dc00d216b,6bfd228dc00d216b,false]
23030 --- [nio-8081-exec-3] ...
  2016-02-02 15:30:58.372 ERROR [bar,6bfd228dc00d216b,6bfd228dc00d216b,false]
23030 --- [nio-8081-exec-3] ...
  2016-02-02 15:31:01.936  INFO [bar,46ab0d418373cbc9,46ab0d418373cbc9,false]
23030 --- [nio-8081-exec-4] ...
```

来自 MDC 的[appname, TraceId, spanId, exportable]意义如下：

- appname：记录应用程序的名称，也就是 application.properties 中 spring.application.name 配置项的值。

- TraceId：全局跟踪 ID，用来标记一次完整服务调用，所以和一次服务调用相关的 Span 中的 TraceId 都是相同的。一个 TraceId 包含多个 spanId。
- spanId：发生特定操作的 ID。
- exportable：表示是否应将日志导出到 Zipkin。

从 2.0.0 版本开始，Spring Cloud Sleuth 使用 Brave 作为跟踪库。Brave 是一个库，用于捕获有关分布式操作的延迟信息并将其报告给 Zipkin。

10.2　Zipkin 简介

Zipkin 是一个开源的分布式追踪系统，用于对服务间的调用链路进行监控追踪。在微服务架构下，用户的一个请求可能涉及很多个后台服务间的调用。Zipkin 可以追踪（trace）调用链路、收集在各个微服务上所花的时间等信息，并上报到 Zipkin 服务器。

Zipkin 提供可插拔数据存储方式：In-Memory、MySQL、Cassandra 以及 Elasticsearch。为了方便，在开发环境直接采用 In-Memory 方式进行存储，生产数据量大的情况则推荐使用 Elasticsearch。

Zipkin 内部主要分为四部分：Collector、Storage、API、UI，如图 10-2 所示。

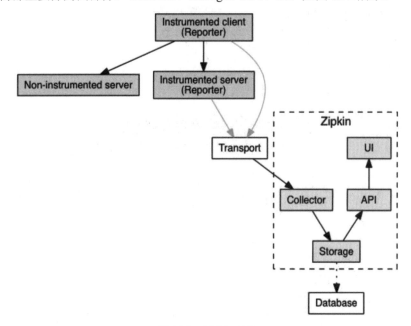

图 10-2　Zipkin 结构

（1）Collector：接收或收集各应用传输的数据。

（2）Storage：存储接收或收集过来的数据，当前支持 Memory、MySQL、Cassandra、ElasticSearch 等，默认存储在内存中。

（3）API（Query）：负责查询 Storage 中存储的数据，提供简单的 JSON API 获取数据，主要提供给 Web UI 使用。

（4）UI：官方默认提供的一个图形用户界面。

（5）Transport：负责运输从 service 收集来的 spans，并把这些 spans 转化为 Zipkin 的通用 Span，将其传递到存储层。这种方法是模块化的，允许任何生产者接收任何类型的数据。Zipkin 配有 HTTP、Kafka、scribe 三种类型的 transport。instrumentations 负责和 transport 进行交互。

Zipkin 以 Trace 结构表示对一次请求的追踪，把每个 Trace 拆分为若干个有依赖关系的 Span。在微服务架构中，一次用户请求可能会由后台若干个服务负责处理，那么每个处理请求的服务就可以理解为一个 Span（可以包括 API 服务、缓存服务、数据库服务以及报表服务等）。这个服务也可以继续请求其他的服务，因此 Span 是一个树形结构，以体现服务之间的调用关系。Zipkin 的用户界面除了可以查看 Span 的依赖关系之外，还以瀑布图的形式显示每个 Span 的耗时情况，可以很清晰地看到各个服务的性能状况。打开每个 Span，还有更详细的数据以键值对的形式呈现，而且这些数据可以在装备应用的时候自行添加。

Zipkin 起到的作用非常类似于 Logstash 和 Kibana 所起到的作用，即完成对 Trace 数据的采集、实现对 Trace 的可视化追踪。

学习 Zipkin 最快的入门方法是获取最新发布的服务器作为独立的可执行 jar。注意，Zipkin 服务器需要最低的 JRE8。Zipkin 的安装步骤如下所示。

步骤 01 在命令行窗口中执行如下命令：

```
### 下载最新可执行的 jar 包
> curl -sSL https://zipkin.io/quickstart.sh | bash -s
### 运行 jar 包
> java -jar zipkin.jar
```

也可以通过 Docker 启动 Zipkin，具体命令如下：

```
> docker run -d -p 9411:9411 openzipkin/zipkin
```

步骤 02 服务器运行后，可以使用 Zipkin UI 在 http://localhost:9411/zipkin/ 上查看跟踪信息。

Zipkin 的简约版启动速度更快，支持内存和 Elasticsearch 存储，但不支持 Kafka 或 RabbitMQ 等消息传递。如果这些约束满足需求，则可以尝试使用简约版。Zipkin 简约版的安装步骤如下所示。

在命令行窗口中执行如下命令：

```
### 下载最新的简约版
> curl -sSL https://zipkin.io/quickstart.sh | bash -s io.zipkin:zipkin-server:LATEST:slim zipkin.jar
### 运行 jar 包
> java -jar zipkin.jar
```

也可以通过 Docker 启动 Zipkin 简约版，具体命令如下：

```
> docker run -d -p 9411:9411 openzipkin/zipkin-slim
```

10.3 Spring Cloud Sleuth 整合 Zipkin

10.3.1 整合 Zipkin

Sleuth 整合 Zipkin 的具体步骤如下。

步骤 01 使用 Intellij IDEA 创建 sleuth-zipkin-demo 项目，包含 sleuth 和 other-service 模块（Spring Boot 项目），具体如图 10-3 所示。

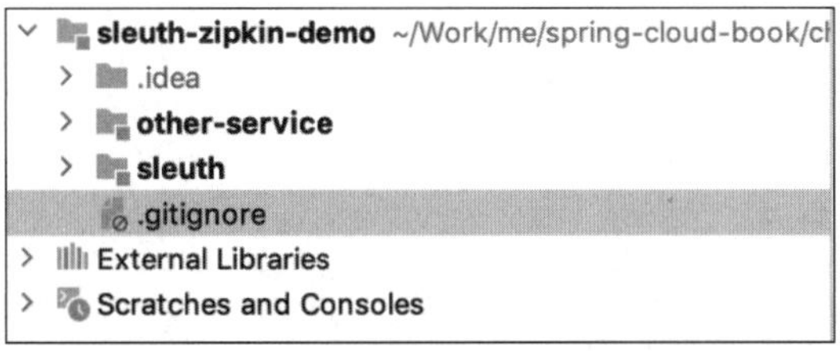

图 10-3 sleuth-zipkin-demo 项目结构

步骤 02 在 sleuth 和 other-service 模块中添加如下依赖：

```
//省略部分代码
  <parent>
      <groupId>org.springframework.boot</groupId>
      <artifactId>spring-boot-starter-parent</artifactId>
      <version>2.3.0.RELEASE</version>
      <relativePath/> <!-- lookup parent from repository -->
  </parent>

  <dependencies>
      <dependency>
          <groupId>org.springframework.cloud</groupId>
          <artifactId>spring-cloud-starter-sleuth</artifactId>
      </dependency>
      <dependency>
          <groupId>org.springframework.cloud</groupId>
          <artifactId>spring-cloud-sleuth-zipkin</artifactId>
      </dependency>
      <dependency>
          <groupId>org.springframework.boot</groupId>
          <artifactId>spring-boot-starter-web</artifactId>
      </dependency>
  </dependencies>
  
  <dependencyManagement>
      <dependencies>
          <dependency>
```

```xml
            <groupId>org.springframework.cloud</groupId>
            <artifactId>spring-cloud-dependencies</artifactId>
            <version>Hoxton.SR6</version>
            <type>pom</type>
            <scope>import</scope>
        </dependency>
      </dependencies>
   </dependencyManagement>
</project>
```

如果仅想使用 Spring Cloud Sleuth 而没有 Zipkin 集成，则将 spring-cloud-starter-sleuth 模块添加到项目中。如果同时需要 Sleuth 和 Zipkin，就添加 spring-cloud-starter-zipkin 依赖项。

步骤 03 在 sleuth 模块的配置文件 application.properties 中添加如下配置：

```
### 应用名称
spring.application.name=sleuth
### 端口
server.port=9987
### zipkin 地址
spring.zipkin.baseUrl=http://localhost:9411
### other-service 端口
other.service.port=9547
spring.zipkin.enabled=true
```

如果 spring-cloud-sleuth-zipkin 位于类路径中，则该应用程序会生成并收集与 Zipkin 兼容的 trace。默认情况下，应用程序通过 HTTP 将 trace 信息发送到本地主机（端口 9411）上的 Zipkin 服务器。可以通过设置 spring.zipkin.baseUrl 来配置服务的位置。

spring.zipkin.enabled 和 spring.sleuth.enabled 两个配置默认都是 true，对客户端而言，只要 pom 添加了依赖，就默认开启了链路跟踪功能。

如果依赖 spring-rabbit，则应用会将跟踪发送到 RabbitMQ 代理。

如果依赖 spring-kafka 并设置 spring.zipkin.sender.type:kafka，则应用会将跟踪信息发送到 Kafka 代理。

> **注　意**
>
> spring-cloud-sleuth-stream 已弃用，不应再使用。

在 other-service 模块的配置文件 application.properties 中添加如下配置：

```
spring.application.name=other-service
server.port=9547
spring.zipkin.baseUrl=http://localhost:9411
spring.zipkin.enabled=true
```

步骤 04 在 sleuth 模块中创建 SleuthController 类，具体代码如下：

```
/**
 * @author ay
 * @since 2020-12-01
```

```java
 */
@RestController
public class SleuthController {

    private static final Logger LOGGER =
LoggerFactory.getLogger(SleuthController.class);

    @Resource
    private RestTemplate restTemplate;

    @Value("${server.port}")
    private int appPort;

    @Value("${other.service.port}")
    private int otherServicePort;

    @RequestMapping("/getRemoteTime")
    public String getRemoteTime() throws URISyntaxException {
        LOGGER.info("Request to /getRemoteTime endpoint");
        ResponseEntity<String> forEntity =
                restTemplate.getForEntity(new URI("http://localhost:" +
otherServicePort + "/getTime"), String.class);
        LOGGER.info("Got response code: {}",
forEntity.getStatusCode().toString());
        return "The remote time is: " + forEntity.getBody();
    }

    @RequestMapping("/getLocalTime")
    public String getLocalTime() throws URISyntaxException {
        LOGGER.info("Request to /getLocalTime endpoint");
        ResponseEntity<String> forEntity =
                restTemplate.getForEntity(new URI("http://localhost:" + appPort
+ "/getTime"), String.class);
        LOGGER.info("Got response code: {}",
forEntity.getStatusCode().toString());
        return "The localtime is: " + forEntity.getBody();
    }
}
```

- getRemoteTime：主要通过 RestTemplate 向 other-service 发起请求。
- getLocalTime：向自己发起请求，获取时间。

步骤 05 在 other-service 模块中创建 OtherServiceController 类，具体代码如下：

```java
/**
 * 描述：
 *
 * @author ay
 * @since 2020/12/3
 */
@RestController
```

```
public class OtherServiceController {

    private static final Logger LOGGER =
LoggerFactory.getLogger(OtherServiceApplication.class);

    @RequestMapping("/getTime")
    public String getTime() {
        LOGGER.info("Request to /getTime endpoint");
        return LocalDateTime.now().toString();
    }
}
```

步骤06 运行 sleuth 和 other-service 模块的 main 方法,启动应用,并在浏览器中输入请求地址 "http://localhost:9987/getRemoteTime"。在 Zipkin 界面的 Dependencies 选项卡中可看到具体请求链路,如图 10-4 所示。

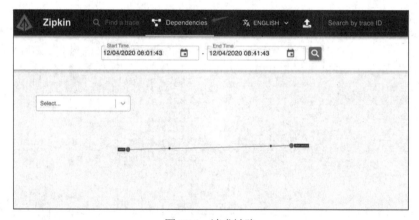

图 10-4 请求链路

在 Find a trace 选项卡中,单击 Run QUERY 按钮,可看到具体的请求信息,如图 10-5 所示。

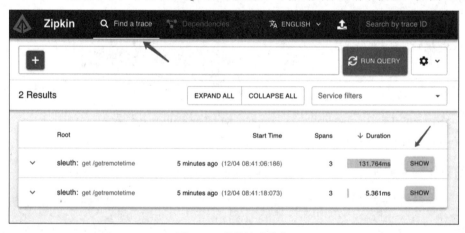

图 10-5 具体请求信息

单击 SHOW 按钮,可以看到更为详细的链路追踪信息,如图 10-6 所示。

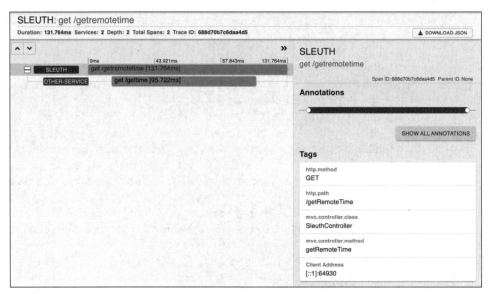

图 10-6　详细链路追踪信息

10.3.2　MySQL 存储链路数据

步骤 01　事先搭建好 MySQL 数据库，新建数据库 zipkin。

步骤 02　执行官网建库脚本，地址为 https://github.com/openzipkin/zipkin/blob/master/zipkin-storage/mysql-v1/src/main/resources/mysql.sql，具体 SQL 脚本如下：

```
    CREATE TABLE IF NOT EXISTS zipkin_spans (
      `trace_id_high` BIGINT NOT NULL DEFAULT 0 COMMENT 'If non zero, this means the trace uses 128 bit traceIds instead of 64 bit',
      `trace_id` BIGINT NOT NULL,
      `id` BIGINT NOT NULL,
      `name` VARCHAR(255) NOT NULL,
      `remote_service_name` VARCHAR(255),
      `parent_id` BIGINT,
      `debug` BIT(1),
      `start_ts` BIGINT COMMENT 'Span.timestamp(): epoch micros used for endTs query and to implement TTL',
      `duration` BIGINT COMMENT 'Span.duration(): micros used for minDuration and maxDuration query',
      PRIMARY KEY (`trace_id_high`, `trace_id`, `id`)
    ) ENGINE=InnoDB ROW_FORMAT=COMPRESSED CHARACTER SET=utf8 COLLATE utf8_general_ci;

    ALTER TABLE zipkin_spans ADD INDEX(`trace_id_high`, `trace_id`) COMMENT 'for getTracesByIds';
    ALTER TABLE zipkin_spans ADD INDEX(`name`) COMMENT 'for getTraces and getSpanNames';
    ALTER TABLE zipkin_spans ADD INDEX(`remote_service_name`) COMMENT 'for getTraces and getRemoteServiceNames';
```

```sql
    ALTER TABLE zipkin_spans ADD INDEX(`start_ts`) COMMENT 'for getTraces ordering and range';

    CREATE TABLE IF NOT EXISTS zipkin_annotations (
      `trace_id_high` BIGINT NOT NULL DEFAULT 0 COMMENT 'If non zero, this means the trace uses 128 bit traceIds instead of 64 bit',
      `trace_id` BIGINT NOT NULL COMMENT 'coincides with zipkin_spans.trace_id',
      `span_id` BIGINT NOT NULL COMMENT 'coincides with zipkin_spans.id',
      `a_key` VARCHAR(255) NOT NULL COMMENT 'BinaryAnnotation.key or Annotation.value if type == -1',
      `a_value` BLOB COMMENT 'BinaryAnnotation.value(), which must be smaller than 64KB',
      `a_type` INT NOT NULL COMMENT 'BinaryAnnotation.type() or -1 if Annotation',
      `a_timestamp` BIGINT COMMENT 'Used to implement TTL; Annotation.timestamp or zipkin_spans.timestamp',
      `endpoint_ipv4` INT COMMENT 'Null when Binary/Annotation.endpoint is null',
      `endpoint_ipv6` BINARY(16) COMMENT 'Null when Binary/Annotation.endpoint is null, or no IPv6 address',
      `endpoint_port` SMALLINT COMMENT 'Null when Binary/Annotation.endpoint is null',
      `endpoint_service_name` VARCHAR(255) COMMENT 'Null when Binary/Annotation.endpoint is null'
    ) ENGINE=InnoDB ROW_FORMAT=COMPRESSED CHARACTER SET=utf8 COLLATE utf8_general_ci;

    ALTER TABLE zipkin_annotations ADD UNIQUE KEY(`trace_id_high`, `trace_id`, `span_id`, `a_key`, `a_timestamp`) COMMENT 'Ignore insert on duplicate';
    ALTER TABLE zipkin_annotations ADD INDEX(`trace_id_high`, `trace_id`, `span_id`) COMMENT 'for joining with zipkin_spans';
    ALTER TABLE zipkin_annotations ADD INDEX(`trace_id_high`, `trace_id`) COMMENT 'for getTraces/ByIds';
    ALTER TABLE zipkin_annotations ADD INDEX(`endpoint_service_name`) COMMENT 'for getTraces and getServiceNames';
    ALTER TABLE zipkin_annotations ADD INDEX(`a_type`) COMMENT 'for getTraces and autocomplete values';
    ALTER TABLE zipkin_annotations ADD INDEX(`a_key`) COMMENT 'for getTraces and autocomplete values';
    ALTER TABLE zipkin_annotations ADD INDEX(`trace_id`, `span_id`, `a_key`) COMMENT 'for dependencies job';

    CREATE TABLE IF NOT EXISTS zipkin_dependencies (
      `day` DATE NOT NULL,
      `parent` VARCHAR(255) NOT NULL,
      `child` VARCHAR(255) NOT NULL,
      `call_count` BIGINT,
      `error_count` BIGINT,
      PRIMARY KEY (`day`, `parent`, `child`)
    ) ENGINE=InnoDB ROW_FORMAT=COMPRESSED CHARACTER SET=utf8 COLLATE utf8_general_ci;
```

步骤03 启动 Zipkin，连接 MySQL。具体启动命令如下：

```
STORAGE_TYPE=mysql MYSQL_USER=root MYSQL_PASS=root MYSQL_HOST=xxx.xx.xx.xx
MYSQL_TCP_PORT=3306 java -jar zipkin-server-2.23.0-exec.jar
```

步骤04 执行服务调用，采集的数据将被保存到数据库中，如图 10-7 所示。

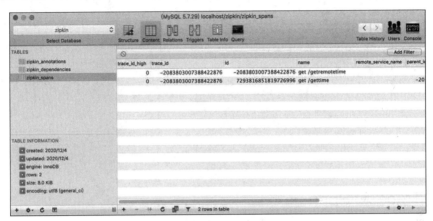

图 10-7　Zipkin 保存数据到 MySQL

10.3.3　Sleuth 抽样采集

在高并发请求下，如果每个请求都像记录日志一样保存起来，就会对性能产生影响。为了防止系统过载，可以设置以下参数：

```
# 跟踪信息收集采样比例，默认为 0.1，若为 1，则是 100%，即收集所有
spring.sleuth.sampler.probability=1
# 每秒速率，即每秒最多能跟踪的请求，rate 优先
spring.sleuth.sampler.rate=50
```

跟踪信息收集默认是 0.1（10%）的采样比例，可通过 probability 属性修改；或可采用每秒速率来控制采集数据，属性是 rate。

默认情况下，sleuth 采用 ProbabilityBasedSampler 实现抽样策略，以请求百分比的方式配置和收集跟踪信息。采样器也可以由 Java Config 设置，如以下示例所示：

```
@Bean
public Sampler defaultSampler() {
    return Sampler.ALWAYS_SAMPLE;
}
```

通过如上方式可以覆盖 ProbabilityBasedSampler 默认抽样策略。

根据操作的不同，可能需要应用不同的策略。例如，不想跟踪对静态资源（例如图像）的请求，或者想跟踪所有对新 API 的请求。

可以使用拦截器来自动化这种策略。以下示例显示了它如何在内部工作：

```
@Autowired Tracer tracer;
@Autowired Sampler fallback;
```

```
Span nextSpan(final Request input) {
//实现Sampler接口
Sampler requestBased = Sampler() {
    @Override public boolean isSampled(long traceId) {
      if (input.url().startsWith("/experimental")) {
        return true;
      //不想跟踪静态资源
      } else if (input.url().startsWith("/static")) {
        return false;
      }
      return fallback.isSampled(traceId);
    }
  };
  return tracer.withSampler(requestBased).nextSpan();
}
```

10.3.4 Trace 和 Span

Brave 提供当前 tracing 组件概念，可通过 Tracing.current()获得实例化的最新 tracing 组件。也可以使用 Tracing.currentTracer()获取 tracer 对象。如果使用这些方法之一，就不要缓存结果，而是在每次需要时查找它们。

Brave 提供当前 Span 概念，可通过 Tracer.currentSpan()将自定义标签添加到 Span。Tracer.nextSpan()创建正在运行的子项。

接下来了解 Span 的生命周期，可以通过 brave.Tracer 在 Span 上执行以下操作：

- start：开始 Span 时，将为其分配名称并记录开始时间戳。
- close：Span 完成（记录跨度的结束时间），如果对跨度进行了采样，则进行收集（例如收集到 Zipkin）。
- continue：创建一个新的 Span 实例，它是 continue 的副本。
- detach：Span 不会停止或关闭，它只会从当前线程中删除。
- create with explicit parent（使用显式父项创建）：可以创建一个新的 Span 并为其设置一个显式父项。

可以使用 Tracer 手动创建 Span，如下例所示：

```
Span newSpan = this.tracer.nextSpan().name("calculateTax");
try (Tracer.SpanInScope ws = this.tracer.withSpanInScope(newSpan.start())) {
    // ...
    // You can tag a span
    newSpan.tag("taxValue", taxValue);
    // ...
    // You can log an event on a span
    newSpan.annotate("taxCalculated");
}
finally {
    // Once done remember to finish the span. This will allow collecting
```

```
    // the span to send it to Zipkin
    newSpan.finish();
}
```

从前面的示例中可以看到如何创建 Span 的新实例。如果此线程中已经有一个跨度，那么它将成为新 Span 的父级。

> **注 意**
>
> 创建 Span 后，记得调用 Span 的 finish()方法。另外，请始终 finish 要发送给 Zipkin 的所有 Span。如果 Span 包含的名称大于 50 个字符，则该名称将被截断为 50 个字符，太长的名称会导致延迟问题，有时甚至会引发异常。同时，名字必须明确。

有时不想创建一个新 Span，而是继续一个 Span，这时可以使用 brave.Tracer，如以下示例所示：

```
//假如我们处在线程 Y 和已经从线程 X 收到 initialSpan
Span continuedSpan = this.tracer.toSpan(newSpan.context());
try {
    // ...
    // You can tag a span
    continuedSpan.tag("taxValue", taxValue);
    // ...
    // You can log an event on a span
    continuedSpan.annotate("taxCalculated");
}
finally {
    // Once done remember to flush the span. That means that
    // it will get reported but the span itself is not yet finished
    continuedSpan.flush();
}
```

有时你可能要开始一个新的 Span 并提供该 Span 的显式父项。假定 Span 的父级在一个线程中，而你想在另一个线程中开始一个新的 Span。在 Brave 中，每当调用 nextSpan()时都会参考当前的 Scope 创建一个 Span。可以将 Span 放入 Scope 中，然后调用 nextSpan()，如以下示例所示：

```
//假如我们处在线程 Y 和已经从线程 X 收到 initialSpan, initialSpan 将是 newSpan 的父 Span
Span newSpan = null;
try (Tracer.SpanInScope ws = this.tracer.withSpanInScope(initialSpan)) {
    newSpan = this.tracer.nextSpan().name("calculateCommission");
    // ...
    // You can tag a span
    newSpan.tag("commissionValue", commissionValue);
    // ...
    // You can log an event on a span
    newSpan.annotate("commissionCalculated");
}
finally {
    // Once done remember to finish the span. This will allow collecting
    // the span to send it to Zipkin. The tags and events set on the
```

```
        // newSpan will not be present on the parent
        if (newSpan != null) {
            //创建 Span 后,必须调用 finish,否则不报告(例如,向 Zipkin 报告)
            newSpan.finish();
        }
    }
```

选择一个 Span 名称不是一件容易的事。Span 名称应描述一个操作名称,该名称应为低基数,因此不应包含标识符。一些 Span 的名称是约定下来的:

- 由方法名称为 controllerMethodName 的 Controller 接收时,使用控制层的方法作为 Span 的名称。
- 用@Scheduled 注释的方法返回类的简单名称,作为 Span 的名称。
- 用于使用包装的 Callable 和 Runnable 接口完成的异步操作,使用 async 作为 Span 的名称。

幸运的是,对于异步处理,可以使用@SpanName 注释显式命名 Span,如以下示例所示:

```
@SpanName("calculateTax")
class TaxCountingRunnable implements Runnable {
    @Override
    public void run() {
        // perform logic
    }
}
```

在这种情况下,按以下方式进行处理时 Span 将被命名为 calculateTax:

```
Runnable runnable = new TraceRunnable(this.tracing, spanNamer,
        new TaxCountingRunnable());
Future<?> future = executorService.submit(runnable);
// ... some additional logic ...
future.get();
```

当然,也可以使用各种注解来管理 Span。使用注解管理 Span 有很多好处,比如减少对用户代码的影响。

如果不想手动创建 Span,则可以使用@NewSpan 注解。另外,可以用@SpanTag 注解以自动方式添加 Tag 标签,示例如下:

```
@NewSpan
void testMethod();
```

在不带任何参数的情况下对方法进行注解会创建一个新的 Span,Span 的名称等于方法名称。

```
@NewSpan("customNameOnTestMethod4")
void testMethod4();
```

在@NewSpan 注解中提供的值(直接或通过设置 name 参数)将作为 Span 的名称。

```
// method declaration
@NewSpan(name = "customNameOnTestMethod5")
void testMethod5(@SpanTag("testTag") String param);
```

```
// and method execution
this.testBean.testMethod5("test");
```

可以同时使用名称和标签。在这种情况下，带注解的方法的参数运行时值的值将成为标记的值。在上面的示例中，标签 key 是 testTag，标记 value 值是 test。

更多注解可参考 Sping Cloud Sleuth 官网：https://spring.io/projects/spring-cloud- sleuth#learn。

10.4 Spring Cloud Sleuth 整合 ELK

Spring Cloud Sleuth 主要记录链路调用数据，本身只支持内存存储。在业务量大的场景下，为了提升系统性能可以通过 HTTP 传输数据，也可以换作 rabbit 或者 kafka 来传输数据。

Zipkin 是 Twitter 开源的分布时追踪系统，可接收数据、存储数据（内存/cassandra/mysql/es）、检索数据、展示数据，本身不会直接在分布式的系统服务中追踪数据，可便捷地使用 sleuth 来收集传输数据。

Sleuth+Zipkin 做链路跟踪时，Zipkin 将 trace 信息存储在内存中，一旦访问量上去，Zipkin 就容易被压崩，从网上搜索资料发现基本都是 Kafka+ZK+Elasticsearch 做解决方案的。

ELK 项目是开源项目 Elasticsearch、Logstash 和 Kibana 的集合，集合中每个项目的职责如下。

- Elasticsearch：一个基于 Lucene 搜索引擎的 NoSQL 数据库。
- Logstash：一个基于管道的处理工具，从不同的数据源接收数据，执行不同的转换，然后发送数据到不同的目标系统。
- Kibana：工作在 Elasticsearch 上，是数据的展示层系统。

这 3 个项目组成的 ELK 系统通常用于现代服务化系统的日志管理。Logstash 用来收集和解析日志，把日志存储到 Elasticsearch 中并建立索引。Kibana 通过可视化的方式把数据展示给使用者。更多信息可查看 Elastic 官网：https://www.elastic.co/。

步骤 01 使用 Docker 快速安装 ELK：

```
### 使用 docker 快速安装启动 ELK
### Elasticsearch 端口：9200
### Kibana 端口：5601
### Logstash 端口：5044
> docker run -p 5601:5601 -p 9200:9200 -p 5044:5044 -e ES_MIN_MEM=128m  -e ES_MAX_MEM=1024m -it --name elk sebp/elk
```

步骤 02 ELK 启动完成后，在浏览器中输入请求地址"http://localhost:5601"，便可进入容器，具体命令如下所示：

```
> docker exec -it elk /bin/bash
```

通过 cd 命令进入目录：

```
> cd etc/logstash/conf.d/
```

修改 02-beats-input.conf 文件中的 input 配置:

```
> vim 02-beats-input.conf

input {
    tcp {
        port => 5044
        codec => json_lines
    }
}
output{
    elasticsearch {
    hosts => ["localhost:9200"]
    }
}
```

退出容器,重启 ELK:

```
> docker restart elk
```

步骤 03 启动 Zipkin,具体命令如下:

```
### STORAGE_TYPE 指定存储类型是 Elasticsearch
### DES_HOSTS 指定 Elasticsearch 端口 9200
> java -jar zipkin-server-2.23.0-exec.jar  --STORAGE_TYPE=elasticsearch
--DES_HOSTS=http://localhost:9200
```

步骤 04 在 sleuth 模块的 pom.xml 文件中添加如下依赖:

```xml
<dependency>
    <groupId>net.logstash.logback</groupId>
    <artifactId>logstash-logback-encoder</artifactId>
    <version>5.2</version>
</dependency>
```

logstash-logback-encoder 是 logstash 和 logback 集成的工具包,向 lostash 进行输送日志。
在 sleuth 模块的 resoures 资源目录下添加 logback.xml 日志配置文件,具体代码如下:

```xml
<?xml version="1.0" encoding="UTF-8"?>
<!--该日志将日志级别不同的 log 信息保存到不同的文件中 -->
<configuration>
    <include
resource="org/springframework/boot/logging/logback/defaults.xml" />

    <springProperty scope="context" name="springAppName"
                source="spring.application.name" />

    <!-- 日志在工程中的输出位置 -->
    <property name="LOG_FILE"
value="${BUILD_FOLDER:-build}/${springAppName}" />

    <!-- 控制台的日志输出样式 -->
```

```xml
<property name="CONSOLE_LOG_PATTERN"
          value="%clr(%d{yyyy-MM-dd HH:mm:ss.SSS}){faint} %clr(${LOG_LEVEL_PATTERN:-%5p}) %clr(${PID:- }){magenta} %clr(---){faint} %clr([%15.15t]){faint} %m%n${LOG_EXCEPTION_CONVERSION_WORD:-%wEx}}" />

<!-- 控制台输出 -->
<appender name="console" class="ch.qos.logback.core.ConsoleAppender">
    <filter class="ch.qos.logback.classic.filter.ThresholdFilter">
        <level>INFO</level>
    </filter>
    <!-- 日志输出编码 -->
    <encoder>
        <pattern>${CONSOLE_LOG_PATTERN}</pattern>
        <charset>utf8</charset>
    </encoder>
</appender>

<!-- 为logstash输出的JSON格式的Appender -->
<appender name="logstash"
          class="net.logstash.logback.appender.LogstashTcpSocketAppender">
    ### 重点，将日志发送到5044端口
    <destination>127.0.0.1:5044</destination>
    <!-- 日志输出编码 -->
    <encoder
          class="net.logstash.logback.encoder.LoggingEventCompositeJsonEncoder">
        <providers>
            <timestamp>
                <timeZone>UTC</timeZone>
            </timestamp>
            <pattern>
                <pattern>
                    {
                    "severity": "%level",
                    "service": "${springAppName:-}",
                    "trace": "%X{X-B3-TraceId:-}",
                    "span": "%X{X-B3-SpanId:-}",
                    "exportable": "%X{X-Span-Export:-}",
                    "pid": "${PID:-}",
                    "thread": "%thread",
                    "class": "%logger{40}",
                    "rest": "%message"
                    }
                </pattern>
            </pattern>
        </providers>
    </encoder>
</appender>
```

```xml
    <!-- 日志输出级别 -->
    <root level="INFO">
        <appender-ref ref="console" />
        <appender-ref ref="logstash" />
    </root>
</configuration>
```

`<destination>`127.0.0.1:5044`</destination>`配置用于将日志发送到 5044 端口。

步骤 05 在 Kibana 中添加 zipkin-*索引，如图 10-8 所示。

图 10-8　在 Kibana 中添加 zipkin-*索引

索引添加完毕后，便可以在 Kibana 上查看链路追踪日志，如图 10-9 所示。

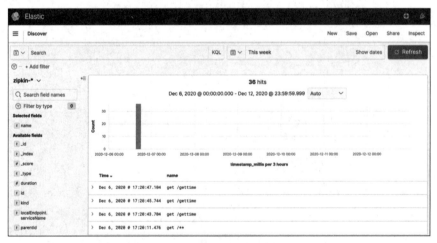

图 10-9　在 Kibana 上查看链路追踪日志

10.5　Sleuth 原理浅析

10.5.1　TraceId 传递

通过 spring-cloud-sleuth-core 的 jar 包结构可以很明显地看出 Sleuth 支持链路追踪的组件（web 下面包括 http、client 和 feign），如图 10-10 所示。

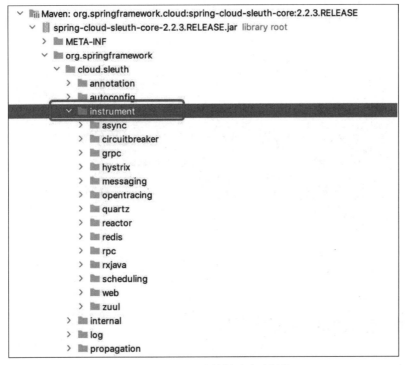

图 10-10　Sleuth 支持链路追踪组件

在微服务中，服务之间的相互调用是如何传递 TraceId 的呢？我们重点关注两个问题：一是网关是如何将 TraceId 传递给微服务的；二是微服务之间的 Feign 调用是如何传递 TraceId 的。

请求一般都是先通过网关再进行转发。例如，Zuul 网关通过 TracePostZuulFilter 将 TraceId 添加到 HTTP 头 X-B3-TraceId 中，以便所转发请求对应的服务能从头中获取到 TraceId；同时还将 TraceId 放到 MDC 中，以便本应用在输出日志时带 TraceId。

服务内部通过 Feign 注解调用另一个服务，由于 Feign 注解的实现是通过生成 Hystrix 的代理类来完成请求处理的，而 Hystrix 在进行请求发送时是通过异步的方式调用 Ribbon 的组件进行负载均衡的，然后通过 Feign.Client 的 execute 方法来进行请求的发送，故此时需要解决以下两个问题：

1. 如何在异步线程中传递 TraceId

Sleuth 通过实现 HystrixConcurrencyStrategy 接口来解决 TraceId 异步传递的问题。Hystrix 在实际调用时会调用 HystrixConcurrencyStrategy 的 wrapCallable 方法。因此，通过实现这个接口，在

wrapCallable 中将 TraceId 存放起来，具体参考 SleuthHystrixConcurrencyStrategy 类。

2. Feign 如何在服务中传递 TraceId

Sleuth 通过实现 Feign.Client，在 execute 前将 TraceId 存放到 X-B3-TraceId 头中，具体参考 TracingFeignClient 类。

服务本身对 TraceId 的处理是通过 Filter 的方式，在 Filter 中对请求头进行处理来实现 TraceId 的追踪。

Sleuth实现的原理如图10-11所示。先判断HTTP头中是否存在"X-B3-TraceId"，不存在则生成新的TraceId；存在则以头X-B3-TraceId的值作为TraceId，最后将X-B3-TraceId的值放到MDC中，以便日志输出中带上X-B3-TraceId。这样使得在本服务中用户请求产生的日志输出都会带有TraceId。

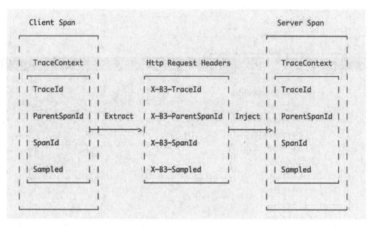

图 10-11　Sleuth 原理

10.5.2　spring.factories 配置文件

Spring Boot 中一个强大的特性是自动配置，只需在 pom.xml 文件中添加依赖，在 application.yml 文件中添加配置，再配合 spring.factories 文件即可。Sleuth 自动配置类都在 sleuth-core.jar 包的 spring.factories 文件中。

spring.factories 代码如下所示：

```
# Auto Configuration
### 下面三个自动配置类属于基础类，在任何场景下都需要执行
org.springframework.boot.autoconfigure.EnableAutoConfiguration=\
  org.springframework.cloud.sleuth.annotation.SleuthAnnotationAutoConfiguration,\
  org.springframework.cloud.sleuth.autoconfig.TraceAutoConfiguration,\
  org.springframework.cloud.sleuth.propagation.SleuthTagPropagationAutoConfiguration,\
### 下面每个自动配置类都是应用于具体框架或者中间件的
  org.springframework.cloud.sleuth.instrument.web.TraceHttpAutoConfiguration,\
  org.springframework.cloud.sleuth.instrument.web.TraceWebServletAutoConfigu
```

```
ration,\
    org.springframework.cloud.sleuth.instrument.web.client.TraceWebClientAutoC
onfiguration,\
    org.springframework.cloud.sleuth.instrument.web.client.TraceWebAsyncClient
AutoConfiguration,\
    org.springframework.cloud.sleuth.instrument.async.AsyncAutoConfiguration,\
    org.springframework.cloud.sleuth.instrument.async.AsyncCustomAutoConfigura
tion,\
    org.springframework.cloud.sleuth.instrument.async.AsyncDefaultAutoConfigur
ation,\
    org.springframework.cloud.sleuth.instrument.scheduling.TraceSchedulingAuto
Configuration,\
    org.springframework.cloud.sleuth.instrument.web.client.feign.TraceFeignCli
entAutoConfiguration,\
    org.springframework.cloud.sleuth.instrument.hystrix.SleuthHystrixAutoConfi
guration,\
    org.springframework.cloud.sleuth.instrument.circuitbreaker.SleuthCircuitBr
eakerAutoConfiguration,\
    org.springframework.cloud.sleuth.instrument.rxjava.RxJavaAutoConfiguration
,\
    org.springframework.cloud.sleuth.instrument.reactor.TraceReactorAutoConfig
uration,\
    org.springframework.cloud.sleuth.instrument.web.TraceWebFluxAutoConfigurat
ion,\
    org.springframework.cloud.sleuth.instrument.zuul.TraceZuulAutoConfiguratio
n,\
    org.springframework.cloud.sleuth.instrument.rpc.TraceRpcAutoConfiguration,
\
    org.springframework.cloud.sleuth.instrument.grpc.TraceGrpcAutoConfiguratio
n,\
    org.springframework.cloud.sleuth.instrument.messaging.SleuthKafkaStreamsCo
nfiguration,\
    org.springframework.cloud.sleuth.instrument.messaging.TraceMessagingAutoCo
nfiguration,\
    org.springframework.cloud.sleuth.instrument.messaging.TraceSpringIntegrati
onAutoConfiguration,\
    org.springframework.cloud.sleuth.instrument.messaging.TraceSpringMessaging
AutoConfiguration,\
    org.springframework.cloud.sleuth.instrument.messaging.websocket.TraceWebSo
cketAutoConfiguration,\
    org.springframework.cloud.sleuth.instrument.opentracing.OpentracingAutoCon
figuration,\
    org.springframework.cloud.sleuth.instrument.redis.TraceRedisAutoConfigurat
ion,\
    org.springframework.cloud.sleuth.instrument.quartz.TraceQuartzAutoConfigur
ation

    # Environment Post Processor
    # TraceEnvironmentPostProcessor 后处理器与日志打印相关
    org.springframework.boot.env.EnvironmentPostProcessor=\
```

org.springframework.cloud.sleuth.autoconfig.TraceEnvironmentPostProcessor

10.5.3 TraceEnvironmentPostProcessor 处理日志

TraceEnvironmentPostProcessor 用于处理日志打印，如果应用程序不设置日志打印格式，那么该类会设置默认的打印格式，具体源码如下：

```java
//在 3.0 版本，该类将被隐藏或者废弃
@Deprecated
public class TraceEnvironmentPostProcessor implements EnvironmentPostProcessor {

    private static final String PROPERTY_SOURCE_NAME = "defaultProperties";

    @Override
    public void postProcessEnvironment(ConfigurableEnvironment environment,
            SpringApplication application) {
        Map<String, Object> map = new HashMap<String, Object>();
        // This doesn't work with all logging systems but it's a useful default
        //so you see traces in logs without having to configure it.
        //将打印的日志格式存入 map 中
        //日志打印四个内容：应用名、trace id、span id、是否发送到 zipkin
        if (Boolean
                .parseBoolean(environment.getProperty("spring.sleuth.enabled", "true"))) {
            map.put("logging.pattern.level", "%5p [${spring.zipkin.service.name:"
                    + "${spring.application.name:}},%X{X-B3-TraceId:-},%X{X-B3-SpanId:-},%X{X-Span-Export:-}]");
        }
        addOrReplace(environment.getPropertySources(), map);
    }

    //addOrReplace 方法将日志打印的格式设置到默认配置中
    //如果应用没有设置打印格式，则使用默认配置
    private void addOrReplace(MutablePropertySources propertySources,
            Map<String, Object> map) {
        MapPropertySource target = null;
        if (propertySources.contains(PROPERTY_SOURCE_NAME)) {
            PropertySource<?> source = propertySources.get(PROPERTY_SOURCE_NAME);
            if (source instanceof MapPropertySource) {
                target = (MapPropertySource) source;
                for (String key : map.keySet()) {
                    if (!target.containsProperty(key)) {
                        target.getSource().put(key, map.get(key));
                    }
                }
            }
```

```
        }
    }
    if (target == null) {
        target = new MapPropertySource(PROPERTY_SOURCE_NAME, map);
    }
    if (!propertySources.contains(PROPERTY_SOURCE_NAME)) {
        propertySources.addLast(target);
    }
}
```

SleuthLogAutoConfiguration 类也与日志打印相关，SleuthLogAutoConfiguration 将 Slf4jScopeDecorator 添加到配置中，可以在日志中打印追踪信息。SleuthLogAutoConfiguration 只与 Slf4j 对接，具体源码如下所示：

```
@Configuration(proxyBeanMethods = false)
@ConditionalOnClass(MDC.class)
@EnableConfigurationProperties(SleuthSlf4jProperties.class)
public static class Slf4jConfiguration {
    @Bean
    @ConditionalOnProperty(value = "spring.sleuth.log.slf4j.enabled",
            matchIfMissing = true)
    static CurrentTraceContext.ScopeDecorator slf4jSpanDecorator(
    SleuthProperties sleuthProperties,
    SleuthSlf4jProperties sleuthSlf4jProperties) {
        return new Slf4jScopeDecorator(sleuthProperties, sleuthSlf4jProperties);
    }
}
```

slf4jSpanDecorator 方法用于创建 Slf4jScopeDecorator 对象并放入 Spring 容器中。Slf4jScopeDecorator 类的 decorateScope 方法将要打印的内容设置到 MDC 里面。

10.5.4 TraceAutoConfiguration

```
@Configuration(proxyBeanMethods = false)
//读取配置文件，默认是 true，因此该类在启动时会加载
@ConditionalOnProperty(value = "spring.sleuth.enabled", matchIfMissing = true)
//读取配置类，并且注入 spring bean
@EnableConfigurationProperties(SleuthProperties.class)
@Import({ SleuthLogAutoConfiguration.class,
TraceBaggageConfiguration.class,
        SamplerAutoConfiguration.class })
// public allows @AutoConfigureAfter(TraceAutoConfiguration)
// for components needing Tracing
public class TraceAutoConfiguration {

    @Bean
    @ConditionalOnMissingBean
```

```
            // NOTE: stable bean name as might be used outside sleuth
        Tracing tracing(@LocalServiceName String serviceName, Propagation.Factory factory,
                CurrentTraceContext currentTraceContext, Sampler sampler,
                ErrorParser errorParser, SleuthProperties sleuthProperties,
                @Nullable List<Reporter<zipkin2.Span>> spanReporters,
                @Nullable List<SpanAdjuster> spanAdjusters,
                @Nullable List<SpanHandler> spanHandlers,
                @Nullable List<TracingCustomizer> tracingCustomizers) {
            if (spanAdjusters == null) {
                spanAdjusters = Collections.emptyList();
            }
            Tracing.Builder builder = Tracing.newBuilder().sampler(sampler)
                    .errorParser(errorParser)
                    .localServiceName(StringUtils.isEmpty(serviceName) ? DEFAULT_SERVICE_NAME
                            : serviceName)

                    .propagationFactory(factory).currentTraceContext(currentTraceContext)
                    .spanReporter(new CompositeReporter(spanAdjusters,
                            spanReporters != null ? spanReporters : Collections.emptyList()))
                    .traceId128Bit(sleuthProperties.isTraceId128())
                    .supportsJoin(sleuthProperties.isSupportsJoin());
            if (spanHandlers != null) {
                for (SpanHandler spanHandlerFactory : spanHandlers) {
                    builder.addSpanHandler(spanHandlerFactory);
                }
            }
            if (tracingCustomizers != null) {
                for (TracingCustomizer customizer : tracingCustomizers) {
                    customizer.customize(builder);
                }
            }

            reorderZipkinHandlersLast(builder);
            return builder.build();
        }
    }
```

通过 Tracing.tracer()可获取 Tracer 类，通过 Traceer 类可进行 span 创建、保存以及更新等。

10.5.5 TracingFilter 过滤器

在 Web 环境下，Spring Boot 会自动创建 TracingFilter（在 TraceWebServletAutoConfiguration 中完成），具体源码如下所示：

```
@Deprecated
@Configuration(proxyBeanMethods = false)
```

```java
@ConditionalOnProperty(value = "spring.sleuth.web.enabled", matchIfMissing = true)
@ConditionalOnWebApplication(type = ConditionalOnWebApplication.Type.SERVLET)
@ConditionalOnBean(HttpTracing.class)
@AutoConfigureAfter(TraceHttpAutoConfiguration.class)
@Import(SpanCustomizingAsyncHandlerInterceptor.class)
public class TraceWebServletAutoConfiguration {
    @Bean
    @ConditionalOnMissingBean
    public TracingFilter tracingFilter(HttpTracing tracing) {
        return (TracingFilter) TracingFilter.create(tracing);
    }
}
```

TracingFilter 实现了 javax.servlet.Filter 接口。Spring 将 TracingFilter 作为 Web 过滤器设置到 Web 容器中,这样 TracingFilter 会对所有的网络请求拦截。TracingFilter.doFilter 的源码如下:

```java
@Override
  public void doFilter(ServletRequest request, ServletResponse response, FilterChain chain)
      throws IOException, ServletException {
    HttpServletRequest req = (HttpServletRequest) request;
    HttpServletResponse res = servlet.httpServletResponse(response);

    // Prevent duplicate spans for the same request
    TraceContext context = (TraceContext) request.getAttribute(TraceContext.class.getName());
    if (context != null) {
      // A forwarded request might end up on another thread, so make sure it
      // is scoped
      Scope scope = currentTraceContext.maybeScope(context);
      try {
        chain.doFilter(request, response);
      } finally {
        scope.close();
      }
      return;
    }

    Span span = handler.handleReceive(new HttpServletRequestWrapper(req));

    // Add attributes for explicit access to customization or span context
    request.setAttribute(SpanCustomizer.class.getName(), span.customizer());
    request.setAttribute(TraceContext.class.getName(), span.context());
    SendHandled sendHandled = new SendHandled();
    request.setAttribute(SendHandled.class.getName(), sendHandled);

    Throwable error = null;
    Scope scope = currentTraceContext.newScope(span.context());
    try {
```

```
            // any downstream code can see Tracer.currentSpan() or use
            // Tracer.currentSpanCustomizer()
            chain.doFilter(req, res);
        } catch (Throwable e) {
            error = e;
            throw e;
        } finally {
            // When async, even if we caught an exception, we don't have the final
response: defer
            if (servlet.isAsync(req)) {
                servlet.handleAsync(handler, req, res, span);
            } else if (sendHandled.compareAndSet(false, true)){
                // we have a synchronous response or error: finish the span
                HttpServerResponse responseWrapper =
HttpServletResponseWrapper.create(req, res, error);
                handler.handleSend(responseWrapper, span);
            }
            scope.close();
        }
    }
```

在 doFilter 方法中创建 Span 对象时,如果是调用链的第一个服务,那么 span id 是一个 long 型的随机数,然后设置 trace id=span id。如果不是第一个服务,则将 header 里面的 trace id、span id、parent span id 直接设置到新建的 Span 对象中。

当将 trace id、span id 设置到 MDC 时,就要调用之前提到的 Slf4jScopeDecorator.decorateScope 方法。

```
    @Override
    public Scope decorateScope(TraceContext context, Scope scope) {
        return LEGACY_IDS.decorateScope(context, delegate.decorateScope(context,
scope));
    }
```

delegate.decorateScope 方法根据 Slf4jScopeDecorator 构造方法的 add 方法添加的属性,从入参 context 里面读取属性值,然后调用 MDC.put 设置到 MDC 中。打印日志时可以将这些信息打印出来。Slf4jScopeDecorator 的构造方法如下:

```
    Slf4jScopeDecorator(SleuthProperties sleuthProperties,
            SleuthSlf4jProperties sleuthSlf4jProperties) {

        CorrelationScopeDecorator.Builder builder =
MDCScopeDecorator.newBuilder().clear()
                    .add(SingleCorrelationField.create(BaggageFields.TRACE_ID))
    .add(SingleCorrelationField.create(BaggageFields.PARENT_ID))
                    .add(SingleCorrelationField.create(BaggageFields.SPAN_ID))

    .add(SingleCorrelationField.newBuilder(BaggageFields.SAMPLED)
                        .name("spanExportable").build());
```

```
        Set<String> whitelist = new TreeSet<>(String.CASE_INSENSITIVE_ORDER);
        whitelist.addAll(sleuthSlf4jProperties.getWhitelistedMdcKeys());

        // Note: we are adding all the keys as-is because correlation context
        //doesn't prefix, only ExtraFieldPropagation does
        Set<String> retained = new LinkedHashSet<>();
        retained.addAll(sleuthProperties.getBaggageKeys());
        retained.addAll(sleuthProperties.getLocalKeys());
        retained.addAll(sleuthProperties.getPropagationKeys());
        retained.retainAll(whitelist);

        // For backwards compatibility set all fields dirty, so that any changes
        //made by MDC directly are reverted.
        for (String name : retained) {
builder.add(SingleCorrelationField.newBuilder(BaggageField.create(name))
                    .dirty().build());
        }

        this.delegate = builder.build();
    }
```

在 MDC 中存放的属性有 X-B3-TraceId、X-B3-SpanId、X-Span-Export、X-B3-ParentSpanId、TraceId、spanId、spanExportable 和 parentId。

10.5.6 TraceWebClientAutoConfiguration

Spring Boot 启动时执行 TraceWebClientAutoConfiguration 自动配置，该类中有一个内部类 TraceRestTemplateBeanPostProcessor，具体代码如下：

```
class TraceRestTemplateBeanPostProcessor implements BeanPostProcessor {
    //Spring 容器
    private final BeanFactory beanFactory;

    TraceRestTemplateBeanPostProcessor(BeanFactory beanFactory) {
        this.beanFactory = beanFactory;
    }

    @Override
    public Object postProcessBeforeInitialization(Object bean, String beanName)
            throws BeansException {
        return bean;
    }
    //bean 对象初始化后要执行该后处理器
    @Override
    public Object postProcessAfterInitialization(Object bean, String beanName)
            throws BeansException {
```

```
            if (bean instanceof RestTemplate) {
                RestTemplate rt = (RestTemplate) bean;
                new RestTemplateInterceptorInjector(interceptor()).inject(rt);
            }
            return bean;
        }
        //interceptor 方法返回拦截器,
        //拦截器会被添加到 RestTemplate 中, 用于对 http 请求拦截
        private LazyTracingClientHttpRequestInterceptor interceptor() {
            return new
LazyTracingClientHttpRequestInterceptor(this.beanFactory);
        }
    }
```

下面是 RestTemplateInterceptorInjector 的 inject 方法:

```
    class RestTemplateInterceptorInjector {

        void inject(RestTemplate restTemplate) {
            if (hasTraceInterceptor(restTemplate)) {
                return;
            }
            List<ClientHttpRequestInterceptor> interceptors = new
ArrayList<ClientHttpRequestInterceptor>(
                    restTemplate.getInterceptors());
            interceptors.add(0, this.interceptor);
            //将拦截器 LazyTracingClientHttpRequestInterceptor 设置到 restTemplate 对象中
            restTemplate.setInterceptors(interceptors);
        }

        private boolean hasTraceInterceptor(RestTemplate restTemplate) {
            for (ClientHttpRequestInterceptor interceptor :
restTemplate.getInterceptors()) {
                if (interceptor instanceof TracingClientHttpRequestInterceptor
                        || interceptor instanceof
LazyTracingClientHttpRequestInterceptor) {
                    return true;
                }
            }
            return false;
        }
    }
```

设置好拦截器后, 当每次 RestTemplate 发起 HTTP 请求时都会被拦截器拦截。LazyTracingClientHttpRequestInterceptor 的源码如下:

```
    class LazyTracingClientHttpRequestInterceptor implements
ClientHttpRequestInterceptor {
        //省略部分代码
        @Override
        //发起请求时, intercept 方法被调用
```

```java
public ClientHttpResponse intercept(HttpRequest request, byte[] body,
        ClientHttpRequestExecution execution) throws IOException {
    //interceptor()方法返回TracingClientHttpRequestInterceptor对象
    return interceptor().intercept(request, body, execution);
}
private TracingClientHttpRequestInterceptor interceptor() {
    if (this.interceptor == null) {
        this.interceptor = this.beanFactory
                .getBean(TracingClientHttpRequestInterceptor.class);
    }
    return this.interceptor;
}
```

TracingClientHttpRequestInterceptor 的 intercept 方法的代码如下：

```java
@Override public ClientHttpResponse intercept(HttpRequest req, byte[] body,
    ClientHttpRequestExecution execution) throws IOException {
  HttpRequestWrapper request = new HttpRequestWrapper(req);
  //创建Span对象，其中span id重新生成，trace id不变
  //同时将span id、trace id等信息添加到http的header中
  Span span = handler.handleSend(request);
  ClientHttpResponse response = null;
  Throwable error = null;
  try (Scope ws = currentTraceContext.newScope(span.context())) {
    return response = execution.execute(req, body);
  } catch (Throwable e) {
    error = e;
    throw e;
  } finally {
    handler.handleReceive(new ClientHttpResponseWrapper(request, response,
error), span);
  }
}
```

执行完 currentTraceContext.newScop 方法后，newScope 方法会更新 MDC 数据。RestTemplate 每次发起请求时，拦截器会在 HTTP 请求的 header 中放入如下信息：

```
### 表示当前调用链的trace id
x-b3-traceid = xxx
### 表示span id
x-b3-spanid = xxx
#调用链中前一个服务的span id
x-b3-parentspanid = xxx
#表示是否取样，1表示要将调用信息发送到zipkin
x-b3-sampled = 1
```

总结：当使用 RestTemplate 发送请求时，RestTemplateInterceptorInjector 拦截器对请求进行拦截，将新生成的 span id、trace id 等信息设置到请求的 header 中。服务端收到请求后就可以从 header 中解析出 Span 信息。

第 11 章

Spring Cloud Commons

本章主要介绍 Spring Cloud Commons 公共包、Spring Cloud Context 功能、Spring Cloud Commons 功能、Spring Cloud LoadBalance 负载均衡、Spring Cloud Circuit Breaker 断路器的核心概念等。

11.1　Spring Cloud Commons 简介

Spring Cloud Commons 提供两个库的功能：Spring Cloud Context 和 Spring Cloud Commons。

- Spring Cloud Context：为 Spring Cloud 应用程序的 ApplicationContext（bootstrap 上下文、加密、刷新范围和环境端点）提供实用程序和特殊服务。
- Spring Cloud Commons：一组用于不同 Spring Cloud 实现的抽象和公共类（比如 Spring Cloud Netflix、Spring Cloud Consul）。

11.2　Spring Cloud Context 功能

11.2.1　bootstrap 应用程序上下文

Spring Boot 有以下两种配置文件：

- bootstrap (.yml 或者 .properties)
- application (.yml 或者 .properties)

Spring Cloud 应用程序通过创建 bootstrap 上下文来运行，该上下文是主应用程序的父上下文。此上下文负责从外部源加载配置属性，并负责解密本地外部配置文件中的属性。这两个上下文共享一个环境（Environment），该环境是任何 Spring 应用程序外部属性的来源。默认情况下，bootstrap

属性具有较高的优先级，因此它们不能被本地配置覆盖。

bootstrap 上下文使用不同于主应用程序上下文的约定来定位外部配置，使用 bootstrap.yml 进行配置。例如：

```
spring:
  application:
    name: foo
  cloud:
    config:
      uri: ${SPRING_CONFIG_URI:http://localhost:8888}
```

要检索特定的配置文件，应该在 bootstrap.yml（或者 properties）中设置 spring.profiles.active 配置。另外，可以通过设置 spring.cloud.bootstrap.enabled = false 来完全禁用 bootstrap。

11.2.2 修改 bootstrap.properties 位置

bootstrap ApplicationContext 寻找的外部属性文件的名字不一定是 bootstrap，可以通过设置 spring.cloud.bootstrap.name 来修改 bootstrap 属性文件的名称（默认值为 bootstrap）。例如：通过系统属性 spring.cloud.bootstrap.name 指定 bootstrap 属性文件的名称是 application，即会在 Classpath 根目录或 config 目录下寻找 application.yml 或 application.properties 文件。

还可以通过系统属性 spring.cloud.bootstrap.location 指定 bootstrap 文件的位置。它们的用法类似于在 Spring Boot 中指定配置文件的 spring.config.name 和 spring.config.location。如果在 bootstrap.yml 中指定了 spring.profiles.active=dev，那么 bootstrap ApplicationContext 寻找配置文件时还会寻找 bootstrap-dev.yml 文件。

11.2.3 覆盖远程属性的值

通过 bootstrap 上下文添加到应用程序中的属性源通常是"远程"的（例如，来自 Spring Cloud Config Server）。默认情况下，不能在本地覆盖它们。如果想让应用程序使用自己的系统属性或配置文件覆盖远程属性，则远程属性源必须通过设置 spring.cloud.config.allowOverride = true 来授予其权限（在本地设置无效）。一旦设置该属性，就有两个更细粒度的设置：

- spring.cloud.config.overrideNone=true：从任何本地属性来源覆盖。
- spring.cloud.config.overrideSystemProperties=false：只有系统属性、命令行参数和环境变量（而不是本地配置文件）才应覆盖远程设置。

11.2.4 自定义 bootstrap 配置

bootstrap context 可以被设置来做任何你想要做的事，只要在/META-INF/spring.factories 文件中配置 org.springframework.cloud.bootstrap.BootstrapConfiguration 的值即可。

它的值是以逗号分隔的@Configuration 类的全限定名，所以任何你想要在 main application context 中注入的 bean 都可以在这里配置。如果你希望控制启动顺序，就在类上添加@Order 注解

（默认顺序为最后）。

11.2.5　刷新范围

RefreshScope（org.springframework.cloud.context.scope.refresh）是 Spring Cloud 提供的一种特殊的 scope 实现，用来实现配置、实例热加载。

> **注　意**
>
> 如果项目使用的是 HikariDataSource 数据源，则无法刷新它，它是 spring.cloud.refresh.never-refreshable 的默认值。如果需要刷新，就选择其他数据源实现。

11.2.6　加密与解密

在微服务架构中，很多配置文件包含大量的敏感信息，比如数据库密码、缓存密码等。将敏感信息以明文的方式存储于微服务应用的配置文件中是非常危险的，针对这个问题，Spring Cloud Config 提供了对属性进行加密解密的功能，以保护配置文件中的信息安全。例如：

```
### 用户名
spring.datasource.username=ay
### 密码
spring.datasource.password={cipher}429de499bd4c2053c05f029e7cfbf143695f5b
```

在 Spring Cloud Config 中，通过在属性值前使用{cipher}前缀来标注该内容是一个加密值，当微服务客户端来加载配置时，配置中心会自动为带有{cipher}前缀的值进行解密。

如果由于"密钥大小非法"而导致异常，就需要在配置中心的运行环境中安装不限长度的 JCE 版本（Unlimited Strength Java Cryptography Extension）。虽然 JCE 功能在 JRE 中自带，但是默认使用的是有长度限制的版本。从 Oracle 的官方网站中下载到它的压缩包，解压后可以看到三个文件：README.txt、local_policy.jar、US_export_policy.jar。

将 local_policy.jar 和 US_export_policy.jar 两个文件复制到$JAVA_HOME/jre/lib/security 目录下，覆盖原来的默认内容。至此，完成加密解密的准备工作。

最后，在配置文件中直接指定密钥信息（对称性密钥），比如：

```
encrypt.key=ayencrypt
```

11.2.7　Endpoints 端点

对于 Spring Boot Actuator 应用程序，可以使用管理端点来管理运用，例如：

- /actuator/env 端点可以更新环境并重新绑定@ConfigurationProperties 和日志级别。
- /actuator/refresh 端点可以重新加载引导上下文并刷新@RefreshScope bean。
- /actuator/restart 端点可以关闭 ApplicationContext 并重新启动（默认情况下禁用）。
- /actuator/pause 和/actuator/resume 用来调用生命周期方法（ApplicationContext 上的 stop()和

start()）。

> **注 意**
>
> 如果禁用/actuator/restart 端点，则/actuator/pause 和/actuator/resume 端点也将被禁用，因为它们只是/actuator/restart 的特例。

11.3 Spring Cloud Commons 功能

Spring Cloud Commons 模块为微服务中的服务注册与发现、负载均衡、熔断器等功能提供一个抽象层代码，该抽象层与具体的实现无关，具体的实现可以采用不同的技术，并做到灵活的更换。

常用的抽象点有@EnableDiscoveryClient、ServiceRegistry 等。

11.3.1 @EnableDiscoveryClient 注解

Spring Cloud Commons 提供了@EnableDiscoveryClient 注释。它通过 META-INF/spring.factories 寻找 DiscoveryClient 和 ReactiveDiscoveryClient 接口的实现。

DiscoveryClient 实现的例子包括 Spring Cloud Netflix Eureka、Spring Cloud Consul Discovery 和 Spring Cloud Zookeeper Discovery。

默认情况下，Spring Cloud 将同时提供阻塞和响应式服务发现客户端。可以通过设置 spring.cloud.discovery.blocking.enabled=false 或 spring.cloud.discovery.reactive.enabled=false 轻松禁用阻塞和响应式客户端。要完全禁用服务发现，只需设置 spring.cloud.discovery.enabled=false 即可。

认情况下，DiscoveryClient 的实现会自动将本地 Spring Boot 服务注册到远程注册中心。可以通过在@EnableDiscoveryClient 中设置 autoRegister=false 来禁用此行为。

> **注 意**
>
> @EnableDiscoveryClient 注解不是必需的，可以将 DiscoveryClient 实现放在类路径上，让 Spring Boot 应用程序向注册中心注册。

DiscoveryClient 接口扩展 Ordered。这在使用多个 Discovery Client 时非常有用，因为它允许定义 Discovery Client 的顺序，类似于如何排序 Spring 应用程序加载的 bean。

默认情况下，任何 DiscoveryClient 的顺序都设置为 0。如果想自定义 DiscoveryClient 实现设置不同的顺序，只需覆盖 getOrder()方法，以便返回合适的值。

除此之外，可以使用属性来设置 Spring Cloud 提供的 DiscoveryClient 实现的顺序，只需要设置 spring.cloud.{clientIdentifier}.discovery.order（或 Eureka 的 eureka.client.order）属性的值即可。

11.3.2 服务注册 ServiceRegistry

Commons 提供了一个 ServiceRegistry 接口，该接口提供了 register(Registration) 和 deregister(Registration)等方法，可以提供自定义注册服务。其中，Registration 是一个接口。

```
/**
 * @author ay
 * @since 2020-12-014
 */
@Configuration
@EnableDiscoveryClient(autoRegister=false)
public class MyConfiguration {
    private ServiceRegistry registry;

    public MyConfiguration(ServiceRegistry registry) {
        this.registry = registry;
    }

    // register 方法可以通过其他程序调用，例如事件或者自定义的端点
    public void register() {
        Registration registration = constructRegistration();
        this.registry.register(registration);
    }
}
```

每个 ServiceRegistry 实现都有自己的 Registry 实现：

- ZookeeperRegistration 与 ZookeeperServiceRegistry 一起使用。
- Eurekaregistry 与 EurekaServiceRegistry 一起使用。
- ConsulRegistration 与 ConsulServiceRegistry 一起使用。

默认情况下，ServiceRegistry 实现自动注册正在运行的服务。要禁用该行为，可以设置 @EnableDiscoveryClient(autoRegister=false) 永久禁用自动注册。通过配置 spring.cloud.service-registry.auto-registration.enabled=false 禁用该行为。

当服务自动注册时，将触发两个事件：第一个事件称为 InstancePreRegisteredEvent，它在注册服务之前被触发；第二个事件称为 InstanceRegisteredEvent，在服务注册后触发。可以注册一个 ApplicationListener 来侦听这些事件并对其做出反应。

如果 spring.cloud.service-registry.auto-registration.enabled 属性为 false，则不会触发这些事件。

Spring Cloud Commons 提供了一个/service-registry 执行器端点，这个端点依赖于 Spring 应用程序上下文中的 Registration bean。使用 GET 调用/service-registry 返回 Registration 的状态。

11.3.3 多个 RestTemplate 实例

如果想要一个没有负载均衡的 RestTemplate，就创建一个 RestTemplate bean 并注入它。要访

问负载平衡的 RestTemplate，就在创建@Bean 时使用@LoadBalanced 注解，如以下示例所示：

```
@Configuration
public class MyConfiguration {
    @LoadBalanced
    @Bean
    RestTemplate loadBalanced() {
        return new RestTemplate();
    }
    @Primary
    @Bean
    RestTemplate restTemplate() {
        return new RestTemplate();
    }
}

public class MyClass {
@Autowired
private RestTemplate restTemplate;

    @Autowired
    @LoadBalanced
    private RestTemplate loadBalanced;

    //省略部分代码
}
```

使用负载均衡的 RestTemplate 时，需要在类路径中有一个负载均衡器实现，推荐的实现是 BlockingLoadBalancerClient。将 Spring Cloud LoadBalancer starter 依赖包添加到项目中：

```
<dependency>
  <groupId>org.springframework.cloud</groupId>
  <artifactId>spring-cloud-starter-loadbalancer</artifactId>
</dependency>
```

虽可使用 RibbonLoadBalancerClient，但它正在维护中，所以不建议将其添加到新项目中。

默认情况下，如果同时具有 RibbonLoadBalancerClient 和 BlockingLoadBalancerClient，则为了保持向后兼容性将使用 RibbonLoadBalancerClient。要覆盖它，可以将 spring.cloud.loadbalancer.ribbon.enabled 属性设置为 false。

11.3.4　多个 WebClient 实例

在项目中，如果想要一个不带有负载均衡的 WebClient 对象，可以创建一个 WebClient bean 对象并注入。要访问带有负载均衡的 WebClient，可以在创建@Bean 时使用@LoadBalanced 注解，如下面的例子所示：

```
@Configuration
public class MyConfiguration {
    //具备负载均衡功能
```

```
    @LoadBalanced
    @Bean
    WebClient.Builder loadBalanced() {
        return WebClient.builder();
    }
    //不具备负载均衡功能
    @Primary
    @Bean
    WebClient.Builder webClient() {
        return WebClient.builder();
    }
}

public class MyClass {
    @Autowired
    private WebClient.Builder webClientBuilder;

    @Autowired
    @LoadBalanced
    private WebClient.Builder loadBalanced;

    //省略部分代码
}
```

如果要使用@LoadBalanced WebClient.Builder，则需要在类路径中具有负载平衡器实现。建议将 Spring Cloud LoadBalancer starter 添加到项目中，然后在下面使用 ReactiveLoadBalancer。另外，此功能也可以与 spring-cloud-starter-netflix-ribbon 一起使用，但是该请求由后台的非响应式 LoadBalancerClient 处理。由于 spring-cloud-starter-netflix-ribbon 已经处于维护模式，因此不建议将其添加到新项目中。如果类路径中同时包含 spring-cloud-starter-loadbalancer 和 spring-cloud-starter-netflix-ribbon，则默认使用 Ribbon。要切换到 Spring Cloud LoadBalancer，可将 spring.cloud.loadbalancer.ribbon.enabled 属性设置为 false。可以将负载平衡的 RestTemplate 配置为重试失败的请求。默认情况下，禁用此逻辑。可以通过将 Spring Retry 添加到应用程序的类路径中来启用它。负载平衡的 RestTemplate 支持某些与重试失败请求有关的功能区配置值。可以使用 client.ribbon.MaxAutoRetries、client.ribbon.MaxAutoRetriesNextServer 和 client.ribbon.OkToRetryOnAllOperations 属性。如果想在类路径上使用 Spring Retry 禁用重试逻辑，则可以设置 spring.cloud.loadbalancer.retry.enabled = false。

如果要在重试中实现 BackOffPolicy（回退策略，具体可参考 Spring Retry 相关文档），则需要创建一个类型为 LoadBalancedRetryFactory 的 bean，并重写 createBackOffPolicy 方法：

```
@Configuration
public class MyConfiguration {
    @Bean
    LoadBalancedRetryFactory retryFactory() {
        return new LoadBalancedRetryFactory() {
            @Override
            public BackOffPolicy createBackOffPolicy(String service) {
                return new ExponentialBackOffPolicy();
```

```
            }
        };
    }
}
```

11.3.5 忽略网卡

分布式应用部署到服务上，由于服务器可能存在多张网卡，因此会造成 IP 地址不准的问题。下面的配置忽略 docker0 接口和所有以 veth 开始的接口。

application.yml：

```
spring:
  cloud:
    inetutils:
      ignoredInterfaces:
        - docker0
        ### 支持正则表达式
        - veth.*
```

也可以通过使用正则表达式列表强制只使用指定的网络地址，如下面的示例所示：

```
spring:
  cloud:
    inetutils:
      preferredNetworks:
        - 192.168
        - 10.0
```

还可以强制只使用网站本地地址，如下面的例子所示：

```
spring:
  cloud:
    inetutils:
      useOnlySiteLocalInterfaces: true
```

11.3.6 HTTP 客户端工厂

Spring Cloud Commons 提供了用于创建 Apache HTTP Client(ApacheHttpClientFactory)和 OK HTTP Client(OkHttpClientFactory)的 bean。

只有当 OK HTTP jar 位于类路径中时才会创建 OkHttpClientFactory 对象。另外，Spring Cloud Commons 提供了用于创建两个客户端连接管理器的对象：ApacheHttpClientConnectionManagerFactory 用于 Apache HTTP 客户端；OkHttpClientConnectionPoolFactory 用于 OK HTTP 客户端。

如果想自定义 HTTP Client，可以实现 Spring Cloud Commons 提供的类。另外，如果提供一个类型为 HttpClientBuilder 或 OkHttpClient.Builder 的对象，默认工厂使用这些 builders 作为返回下游项目的构建器基础。

通过设置 spring.cloud.httpclientfactories.apache.enabled 或 spring.cloud.httpclientfactories.ok.

enabled 为 false 来禁用这些 bean 的创建。

11.3.7 启用功能特性

Spring Cloud Commons 提供了一个执行器端点。这个端点返回类路径上可用的特性以及它们是否被启用，返回的信息包括特性类型、名称、版本和供应商。

开启/features 端点，需在 pom.xml 配置文件中添加如下依赖：

```xml
<dependency>
        <groupId>org.springframework.boot</groupId>
        <artifactId>spring-boot-starter-actuator</artifactId>
</dependency>
```

同时在配置文件 application.properties 中添加如下配置：

```
#指定访问这些监控方法的端口
management.server.port=8099
management.endpoint.health.show-details=always
management.endpoints.web.exposure.include=*
```

有两种类型的 features：abstract 和 named。

- abstract 特性：定义接口或抽象类并由实现创建的特性，如 DiscoveryClient、LoadBalancerClient 或 LockService。抽象类或接口用于在上下文中查找该类型的 bean。显示的版本是 bean.getClass().getPackage().getImplementationVersion()。
- named 特性：没有实现特定类的特性，包括"断路器""API 网关""Spring Cloud Bus"等。这些特性需要一个名称和一个 bean 类型。

任何模块都可以声明任意数量的 HasFeature Bean，如以下示例所示：

```java
@Bean
public HasFeatures commonsFeatures() {
  return HasFeatures.abstractFeatures(DiscoveryClient.class,
LoadBalancerClient.class);
}

@Bean
public HasFeatures consulFeatures() {
  return HasFeatures.namedFeatures(
    new NamedFeature("Spring Cloud Bus", ConsulBusAutoConfiguration.class),
    new NamedFeature("Circuit Breaker", HystrixCommandAspect.class));
}
```

> **提 示**
>
> 记得在类上添加@Configuration 注解。

11.3.8 Spring Cloud 兼容性验证

一些用户在设置 Spring Cloud 应用时存在问题，因此 Spring Cloud 添加了一个兼容性验证机制。如果当前设置与 Spring Cloud 的要求不兼容，那么失败时会附带一个报告，显示出究竟哪里出了问题。

报告示例如下：

```
***************************
APPLICATION FAILED TO START
***************************

Description:

Your project setup is incompatible with our requirements due to following
reasons:

- Spring Boot [2.1.0.RELEASE] is not compatible with this Spring Cloud release
train


Action:

Consider applying the following actions:

- Change Spring Boot version to one of the following versions [1.2.x, 1.3.x] .
You can find the latest Spring Boot versions here
[https://spring.io/projects/spring-boot#learn].
If you want to learn more about the Spring Cloud Release train compatibility,
you can visit this page [
```

如果想要禁用该特性，就将 spring.cloud.compatibility-verifier.enabled 设置为 false。

11.4 Spring Cloud LoadBalancer

11.4.1 LoadBalancer 简介

Spring Cloud Load Balancer 不是一个独立的项目，而是 spring-cloud-commons 中的一个模块。Spring Cloud 提供了自己的客户端负载均衡器抽象和实现，对于负载均衡机制，已添加 ReactiveLoadBalancer 接口，并为其提供了基于 Round-Robin 的实现。

Spring Cloud LoadBalancer 提供了可与 WebClient 一起使用的 ReactorLoadBalancerExchangeFilterFunction 和与 RestTemplate 一起使用的 BlockingLoadBalancerClient。

Spring Cloud LoadBalancer 全局只有一个 BlockingLoadBalancerClient，负责执行所有的负载均衡请求。BlockingLoadBalancerClient 从 LoadBalancerClientFactory 里面加载对应微服务的负载均衡

配置。每个微服务下都有独自的 LoadBalancer，LoadBalancer 里面包含负载均衡的算法，例如 RoundRobin。根据算法，从 ServiceInstanceListSupplier 返回的实例列表中选择一个实例返回。

11.4.2　Spring Cloud LoadBalancer 缓存

Spring Cloud LoadBalancer 提供了两个缓存实现：Caffeine 支持的 load 均衡器缓存实现和默认 load 均衡器缓存实现。

如果类路径中有 com.github.ben-manes.caffeine:caffeine，那么将使用基于 Caffeine 的实现。如果类路径中没有 Caffeine，就将使用 spring-cloud-starter-loadbalancer 自动提供的 DefaultLoadBalancerCache。要使用 Caffeine，而不是默认缓存，可将 com.github.ben-manes.caffeine:caffeine 依赖项添加到类路径，例如：

```xml
<dependency>
    <groupId>com.github.ben-manes.caffeine</groupId>
    <artifactId>caffeine</artifactId>
</dependency>
```

缓存的 ttl 可以通过 spring.cloud.loadbalancer.cache.ttl 属性设置。可以通过设置 spring.cloud.loadbalancer.cache.capacity 属性的值来设置自己的 LoadBalancer 缓存初始容量。

默认 ttl 设置为 35 秒，默认初始容量为 256s。

还可以通过将 spring.cloud.loadbalancer.cache.enabled 的值设置为 false 来禁用 loadBalancer 缓存。

11.4.3　Spring Cloud LoadBalancer Starter

Spring Cloud 提供了一个 starter 组件，可以轻松地在 Spring Boot 应用程序中添加 Spring Cloud LoadBalancer。只需将 spring-cloud-starter-loadbalancer 添加到依赖项中即可使用 starter 组件：

```xml
<dependency>
  <groupId>org.springframework.cloud</groupId>
  <artifactId>spring-cloud-starter-loadbalancer</artifactId>
</dependency>
```

> **注　意**
>
> Spring Cloud LoadBalancer starter 包括 Spring Boot Caching 和 Evictor。

如果类路径中同时具有 Ribbon 和 Spring Cloud LoadBalancer，则为了保持向后兼容性，默认情况下将使用基于 Ribbon 的实现。如果想要切换到 Spring Cloud LoadBalancer，就将属性 spring.cloud.loadbalancer.ribbon.enabled 设置为 false。

11.4.4 自定义 Spring Cloud LoadBalancer 配置

可以使用@LoadBalancerClient 注解设置自己的负载均衡器配置，如下所示：

```
@Configuration
//value 值指定服务的 ID
@LoadBalancerClient(value = "stores", configuration =
CustomLoadBalancerConfiguration.class)
public class MyConfiguration {

    @Bean
    @LoadBalanced
    public WebClient.Builder loadBalancedWebClientBuilder() {
        return WebClient.builder();
    }
}
```

还可以通过@LoadBalancerClients 注解传递多种配置（用于多个负载均衡器客户端），如以下示例所示：

```
@Configuration
@LoadBalancerClients({@LoadBalancerClient(value = "stores", configuration =
StoresLoadBalancerClientConfiguration.class), @LoadBalancerClient(value =
"customers", configuration = CustomersLoadBalancerClientConfiguration.class)})
public class MyConfiguration {

    @Bean
    @LoadBalanced
    public WebClient.Builder loadBalancedWebClientBuilder() {
        return WebClient.builder();
    }
}
```

11.5 Spring Cloud Circuit Breaker

11.5.1 Circuit Breaker 介绍

在 Honxton 版本中，Spring Cloud Commons 提供了熔断的统一接口：CircuitBreakerFactory 和 ReactiveCircuitBreakerFactory。这两个接口分别用于非响应式和响应式编程。Spring Cloud 支持以下断路器实现：

- Netflix Hystrix。
- Resilience4J（官方推荐，比 Hystrix 轻量）。
- Sentinel。

- Spring Retry。

Netflix Hystrix 和 Resilience4J 依赖如下所示：

```
<dependency>
    <groupId>org.springframework.cloud</groupId>
    <artifactId>spring-cloud-starter-circuitbreaker-resilience4j</artifactId>
</dependency>
<dependency>
    <groupId>org.springframework.cloud</groupId>
    <artifactId>spring-cloud-starter-netflix-hystrix</artifactId>
</dependency>
```

11.5.2 核心概念

要在代码中创建断路器，可以使用 CircuitBreakerFactory API。当类路径中包含 Spring Cloud Breaker Starter 时，实现此 CircuitBreakerFactory 的 bean 会自动创建，以下示例显示了如何使用此 API 的简单示例：

```
// 基本用法
@Autowired
private CircuitBreakerFactory cbFactory;
@Autowired
private RestTemplate rest;

public String slow() {
    return cbFactory.create("slow").run(() -> rest.getForObject("/slow", String.class), throwable -> "fallback");
}
```

cbFactory.create 是一个工厂方法，接口的参数是一个 id，这个 id 是用来获取对应配置的。下面给出是 hystrix 和 resilience4j 的 create 方法：

```
//HystrixCircuitBreakerFactory#create
public HystrixCircuitBreaker create(String id) {
    Assert.hasText(id, "A CircuitBreaker must have an id.");
    HystrixCommand.Setter setter = getConfigurations().computeIfAbsent(id,
            defaultConfiguration);
    return new HystrixCircuitBreaker(setter);
}

//Resilience4JCircuitBreakerFactory#create
@Override
public Resilience4JCircuitBreaker create(String id) {
    Assert.hasText(id, "A CircuitBreaker must have an id.");
    Resilience4JConfigBuilder.Resilience4JCircuitBreakerConfiguration config = getConfigurations()
            .computeIfAbsent(id, defaultConfiguration);
    return new Resilience4JCircuitBreaker(id,
config.getCircuitBreakerConfig(),
```

```
            config.getTimeLimiterConfig(), circuitBreakerRegistry,
executorService,
            Optional.ofNullable(circuitBreakerCustomizers.get(id)));
    }
```

run 方法的第一个参数是原逻辑执行，第二个参数是降级逻辑。

11.5.3 配置断路器

可以通过创建 Customizer 类型的 bean 来配置断路器。Customizer 接口只有一个方法，该方法采用 Object 进行自定义。

Spring Cloud 断路器允许你为所有断路器提供默认配置以及特定断路器的配置。例如，要在使用 Resilience4J 时为所有断路器提供默认配置，可以将以下 bean 添加到配置类：

```
CircuitBreakerConfig circuitBreakerConfig = CircuitBreakerConfig.custom()
        .failureRateThreshold(50)
        .waitDurationInOpenState(Duration.ofMillis(1000))
        .slidingWindowSize(2)
        .build();

TimeLimiterConfig timeLimiterConfig = TimeLimiterConfig.custom()
        .timeoutDuration(Duration.ofSeconds(4))
        .build();

@Bean
public Customizer<Resilience4JCircuitBreakerFactory> defaultCustomizer() {
    return factory -> factory.configureDefault(
            id -> new Resilience4JConfigBuilder(id)
                    .timeLimiterConfig(timeLimiterConfig)
                    .circuitBreakerConfig(circuitBreakerConfig)
                    .build());
```

当然，我们的应用中可以有多个断路器。因此，在某些情况下需要为每个断路器配备特定的配置。

```
@Bean
public Customizer<Resilience4JCircuitBreakerFactory>
specificCustomConfiguration1() {
    return factory -> factory.configure(
        builder -> builder
            .circuitBreakerConfig(circuitBreakerConfig)
            .timeLimiterConfig(timeLimiterConfig)
            .build(),"circuitBreaker");
}
```

可以通过提供相同方法的断路器 ID 列表来设置具有相同配置的多个断路器：

```
@Bean
public Customizer<Resilience4JCircuitBreakerFactory>
specificCustomConfiguration2() {
```

```
        return factory -> factory.configure(
            builder -> builder
                .circuitBreakerConfig(circuitBreakerConfig)
                .timeLimiterConfig(timeLimiterConfig).build(),
        "circuitBreaker1","circuitBreaker2","circuitBreaker3");
}
```

11.6 具备缓存功能随机数

Spring Cloud Context 提供了一个 PropertySource，它根据键缓存随机值。在缓存功能之外，它的工作原理与 Spring Boot 的 RandomValuePropertySource 相同。当你希望在 Spring 应用程序上下文启动后保持一致的随机值时，这个随机值可能非常有用。属性值采用 cachedrandom.[yourkey].[type] 的形式，yourkey 是缓存中的密钥，type 可以是 Spring Boot 的 RandomValuePropertySource 支持的任何类型，例如（application.properties 配置文件）：

```
###myrandom 和 myrandom2 获取的随机值是一致的，key 应该相同
myrandom=${cachedrandom.keyOne.uuid}
myrandom2=${cachedrandom.keyOne.uuid}
### keySecond 获取的随机值和 myrandom、myrandom2 不一致
keySecond=${cachedrandom.keySecond.uuid}
```

注意，需要在 pom.xml 配置文件中添加如下依赖：

```
<dependency>
    <groupId>org.springframework.cloud</groupId>
    <artifactId>spring-cloud-context</artifactId>
    <version>2.2.6.RELEASE</version>
</dependency>
```

第 12 章

Spring Cloud OAuth 2.0 保护 API 安全

本章主要介绍 OAuth 2.0 的核心概念、OAuth 2.0 协议流程、OAuth 2.0 的 4 种授权方式、快速搭建 OAuth 2.0 服务、授权码模式实现、JWT 结构和应用，以及如何结合 Spring Security + OAuth 2.0 + JWT 开发具体案例。

12.1 使用 OAuth 2.0 进行授权

12.1.1 OAuth 2.0 简介

OAuth 2.0 是 OAuth 协议的延续版本，但不向前兼容 OAuth 1.0。OAuth 2.0 是一个授权的开放标准，通过认证用户身份并颁发 token（令牌），使得第三方应用可以在限定时间、限定范围使用该令牌访问指定资源。OAuth 2.0 主要涉及的 RFC 规范有 RFC6749（整体授权框架）、RFC6750（令牌使用）、RFC6819（威胁模型）等。

客户端必须得到用户的授权（Authorization Grant）才能获得令牌（Access Token）。OAuth 2.0 定义了 4 种授权方式：授权码模式（Authorization Code）、简化模式（Implicit）、密码模式（Resource Owner Password Credentials）和客户端模式（Client Credentials）。

OAuth 2.0 定义的角色如下所示：

- 资源服务器（Resource Server）：服务提供商存放用户生成的资源的服务器。
- 认证服务器（Authorization Server）：用来进行用户认证并颁发 Token。
- 第三方应用程序（Third-party Application）：又称客户端。
- 用户代理（User-agent）：一般指浏览器。
- 资源所有者（Resource Owner）：一般指用户。

OAuth 2.0 的作用是让"客户端"安全可控地获取"用户"授权，与"服务提供商"进行互动。

12.1.2 OAuth 2.0 协议流程

OAuth 2.0 协议流程如图 12-1 所示。

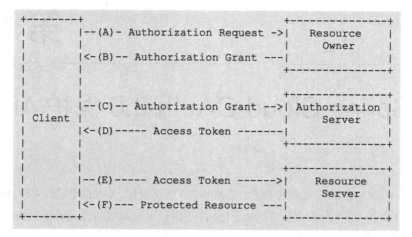

图 12-1　OAuth 2.0 抽象协议流程

图 12-1 所示的 OAuth 2.0 抽象流程描述了角色之间的交互，包括以下步骤：

（A）用户打开客户端以后，客户端要求用户给予授权。
（B）用户同意给予客户端授权。
（C）客户端使用步骤（B）获得的授权，向认证服务器申请令牌（token）。
（D）认证服务器对客户端认证无误后，同意发放令牌（token）。
（E）客户端使用令牌（token）向资源服务器申请获取资源。
（F）资源服务器确认令牌（token）无误，同意向客户端开放资源。

12.1.3　认证与授权

应用访问安全性基本都是围绕着认证（Authentication）和授权（Authorization）这两个核心概念展开的。虽然这两个术语经常混用，但是它们的含义完全不同。首先需要确定用户身份（对用户身份进行认证），确认身份后再确定用户是否有访问指定资源的权限。简单来说，身份认证是验证身份的过程，而授权是验证是否有权访问的过程。

例如，很多人都坐过飞机，登机之前需要身份证和机票：身份证是为了证明张三确实是张三，这就是 Authentication；机票是为了证明张三确实买了票，可以上飞机，这就是 Authorization。

主流认证的解决方案有 CAS、SAML2、OAuth 2.0 等。授权的解决方案有 Spring Security 和 Shiro。Shiro 框架相对比较轻量级，而 Spring Security 架构比较复杂。

12.1.4　OAuth 2.0 的授权方式

客户端（第三方应用）需要得到用户的授权（Authorization Grant）才能获得令牌（Access Token）。

1. 授权码模式

授权码模式如图 12-2 所示。

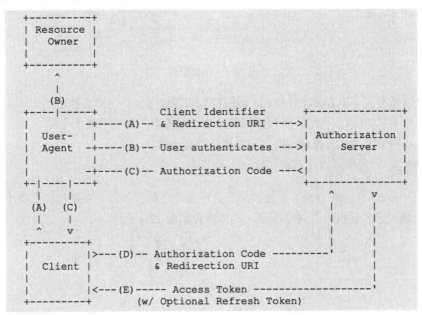

图 12-2　授权码模式

图 12-2 中各步骤的含义如下：

（A）资源所有者（Resource Owner，即用户）通过用户代理（User-Agent）访问客户端（Client），客户端将用户导向认证服务器（Authorization Server）。

（B）用户选中是否给予客户端（Client）授权。

（C）假如用户选择给予授权，认证服务器将用户导向客户端（Client）事先指定的重定向 URI（redirection URI），同时附上授权码（Authorization Code）。

（D）客户端（Client）收到授权码（Authorization Code），加上早先的"重定向 URI"，向认证服务器申请令牌。这一步是在客户端（Client）后台的服务器上完成的，对用户不可见。

（E）认证服务器核对了授权码和重定向 URI，确认无误后，向客户端发送访问令牌（Access Token）和更新令牌（Refresh Token）。

2. 密码模式

密码模式的前提是，用户需要把自己的密码给客户端（Client），具体步骤如图 12-3 所示。

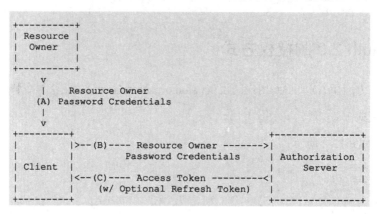

图 12-3　密码模式

（A）资源所有者（Resource Owner，即用户）向客户端提供用户名和密码。
（B）客户端（Client）将用户名和密码发给认证服务器，向后者请求令牌。
（C）认证服务器确认无误后，向客户端提供访问令牌。

3. 简化模式

简化模式（implicit grant type）不通过第三方应用程序的服务器，直接在浏览器中向认证服务器申请令牌，跳过了"授权码"这个步骤，具体步骤如图 12-4 所示。

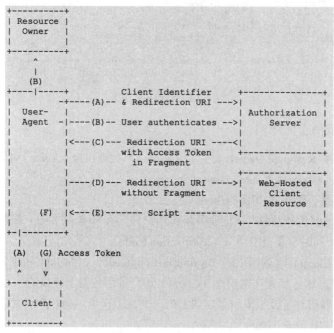

图 12-4　简化模式

（A）客户端（Client）将资源所有者（Resource Owner，即用户）导向认证服务器。
（B）资源所有者（Resource Owner，即用户）决定是否给予客户端授权。
（C）如果用户给予授权，认证服务器将用户导向客户端指定的"重定向 URI"，并在 URI 的 Hash 部分包含访问令牌。

4. 客户端模式

客户端模式的具体步骤如图 12-5 所示。

图 12-5 客户端模式

（A）客户端（Client）向认证服务器（Authorization Server）进行身份认证，并要求一个访问令牌（Token）。

（B）认证服务器必须以某种方式确认无误后向客户端提供访问令牌。

12.1.5 Spring Cloud Security OAuth 2.0 认证流程

以具体实例讲解，现在流行手游王者荣耀，用户首次安装会提示使用"与微信好友玩"还是"与 QQ 好友玩"，具体如图 12-6 和图 12-7 所示。

图 12-6 王者荣耀登录首页　　　　图 12-7 QQ 授权登录界面

单击"与 QQ 好友玩"按钮时，弹出 QQ 授权登录界面，单击"授权"按钮后，王者荣耀即可获取所有的 QQ 好友列表并快速建立社交关系。具体授权流程如下所示：

（1）用户登录王者荣耀，单击"与 QQ 好友玩"按钮，王者荣耀试图访问 QQ 上的好友列表。

（2）王者荣耀发现该资源是 QQ 的受保护资源，于是返回 302 将用户重定向到"QQ 授权登录"界面。

（3）用户完成授权后，QQ 提示用户是否将好友资源列表授权给王者荣耀使用。

（4）用户确认后，王者荣耀通过授权码模式获取 QQ 颁发的 Access_Token。

（5）王者荣耀携带该 Token 访问 QQ 的获取用户接口，QQ 验证 Token 无误后返回与该 Token 绑定的用户信息。

（6）王者荣耀的 Spring Security 安全框架根据返回的用户信息构造 Principal 对象并保存在 Session 中。

（7）王者荣耀再次携带该 Token 访问 QQ 的好友列表，QQ 根据该 Token 对应的用户返回用户的好友列表信息。

（8）该用户后续在王者荣耀发起的访问 QQ 的资源，只要在 Token 有效期以及权限范围内均可正常获取。

上述实例中，王者荣耀是第三方应用（Client），QQ 既是认证服务器又是资源服务器。

RFC6749 规范本身不关心用户身份部分，只关心 token 如何颁发、如何续签、如何用 Token 访问被保护资源。因此，（1）、（2）、（3）、（4）、（7）这几步完成了被保护资源访问的整个过程。Spring Security 作为一套完整的安全框架，还必须关注用户身份，因此有（5）、（6）这两步。

OAuth 2.0 除了进行被保护资源的访问外，还经常被用作单点登录（SSO）。在单点登录中，用户身份是最核心的，但是 Token 不携带用户信息，王者荣耀就无法知道认证服务器颁发的 Token 对应是哪个用户。王者荣耀直接使用 QQ 或者微信的用户体系，其用户和 QQ 或者微信的用户一一对应，如果 OAuth 2.0 认证服务预留一个用户信息获取接口，王者荣耀就可以通过 Token 接口获取用户信息，完成单点登录的整个过程。

12.2 搭建 OAuth 2.0 服务

12.2.1 快速搭建 OAuth 2.0 服务

快速搭建 OAuth 2.0 服务，具体步骤如下：

步骤 01 启动安装并 Nacos 系统。

步骤 02 使用 Intellij IDEA 快速创建 Spring Boot 项目，项目名称为 spring-boot-auth2。在创建过程中，需要勾选 Spring Web、Cloud OAuth2、Nacos Service Discovery 等依赖，如图 12-8 所示。

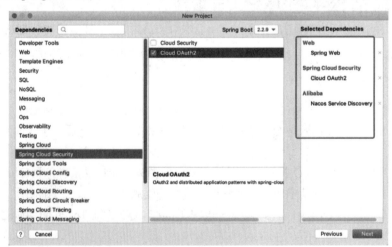

图 12-8　勾选依赖

项目创建完成后，可在 pom.xml 文件中看到如下依赖：

```xml
<dependencies>
  <dependency>
    <groupId>com.alibaba.cloud</groupId>
    <artifactId>spring-cloud-starter-alibaba-nacos-discovery</artifactId>
  </dependency>
  <dependency>
    <groupId>org.springframework.cloud</groupId>
    <artifactId>spring-cloud-starter-oauth2</artifactId>
  </dependency>
  <dependency>
    <groupId>org.springframework.boot</groupId>
    <artifactId>spring-boot-starter-web</artifactId>
  </dependency>
</dependencies>
```

- spring-cloud-starter-alibaba-Nacos-discovery 依赖：服务注册所需依赖。
- spring-cloud-starter-oauth2 依赖：对 spring-cloud-starter-security、spring-security-oauth2、spring-security-jwt 这 3 个依赖的整合。

步骤 03 在 application.yml 配置文件中添加如下配置：

```yml
### 服务端口
server:
  port: 18083
### 应用名称
spring:
  application:
    name: ay-auth
### nacos 注册中心地址
  cloud:
    nacos:
      discovery:
        server-addr: localhost:8848
### 暴露端点
management:
  endpoints:
    web:
      exposure:
        include: '*'
```

> **提 示**
>
> 将服务注入 Nacos 配置和注册中心不是必要的。

步骤 04 在 main 方法中添加 @EnableDiscoveryClient 注解，具体代码如下：

```
@SpringBootApplication
@EnableDiscoveryClient
public class DemoApplication {
```

```
    public static void main(String[] args) {
        SpringApplication.run(DemoApplication.class, args);
    }
}
```

步骤 05 运行 main 方法，启动项目。在浏览器中输入访问地址"http://localhost:18083/login"，如果看到提示输入用户名和密码的登录对话框，就代表 OAuth 服务启动成功，同时从 Nacos 系统中可以看到 OAuth 2.0 服务已注入，如图 12-9 所示。

图 12-9　OAuth 2.0 登录界面

12.2.2　授权码模式实现

在介绍授权码模式之前，这里用 Authentication 和 Authorization 这两个词做一个简单的解释。Authentication 称为鉴权，用户认证，通常是根据某些信息（比如用户名、密码、令牌等）来辨别用户的过程。Authorization 可以理解为授权、访问控制，根据用户的角色、权限等信息来判断目标行为、资源是不是可以被对应的用户使用、消费。简单地说，鉴权就是用户登录、授权就是权限控制。

实现基于内存用户认证和授权，需要三个关键步骤，即配置认证服务器、配置资源服务、配置 Spring Security。

步骤 01 配置认证服务器，使用注解@EnableAuthorizationServer 并扩展 AuthorizationServerConfigurerAdapter 类，具体代码如下：

```
/**
 * 认证服务器
 * @author ay
 * @since 2020-10-19
 */
@Configuration
@EnableAuthorizationServer
public class OAuth2AuthorizationServer extends
AuthorizationServerConfigurerAdapter {

    @Resource
    private BCryptPasswordEncoder passwordEncoder;

    @Override
    public void configure(AuthorizationServerSecurityConfigurer security)
```

```
throws Exception {
        security
                .tokenKeyAccess("permitAll()")
                .checkTokenAccess("isAuthenticated()")
                // 允许表单认证
                .allowFormAuthenticationForClients();
    }

    @Override
    public void configure(ClientDetailsServiceConfigurer clients) throws Exception {
        clients
                .inMemory()
                .withClient("ay-client").secret(passwordEncoder.encode("123456"))
                .authorizedGrantTypes("authorization_code", "refresh_token")
                .authorities("READ_ONLY_CLIENT")
                .scopes("read_user_info")
                .resourceIds("oauth2-resource")
                //重定向url
                .redirectUris("http://localhost:18087/login")
                .accessTokenValiditySeconds(1200)
                .refreshTokenValiditySeconds(240000);
    }
}
```

开发 OAuth2AuthorizationServer 类继承 AuthorizationServerConfigurerAdapter 适配器类，AuthorizationServerConfigurerAdapter 类的源码如下所示：

```
public class AuthorizationServerConfigurerAdapter implements AuthorizationServerConfigurer {

    @Override
    public void configure(AuthorizationServerSecurityConfigurer security) throws Exception {
    }
    @Override
    public void configure(ClientDetailsServiceConfigurer clients) throws Exception {
    }
    @Override
    public void configure(AuthorizationServerEndpointsConfigurer endpoints) throws Exception {
    }
}
```

由源码可知，AuthorizationServerConfigurerAdapter 提供了以下 3 个重载的 configure 方法：

- AuthorizationServerSecurityConfigurer：用来配置令牌端点（Token Endpoint）的安全约束。
- ClientDetailsServiceConfigurer：用来配置客户端详情服务（ClientDetailsService）。客户端详

情信息在这里进行初始化，既可以先通过代码写死，也可以通过查询数据库获取。
- AuthorizationServerEndpointsConfigurer：用来配置授权（Authorization）以及令牌（Token）的访问端点和令牌服务（Token Services），以及 Token 的存储方式（tokenStore）。

Security OAuth 2 公开两个端点用于检查令牌和获取令牌(/oauth/check_token 和/oauth/token_key)，这些端点默认受保护 denyAll()。在 OAuth2AuthorizationServer 类中，通过 AuthorizationServerSecurityConfigurer 配置类的 tokenKeyAccess()和 checkTokenAccess()方法打开这些端点以供使用。

通过 ClientDetailsServiceConfigurer 定义客户端详细的服务信息。在上面的实例中使用内存实现（inMemory）。它具有以下重要属性：

- **clientId**：客户端 ID，必需。
- **secret**：客户端密钥，可选。
- **scope**：客户受限的范围，如果范围未定义或为空（默认值），则客户端不受范围限制。
- **authorizedGrantTypes**：授权客户端使用的授权类型，默认值为空。"authorization_code"为授权码模式，"client_credentials"为客户端模式，"password"为密码模式，"implicit"为简化模式。
- **authorities**：授予客户的权限（常规 Spring Security 权限）。
- **redirectUris**：将用户代理重定向到客户端的重定向端点，必须是绝对 URL。

步骤02 配置资源服务器，使用 @EnableResourceServer 注解并扩展 ResourceServerConfigurerAdapter 类，具体代码如下：

```java
/**
 * 描述：资源服务配置（Resource Server）
 *
 * @author ay
 * @since 2020-04-21
 */
@Configuration
@EnableResourceServer
public class OAuth2ResourceServer extends ResourceServerConfigurerAdapter {

    @Override
    public void configure(HttpSecurity http) throws Exception {
        http
            .authorizeRequests()
            //保护所有/api/**
            .antMatchers("/api/**").authenticated()
            .antMatchers("/").permitAll();
    }
}
```

以上配置启用/api 下所有端点的保护，但可以自由访问其他端点。

步骤03 配置 Spring Security，具体代码如下：

```java
/**
 * @author ay
```

```
 * @since 2020-10-19
 */
@Configuration
@Order(1)
public class SecurityConfig extends WebSecurityConfigurerAdapter {

    @Override
    protected void configure(HttpSecurity http) throws Exception {
        http
                .antMatcher("/**")
                .authorizeRequests()
                .antMatchers("/oauth/authorize**", "/login**", "/error**")
                .permitAll()
                .and()
                .authorizeRequests()
                .anyRequest().authenticated()
                .and()
                .formLogin().permitAll();
    }

    @Override
    protected void configure(AuthenticationManagerBuilder auth) throws Exception {
        auth.inMemoryAuthentication()
                .withUser("ay")
                .password(passwordEncoder().encode("123456"))
                .roles("USER");
    }

    @Bean
    public BCryptPasswordEncoder passwordEncoder() {
        return new BCryptPasswordEncoder();
    }
}
```

步骤 04 创建一个 RESTful API，返回登录用户的姓名和电子邮件，具体代码如下：

```
/**
 * @author ay
 * @since 2020-10-18
 */
@Controller
public class UserController {
    @RequestMapping("/api/users")
    public ResponseEntity<UserDO> getUser() {
        User user = (User) SecurityContextHolder.getContext().getAuthentication().getPrincipal();
        String email = user.getUsername() + "@163.com";
        UserDO userDO = new UserDO();
        userDO.setName(user.getUsername());
        userDO.setEmail(email);
```

```
        return ResponseEntity.ok(userDO);
    }
}
```

UserDO 实体类的代码如下：

```
public class UserDO {
    private String name;
    private String email;
    //省略 set、get 方法
}
```

所有代码开发完成之后，重新启动应用，在浏览器中访问 http://localhost:18087/oauth/authorize?client_id=ay-client&response_type=code&scope=read_user_info，跳转到登录页面（http://localhost:18087/login），如图 12-10 所示。

图 12-10　登录界面

用户输入用户名和密码登录之后，将被重定向到授予访问页面，可以在其中选择授予对第三方应用程序的访问权限，选中 Approve，单击 Authorize 按钮，会重定向到 http://localhost:18087/login?code=K2WZuC，如图 12-11 所示。在实际生产中，这一步是在客户端（Client）后台的服务器上完成的，对用户不可见。K2WZuC 是第三方应用程序的授权代码。

图 12-11　OAuth 2.0 登录界面

使用授权码来获取访问令牌，可通过发起 curl 命令或者 Postman 软件进行获取。curl 命令如下：

```
curl -X POST --user ay-client:123456 http://localhost:18087/oauth/token -H
"content-type: application/x-www-form-urlencoded" -d
"code=8bVFTS&grant_type=authorization_code&redirect_uri=http://localhost:18087
/login&scope=read_user_info"
```

curl 请求返回结果如下：

```
{
    "access_token":"e81ed62c-fdfc-4167-ba44-099f855b2919",
    "token_type":"bearer",
    "refresh_token":"42f3ed79-d891-4e4c-978c-00054b035a88",
    "expires_in":1199,
    "scope":"read_user_info"
}
```

有了访问令牌 Token，就可以转到资源服务器来获取受保护的用户数据：

```
curl -X GET http://localhost:18087/api/user -H "authorization: Bearer e81ed62c-fdfc-4167-ba44-099f855b2919"
```

获得资源响应：

```
{
    "name":"ay",
    "email":"ay@163.com"
}
```

12.3　JWT 简介

JWT（JSON Web Token）是目前流行的跨域认证解决方案，是一个开放标准（RFC 7519），它定义了一种紧凑的、自包含的方式，用于作为 JSON 对象在各方之间安全地传输信息。该信息可以被验证和信任，因为它是数字签名的。

12.3.1　JWT 的结构

JWT 是由头部（header）、载荷（payload）、签证（signature）三段信息构成的，将三段信息文本用"."链接在一起就构成了 JWT 字符串。

一个典型的 JWT 格式如下所示：

```
xxxxx.yyyyy.zzzzz
```

例如：

```
eyJ0eXAiOiJKV1QiLCJhbGciOiJIUzI1NiJ9.
eyJzdWIiOiJVU0VSVEVTVCIsImV4cCI6MTU4NDEwNTc5MDcwMywiaWF0IjoxNTg0MTA1OTQ4Mz
cyfQ.
1HbleXbvJ_2SW8ry30cXOBGR9FW4oSWBd3PWaWKsEXE
```

使用在线校验工具（https://jwt.io/）将上述 Token 进行解码就可以看到数据，如图 12-12 所示。

图 12-12　https://jwt.io/在线校验工具

1. 头部（header）

JWT 的头部承载两部分信息：

（1）声明类型：这里主要是 JWT。

（2）声明加密算法：通常直接使用 HMAC SHA256。

例如：

```json
{"alg":"HS256","typ":"JWT"}
```

alg 属性表示签名所使用的算法；JWT 签名默认的算法为 HMAC SHA256；alg 属性值 HS256 就是 HMAC SHA256 算法；type 属性表示令牌类型，这里是 JWT。

2. 载荷（payload）

载荷是 JWT 的主体，同样也是一个 JSON 对象。载荷包含三个部分：

（1）标准中注册的声明（Registered Claims）：一组预定义的声明，不是强制的，但是推荐。

- iss（issuer）：JWT 签发者。
- sub（subject）：JWT 所面向的用户。
- aud（audience）：接收 JWT 的一方。
- exp（expiration）：JWT 的过期时间，必须要大于签发时间。
- nbf（not before）：定义了在什么时间之前该 JWT 都是不可用的。
- iat（issued at）：JWT 的发布时间，UNIX 时间戳。
- jti（JWT ID）：JWT 的唯一 ID 编号。

（2）公共的声明：可以添加任意信息，一般添加用户的相关信息或其他业务需要的必要信息，但不建议添加敏感信息。

（3）私有的声明：提供者和消费者所共同定义的声明，一般不建议存放敏感信息。

3. 签证 (signature)

JWT 的第三部分是一个签证信息，由三部分组成：header（base64 后的）、payload（base64 后的）、secret（密钥，需要保存好）。

例如：

```
HMACSHA256(base64UrlEncode(header) + "." + base64UrlEncode(payload), secret)
```

签名用于验证消息在传递过程中有没有被更改，并且对于使用私钥签名的 Token 还可以验证 JWT 的发送方是否为它所称的发送方。

secret 是保存在服务器端的，JWT 的签发生成也是在服务器端的，secret 就是用来进行 JWT 的签发和验证的，所以 secret 是服务端的私钥，在任何场景都不应该流露出去。

12.3.2 JWT 的应用

OAuth 2.0 最大的痛点是不携带用户信息，且资源服务器无法进行本地验证，每次对于资源的访问，资源服务器都需要向认证服务器发起请求，以验证 Token 的有效性、获取 Token 对应的用户信息。

这在分布式架构下是很要命的。因为如果有大量相关请求，处理效率是很低的，并且认证服务器会变成一个中心节点，对于 SLA 和处理性能等均有很高的要求。

JWT 就是为解决这些问题而诞生的，从本质上来说，JWT 就是一种特殊格式的 Token。普通的 OAuth 2.0 颁发的就是一串随机 hash 字符串，本身无意义，而 JWT 格式的 Token 是有特定含义的。

JWT 相对于传统的 Token 来说解决了以下两个痛点：

- 通过验证签名，Token 的验证可以直接在本地完成，不需要连接认证服务器。
- 在 payload 中可以定义用户相关信息，这样就轻松实现了 Token 和用户信息的绑定。

在认证的时候，当用户用他们的凭证成功登录以后，一个 JSON Web Token 将会被返回。此后，Token 就是用户凭证了。无论何时用户想要访问受保护的路由或者资源，用户代理（通常是浏览器）都应该带上 JWT，通常用 Bearer schema 放在 Authorization header 中，例如：

```
'Authorization': 'Bearer ' + token
```

12.3.3 Spring Security+OAuth 2.0+JWT 应用

本小节主要演示如何通过 Spring Security 来一步一步地搭建一套基于 JWT 的 OAuth 2.0 授权体系，具体步骤如下所示。

步骤 01 创建一个项目 spring-security-oauth-demo，包含以下三个模块：

- oauth2-client：客户端模块。
- oauth2-server：授权服务器模块。
- oauth2-user-service：用户服务，资源提供者模块。

父 pom.xml 配置文件如下所示:

```xml
<?xml version="1.0" encoding="UTF-8"?>
<project xmlns:xsi="http://www.w3.org/2001/XMLSchema-instance"
    xmlns="http://maven.apache.org/POM/4.0.0"
    xsi:schemaLocation="http://maven.apache.org/POM/4.0.0 http://maven.apache.org/xsd/maven-4.0.0.xsd">
    <modelVersion>4.0.0</modelVersion>

    <groupId>com.ay</groupId>
    <artifactId>spring-security-demo</artifactId>
    <packaging>pom</packaging>
    <version>1.0-SNAPSHOT</version>

    <parent>
        <groupId>org.springframework.boot</groupId>
        <artifactId>spring-boot-starter-parent</artifactId>
        <version>2.2.6.RELEASE</version>
        <relativePath/>
    </parent>

    <modules>
        <module>oauth2-client</module>
        <module>oauth2-server</module>
        <module>oauth2-user-service</module>
    </modules>

    <properties>
        <project.build.sourceEncoding>UTF-8</project.build.sourceEncoding>
        <project.reporting.outputEncoding>UTF-8</project.reporting.outputEncoding>
        <java.version>1.8</java.version>
    </properties>

    <dependencies>
        <!-- lombok -->
        <dependency>
            <groupId>org.projectlombok</groupId>
            <artifactId>lombok</artifactId>
            <optional>true</optional>
        </dependency>
    </dependencies>

    <dependencyManagement>
        <dependencies>
            <dependency>
                <groupId>org.springframework.cloud</groupId>
                <artifactId>spring-cloud-dependencies</artifactId>
                <version>Hoxton.SR3</version>
                <type>pom</type>
```

```xml
                <scope>import</scope>
            </dependency>
        </dependencies>
    </dependencyManagement>
    <build>
        <plugins>
            <plugin>
                <groupId>org.springframework.boot</groupId>
                <artifactId>spring-boot-maven-plugin</artifactId>
            </plugin>
        </plugins>
    </build>
</project>
```

步骤 02 创建 auth-2.0 数据库，初始化以下库表：

- users 表：存储用户数据。
- authorities 表：存储账号的权限数据。
- oauth_approvals 表：存储授权批准的状态。
- oauth_client_details 表：存储 OAuth 的客户端。
- oauth_code 表：存储授权码。

具体 SQL 如下所示：

```sql
SET NAMES utf8mb4;
SET FOREIGN_KEY_CHECKS = 0;
-- ----------------------------
-- Table structure for users
-- ----------------------------
DROP TABLE IF EXISTS `users`;
CREATE TABLE `users` (
  `username` varchar(50) NOT NULL,
  `password` varchar(100) NOT NULL,
  `enabled` tinyint(1) NOT NULL,
  PRIMARY KEY (`username`)
) ENGINE=InnoDB DEFAULT CHARSET=utf8mb4;
-- ----------------------------
-- Table structure for authorities
-- ----------------------------
DROP TABLE IF EXISTS `authorities`;
CREATE TABLE `authorities` (
  `username` varchar(50) NOT NULL,
  `authority` varchar(50) NOT NULL,
  UNIQUE KEY `ix_auth_username` (`username`,`authority`),
  CONSTRAINT `fk_authorities_users` FOREIGN KEY (`username`) REFERENCES `users` (`username`)
) ENGINE=InnoDB DEFAULT CHARSET=utf8mb4;
-- ----------------------------
-- Table structure for oauth_approvals
-- ----------------------------
```

```sql
DROP TABLE IF EXISTS `oauth_approvals`;
CREATE TABLE `oauth_approvals` (
  `userId` varchar(256) DEFAULT NULL,
  `clientId` varchar(256) DEFAULT NULL,
  `partnerKey` varchar(32) DEFAULT NULL,
  `scope` varchar(256) DEFAULT NULL,
  `status` varchar(10) DEFAULT NULL,
  `expiresAt` datetime DEFAULT NULL,
  `lastModifiedAt` datetime DEFAULT NULL
) ENGINE=InnoDB DEFAULT CHARSET=utf8;
-- ----------------------------
-- Table structure for oauth_client_details
-- ----------------------------
DROP TABLE IF EXISTS `oauth_client_details`;
CREATE TABLE `oauth_client_details` (
  `client_id` varchar(255) NOT NULL,
  `resource_ids` varchar(255) DEFAULT NULL,
  `client_secret` varchar(255) DEFAULT NULL,
  `scope` varchar(255) DEFAULT NULL,
  `authorized_grant_types` varchar(255) DEFAULT NULL,
  `web_server_redirect_uri` varchar(255) DEFAULT NULL,
  `authorities` varchar(255) DEFAULT NULL,
  `access_token_validity` int(11) DEFAULT NULL,
  `refresh_token_validity` int(11) DEFAULT NULL,
  `additional_information` varchar(4096) DEFAULT NULL,
  `autoapprove` varchar(255) DEFAULT NULL,
  PRIMARY KEY (`client_id`)
) ENGINE=InnoDB DEFAULT CHARSET=utf8mb4;
-- ----------------------------
-- Table structure for oauth_code
-- ----------------------------
DROP TABLE IF EXISTS `oauth_code`;
CREATE TABLE `oauth_code` (
  `code` varchar(255) DEFAULT NULL,
  `authentication` blob
) ENGINE=InnoDB DEFAULT CHARSET=utf8mb4;
SET FOREIGN_KEY_CHECKS = 1;
```

后续会使用 JWT 来传输令牌信息，进行本地校验。因此，不需要在数据库中创建相应的表来存放访问令牌和刷新令牌。上述 5 个表都可以自己扩展，只需要继承实现相关的 Spring 类即可。

步骤 03 搭建授权服务器模块。快速创建 Spring Boot 模块，在 pom.xml 文件中添加如下配置：

```xml
<?xml version="1.0" encoding="UTF-8"?>
<project xmlns:xsi="http://www.w3.org/2001/XMLSchema-instance"
    xmlns="http://maven.apache.org/POM/4.0.0"
    xsi:schemaLocation="http://maven.apache.org/POM/4.0.0 http://maven.apache.org/xsd/maven-4.0.0.xsd">
    <parent>
        <artifactId>spring-security-demo</artifactId>
        <groupId>com.ay</groupId>
```

```xml
        <version>1.0-SNAPSHOT</version>
    </parent>
    <modelVersion>4.0.0</modelVersion>

    <artifactId>oauth2-server</artifactId>

    <dependencies>
        <dependency>
            <groupId>org.springframework.cloud</groupId>
            <artifactId>spring-cloud-starter-oauth2</artifactId>
        </dependency>
        <dependency>
            <groupId>org.springframework.boot</groupId>
            <artifactId>spring-boot-starter-jdbc</artifactId>
        </dependency>
        <dependency>
            <groupId>mysql</groupId>
            <artifactId>mysql-connector-java</artifactId>
        </dependency>
        <dependency>
            <groupId>org.springframework.boot</groupId>
            <artifactId>spring-boot-starter-web</artifactId>
        </dependency>
        <dependency>
            <groupId>org.springframework.boot</groupId>
            <artifactId>spring-boot-starter-thymeleaf</artifactId>
        </dependency>
    </dependencies>
</project>
```

- spring-cloud-starter-oauth2：该依赖包是 Spring Cloud 按照 OAuth 2.0 的标准并结合 spring-security 封装好的一个具体实现，并做了自动化配置的工作，使用起来更方便。
- spring-boot-starter-jdbc、mysql-connector-java：数据访问相关依赖。
- spring-boot-starter-web：Web 相关依赖。
- spring-boot-starter-thymeleaf：模板引擎依赖。

在 application.yml 文件中添加如下配置：

```
server:
  port: 8080

spring:
  application:
    name: oauth2-server
  datasource:
    url: jdbc:mysql://localhost:3306/auth-2.0?useSSL=false
    username: root
    password: 123456
```

配置文件定义授权服务器的监听端口 8080，配置 Oauth 数据库连接信息。使用 keytool 工具生成密钥对,把密钥文件 jks 保存到资源目录下,并导出一个公钥留作以后使用。

```java
/**
 * 描述: 授权服务器
 * @author ay
 * @since 2020-10-25
 */
@Configuration
//开启授权服务器
@EnableAuthorizationServer
public class OAuthServerConfiguration extends AuthorizationServerConfigurerAdapter {
    @Resource
    private DataSource dataSource;
    @Resource
    private AuthenticationManager authenticationManager;

    /**
     * 配置使用数据库来维护客户端信息
     * @param clients
     * @throws Exception
     */
    @Override
    public void configure(ClientDetailsServiceConfigurer clients) throws Exception {
        clients.jdbc(dataSource);
    }

    /**
     * @param security
     * @throws Exception
     */
    @Override
    public void configure(AuthorizationServerSecurityConfigurer security) throws Exception {
        //打开验证 Token 的访问权限
        security.checkTokenAccess("permitAll()")
                .allowFormAuthenticationForClients()
                //允许 ClientSecret 明文方式保存并且可以通过表单提交
                .passwordEncoder(NoOpPasswordEncoder.getInstance());
    }

    /**
     *
     * 配置 Token 存放方式: JWT 方式
     * @param endpoints
     * @throws Exception
     */
    @Override
    public void configure(AuthorizationServerEndpointsConfigurer endpoints) {
```

```java
        TokenEnhancerChain tokenEnhancerChain = new TokenEnhancerChain();
        //配置自定义的Token增强器,将更多信息放入Token中
        tokenEnhancerChain.setTokenEnhancers(
                Arrays.asList(tokenEnhancer(), jwtTokenEnhancer()));

        endpoints.approvalStore(approvalStore())
                .authorizationCodeServices(authorizationCodeServices())
                .tokenStore(tokenStore())
                .tokenEnhancer(tokenEnhancerChain)
                .authenticationManager(authenticationManager);
}

/**
 * 使用JDBC数据库方式来保存授权码
 *
 * @return
 */
@Bean
public AuthorizationCodeServices authorizationCodeServices() {
    return new JdbcAuthorizationCodeServices(dataSource);
}

/**
 * 使用JWT令牌存储
 * @return
 */
@Bean
public TokenStore tokenStore() {
    return new JwtTokenStore(jwtTokenEnhancer());
}

/**
 * 使用JDBC数据库方式来保存用户的授权批准记录
 * @return
 */
@Bean
public JdbcApprovalStore approvalStore() {
    return new JdbcApprovalStore(dataSource);
}

/**
 * 自定义的Token增强器,把更多信息放入Token中
 * @return
 */
@Bean
public TokenEnhancer tokenEnhancer() {
    return new CustomTokenEnhancer();
}

/**
```

```java
     * 配置JWT令牌，使用非对称加密方式来验证
     * @return
     */
    @Bean
    protected JwtAccessTokenConverter jwtTokenEnhancer() {
        KeyStoreKeyFactory keyStoreKeyFactory = new KeyStoreKeyFactory(new ClassPathResource("jwt.jks"), "mySecretKey".toCharArray());
        JwtAccessTokenConverter converter = new JwtAccessTokenConverter();
        converter.setKeyPair(keyStoreKeyFactory.getKeyPair("jwt"));
        return converter;
    }

    /**
     * 配置登录页面的视图信息
     */
    @Configuration
    static class MvcConfig implements WebMvcConfigurer {
        @Override
        public void addViewControllers(ViewControllerRegistry registry) {
            registry.addViewController("login").setViewName("login");
        }
    }
}
```

默认情况下，Token 中只会有用户名等基本信息，实际应用中需要把关于用户的更多信息返回给客户端。这时，可以自定义增强器来丰富 Token 的内容。自定义增强器 CustomTokenEnhancer 的具体代码如下所示：

```java
/**
 * 描述：自定义Token增强器
 * @author ay
 * @since 2020-10-25
 */
public class CustomTokenEnhancer implements TokenEnhancer {

    @Override
    public OAuth2AccessToken enhance(OAuth2AccessToken accessToken, OAuth2Authentication authentication) {
        Authentication userAuthentication = authentication.getUserAuthentication();
        if (userAuthentication != null) {
            Object principal = authentication.getUserAuthentication().getPrincipal();
            //把用户标识以userDetails这个Key加入到JWT的额外信息中去
            Map<String, Object> additionalInfo = new HashMap<>();
            additionalInfo.put("userDetails", principal);
            ((DefaultOAuth2AccessToken) accessToken).setAdditionalInformation(additionalInfo);
        }
        return accessToken;
```

        }
    }

WebSecurityConfig 的具体代码如下所示：

```java
/**
 * 描述：WebSecurityConfig
 * @author ay
 * @since 2020-10-25
 */
@Configuration
public class WebSecurityConfig extends WebSecurityConfigurerAdapter {
    @Resource
    private DataSource dataSource;

    @Override
    @Bean
    public AuthenticationManager authenticationManagerBean() throws Exception {
        return super.authenticationManagerBean();
    }

    @Override
    protected void configure(AuthenticationManagerBuilder auth) throws Exception {
        auth.jdbcAuthentication()
                .dataSource(dataSource)
                //配置使用 BCryptPasswordEncoder 哈希来保存用户的密码
                .passwordEncoder(new BCryptPasswordEncoder());
    }

    @Override
    protected void configure(HttpSecurity http) throws Exception {
        http.authorizeRequests()
                //开放/login（用于登录）和/oauth/authorize（用于换授权码）路径
                .antMatchers("/login", "/oauth/authorize")
                .permitAll()
                .anyRequest().authenticated()
                .and()
                //设置/login 使用表单验证进行登录
                .formLogin().loginPage("/login");
    }
}
```

最后在资源目录下创建 templates 文件夹，然后创建 login.html 登录页，具体代码如下：

```html
<!DOCTYPE html>
<html xmlns:th="http://www.thymeleaf.org" class="uk-height-1-1">
<head>
    <meta charset="UTF-8"/>
    <title>OAuth2 Demo</title>
```

```html
        <link rel="stylesheet" href="https://cdnjs.cloudflare.com/ajax/libs/uikit/2.26.3/css/uikit.gradient.min.css"/>
    </head>

    <body class="uk-height-1-1">

    <div class="uk-vertical-align uk-text-center uk-height-1-1">
        <div class="uk-vertical-align-middle" style="width: 250px;">
            <h1>Login Form</h1>
            <p class="uk-text-danger" th:if="${param.error}">
                用户名或密码错误...
            </p>
            <form class="uk-panel uk-panel-box uk-form" method="post" th:action="@{/login}">
                <div class="uk-form-row">
                    <input class="uk-width-1-1 uk-form-large" type="text" placeholder="Username" name="username"
                           value="reader"/>
                </div>
                <div class="uk-form-row">
                    <input class="uk-width-1-1 uk-form-large" type="password" placeholder="Password" name="password"
                           value="reader"/>
                </div>
                <div class="uk-form-row">
                    <button class="uk-width-1-1 uk-button uk-button-primary uk-button-large">Login</button>
                </div>
            </form>

        </div>
    </div>
    </body>
    </html>
```

步骤04 搭建用户服务模块，充当资源服务器，模块名称为 oauth2-user-service（可参考 1.3.1 小节内容快速创建 Spring Boot 模块），项目依赖 spring-cloud-starter-oauth2、spring-boot-starter-web，资源服务器的端口是 8081。在 application.properties 配置文件中添加如下配置：

```
server:
  port: 8081
```

将之前生成的密钥对的公钥命名为 public.cert，并放到资源文件下。这样，资源服务器可以在本地校验 JWT 的合法性。

创建 HelloController 类，其中 hello 方法无须登录就可以被访问，具体代码如下：

```
@RestController
public class HelloController {
    @GetMapping("hello")
    public String hello() {
```

```
        return "Hello Ay";
    }
}
```

创建 UserController 类，具体代码如下：

```
/**
 * 描述：UserController
 * @author ay
 * @since 2020-10-25
 */
@RestController
@RequestMapping("user")
public class UserController {

    @Resource
    private TokenStore tokenStore;

    /***
     * 描述：获取登录用户名（读权限或写权限可访问）
     * @param authentication
     * @return
     */
    @PreAuthorize("hasAuthority('READ') or hasAuthority('WRITE')")
    @GetMapping("name")
    public String getName(OAuth2Authentication authentication) {
        return authentication.getName();
    }

    /**
     * 描述：获取用户信息（读权限或写权限可访问）
     *
     * @param authentication
     * @return
     */
    @PreAuthorize("hasAuthority('READ') or hasAuthority('WRITE')")
    @GetMapping
    public OAuth2Authentication read(OAuth2Authentication authentication) {
        return authentication;
    }

    /**
     * 描述：获取访问令牌中的额外信息（只有写权限可以访问）
     * @param authentication
     * @return
     */
    @PreAuthorize("hasAuthority('WRITE')")
    @PostMapping
    public Object write(OAuth2Authentication authentication) {
        OAuth2AuthenticationDetails details = (OAuth2AuthenticationDetails)
authentication.getDetails();
```

```
        OAuth2AccessToken accessToken = 
tokenStore.readAccessToken(details.getTokenValue());
        return 
accessToken.getAdditionalInformation().getOrDefault("userDetails", null);
    }
}
```

上述代码中创建了 3 个接口，通过@PreAuthorize 在方法执行前进行权限控制。接下来创建核心的资源服务器配置类 ResourceServerConfiguration，具体代码如下：

```
/**
 * @author ay
 * @since 2020-10-25
 */
@Configuration
//启用资源服务器
@EnableResourceServer
//启用方法注解方式来进行权限控制
@EnableGlobalMethodSecurity(prePostEnabled = true)
public class ResourceServerConfiguration extends ResourceServerConfigurerAdapter {

    @Override
    public void configure(ResourceServerSecurityConfigurer resources) {
        //声明资源服务器的 ID 是 user-resource-id
        resources.resourceId("user-resource-id")
            //声明了资源服务器的 TokenStore 是 JWT
            .tokenStore(tokenStore());
    }

    /**
     * 配置 TokenStore
     * @return
     */
    @Bean
    public TokenStore tokenStore() {
        return new JwtTokenStore(jwtAccessTokenConverter());
    }

    /**
     * 配置公钥
     * @return
     */
    @Bean
    protected JwtAccessTokenConverter jwtAccessTokenConverter() {
        JwtAccessTokenConverter converter = new JwtAccessTokenConverter();
        Resource resource = new ClassPathResource("public.cert");
        String publicKey = null;
        try {
            publicKey = new
```

```java
String(FileCopyUtils.copyToByteArray(resource.getInputStream()));
        } catch (IOException e) {
            e.printStackTrace();
        }
        converter.setVerifierKey(publicKey);
        return converter;
    }

    /**
     * 配置除/user 路径之外的请求可以匿名访问
     * @param security
     * @throws Exception
     */
    @Override
    public void configure(HttpSecurity security) throws Exception {
        security.authorizeRequests()
                .antMatchers("/user/**").authenticated()
                .anyRequest().permitAll();
    }
}
```

在上述代码中，为了支持通过本地公钥进行验证，jwtAccessTokenConverter 配置公钥的路径。

步骤 05 完成授权服务器和资源服务器搭建后，接下来开始初始化 Oauth 数据库的数据。首先，配置两个用户：读用户 reader 具有读权限，密码为 reader；写用户 writer 具有读写权限，密码为 writer。具体 SQL 如下：

```sql
INSERT INTO `users` VALUES ('reader',
'$2a$04$C6pPJvC1v6.enW6ZZxX.luTdpSI/1gcgTVN7LhvQV6l/AfmzNU/3i', 1);
INSERT INTO `users` VALUES ('writer',
'$2a$04$M9t2oVs3/VIreBMocOujqOaB/oziWL0SnlWdt8hV4YnlhQrORA0fS', 1);
```

密码使用 BCryptPasswordEncoder 进行哈希，配置 reader 用户具有读权限、writer 用户具有写权限：

```sql
INSERT INTO `authorities` VALUES ('reader', 'READ');
INSERT INTO `authorities` VALUES ('writer', 'READ,WRITE');
```

配置 3 个客户端，其中客户端 client-1 使用密码模式，客户端 client-2 使用客户端模式，客户端 client-3 使用授权码模式，具体 SQL 如下：

```sql
INSERT INTO `oauth_client_details` VALUES ('client-1', 'user-resource-id',
'123456', 'server', 'password,refresh_token', '', 'READ,WRITE', 7200, NULL, NULL,
'true');
INSERT INTO `oauth_client_details` VALUES ('client-2', 'user-resource-id',
'123456', 'server', 'client_credentials,refresh_token', '', 'READ,WRITE', 7200,
NULL, NULL, 'true');
INSERT INTO `oauth_client_details` VALUES ('client-3', 'user-resource-id',
'123456', 'server', 'authorization_code,refresh_token',
'https://baidu.com,http://localhost:8082/ui/login,http://localhost:8083/ui/log
in,http://localhost:8082/ui/remoteCall', 'READ,WRITE', 7200, NULL, NULL,
```

'false');

客户端账号使用的资源 ID 与受保护资源服务器配置的资源 ID 需要保持一致，即 user-resource-id，客户端账号的密码都是 123456，授权范围都是 server，grant_types 字段配置支持不同的授权许可类型。给三个客户端账号各自配置一种授权许可类型。在实际业务场景中，可以为同一个客户端配置支持 OAuth 2.0 的 4 种授权许可类型。

步骤 06 启动 oauth2-server 授权服务器。首先，验证密码模式，通过 Postman 或者 curl 命令发起 POST 请求：

```
> curl -X POST -d "grant_type=password&client_id=client-1&client_secret=123456&username=writer&password=writer" http://localhost:8080/oauth/token
```

响应数据如下所示：

```
{
    "access_token": "eyJhbGciOiJSUzI1NiIsInR5cCI6IkpXVCJ9.eyJhdWQiOlsidXNlci1yZXNvdXJjZS1pZCJdLCJ1c2VyX25hbWUiOiJ3cml0ZXIiLCJzY29wZSI6WyJzZXJ2ZXIiXSwiZXhwIjoxNjA0MjA3NDU4LCJ1c2VyRGV0YWlscyI6eyJwYXNzd29yZCI6bnVsbCwidXNlcm5hbWUiOiJ3cml0ZXIiLCJhdXRob3JpdGllcyI6W3siYXV0aG9yaXR5IjoiUkVBRCxXUklURSJ9XSwiYWNjb3VudE5vbkV4cGlyZWQiOnRydWUsImFjY291bnROb25Mb2NrZWQiOnRydWUsImNyZWRlbnRpYWxzTm9uRXhwaXJlZCI6dHJ1ZSwiZW5hYmxlZCI6dHJ1ZX0sImF1dGhvcml0aWVzIjpbIlJFQUQsV1JJVEUiXSwianRpIjoiMTgwYjc4MzAtYmVjMC00NWVhLTlOTAtYzRjZWYwNWM4MTNlIiwiY2xpZW50X2lkIjoiY2xpZW50LTEifQ.FnKcWNq6yyeJtaz8D0M_t1J7XXS1tvpnOkXxe4vUwZAhAzyQmlmWEXfyboocAXjVF0vpz-1qzub_XX1eUdyPaeF8kNviOiOevkVX61X-QNgtdfOH3k1hPkOi1qffxri7i4zIUUXNPL9Ht1xvUADY_wrv7RqJ8MuDpwfvi58KfT9e_FHJD3HDu3iY25W3n-kFs1kKYyC5WE9d2LqJuodYTVpVz8zeF_V8JVIoIUKydkekxDs71MP7TnqLo0JXUBy-bMJebX1XqvZjVIgVlaQHNhJPXJKz8KxIjYt6rwc58pEEUDOIEyLCpM2JgwQsdZYqbjsJOOIjc4G2P5c7sYRtJQ",
    "token_type": "bearer",
    "refresh_token": "eyJhbGciOiJSUzI1NiIsInR5cCI6IkpXVCJ9.eyJhdWQiOlsidXNlci1yZXNvdXJjZS1pZCJdLCJ1c2VyX25hbWUiOiJ3cml0ZXIiLCJzY29wZSI6WyJzZXJ2ZXIiXSwiYXRpIjoiMTgwYjc4MzAtYmVjMC00NWVhLTlOTAtYzRjZWYwNWM4MTNlIiwiZXhwIjoxNjA2NzkyMjU4LCJ1c2VyRGV0YWlscyI6eyJwYXNzd29yZCI6bnVsbCwidXNlcm5hbWUiOiJ3cml0ZXIiLCJhdXRob3JpdGllcyI6W3siYXV0aG9yaXR5IjoiUkVBRCxXUklURSJ9XSwiYWNjb3VudE5vbkV4cGlyZWQiOnRydWUsImFjY291bnROb25Mb2NrZWQiOnRydWUsImNyZWRlbnRpYWxzTm9uRXhwaXJlZCI6dHJ1ZSwiZW5hYmxlZCI6dHJ1ZX0sImF1dGhvcml0aWVzIjpbIlJFQUQsV1JJVEUiXSwianRpIjoiNmE0ZDYyYzYtMmZhNi00MTgyLTljYTUtOWYzMmE3NDUwYTFlIiwiY2xpZW50X2lkIjoiY2xpZW50LTEifQ.T6w2b9EJ4V8Dfj4oxDQf-dwk6qfNQcYHy6M4Bg83_HGCwB40sEU4urBZ3QU2tx1CsUar1LLrdiJoJ0JgTb0kKvjSuEyxTGKXckYA81itDX7UeWaps3jetIChe34W7FNfS7n3EIe3akfdgdLyr68ODIq3pyy6WqZfRV175naiKSENvbF3mqB5l36J8Wv8SA0IJxYD0fQhG-i43FTXgDcGfO7yvVLVeiSpNZdbc1ygtzmSAQ3RDvwPRqv0ARtWAirTgfWUCR8Wq2SkHqiUZjjBsEN3K47pZp4xZquiZX_hAZ54Kx4LnxViTf1Yn6gOfT6cUKDFZfyjOyc_0Wn7RsB2-Q",
    "expires_in": 7199,
    "scope": "server",
    "userDetails": {
        "password": null,
        "username": "writer",
        "authorities": [
            {
```

```
          "authority": "READ,WRITE"
        }
      ],
      "accountNonExpired": true,
      "accountNonLocked": true,
      "credentialsNonExpired": true,
      "enabled": true
  },
  "jti": "180b7830-bec0-45ea-9e90-c4cef05c813e"
}
```

访问 https://jwt.io/ 地址进行 token 校验，如图 12-13 所示。

图 12-13　JWT 解析

从图 12-13 左侧信息可以看到 Token 增强器加入的 userDetails 自定义信息。将公钥信息粘贴到页面，可以看到 JWT 校验成功，如图 12-14 所示。

图 12-14　JWK 校验

接着，通过 Postman 或者 curl 命令验证客户端模式：

```
> curl -X POST -d
"grant_type=client_credentials&client_id=client-2&client_secret=123456"
http://localhost:8080/oauth/token
```

响应数据如下所示：

```
{
    "access_token": "eyJhbGciOiJSUzI1NiIsInR5cCI6IkpXVCJ9.eyJhdWQiOlsidXNl
ciIyZXNvdXJjZS1pZCJdLCJzY29wZSI6WyJzZXJ2ZXIiXSwiZXhwIjoxNjA0MjA4ODQ3LCJhdXRob3J
pdGllcyI6WyJSRUFEIiwiV1JJVEUiXSwianRpIjoiZDU4YTg3NjctMDE2Zi00NTMyLTkxOWItMDZl
mYxNDAwOTg2IiwiY2xpZW50X2lkIjoiY2xpZW50LTIifQ.BN_tgH8O5O1-avFXK3K96fkJ8HP_KoA6
Mem9J_24a7TTREQHTljotTH62G4lFQa0RabN6f6XutQzioT-8ZbJ_K0K_ox1pRRQHKjC91-B2-_Hlc
06q9jHSucSAvtaWiEJcDCw-T1Ic3cySbvz7JxIDbYjNygYPvd6cOE1xrixXRrzwBUTpuKFNpLjqUy4
Oo4DUrSD0YYnl0pX6YiUOKK3L7Kskobmq-KCj6xvfmON5xMvFQW8mJLesr8bj5ljgBRgffAfQFk32s
Kx026gHwGJtr2mWgoKyN9Y96nJS5AOrUPL70BliqWQh2hkMLJYRzSD1336U5F8lh9MwtFPLUlyng",
    "token_type": "bearer",
    "expires_in": 7199,
    "scope": "server",
    "jti": "d58a8767-016f-4532-919b-06eff1400986"
}
```

客户端模式没有用户的概念，所以没有刷新令牌。刷新令牌主要用于避免访问令牌失效后需要用户再次登录的问题。

最后，验证授权码模式，在浏览器中输入请求地址"http://localhost:8080/oauth/authorize?response_type=code&client_id=client-3&redirect_uri=https://baidu.com"，访问后页面直接跳转到登录界面，如图 12-15 所示。

输入用户名和密码"reader/reader"，由于数据库设置禁用自动批准授权的模式，因此登录后跳转到批准界面，如图 12-16 所示。选择 Approve 选项，单击 Authorize 按钮同意后数据库 oauth_approvals 表产生授权通过记录。浏览器跳转到百度并且提供授权码（https://www.baidu.com/?code=fWdrBb）。oauth_code 表也有授权码记录。

图 12-15　登录界面　　　　　　图 12-16　批准界面

通过 Postman 或者 curl 命令发起请求，code 等于刚刚获取到的授权码：

```
curl -X POST -d
"grant_type=authorization_code&client_id=client-3&client_secret=123456&code=fW
drBb&redirect_uri=https://baidu.com" http://localhost:8080/oauth/token
```

响应数据如下所示：

```
{
    "access_token": "eyJhbGciOiJSUzI1NiIsInR5cCI6IkpXVCJ9.eyJhdWQiOlsidXNlci1yZXNvdXJjZS1pZCJdLCJ1c2VyX25hbWUiOiJyZWFkZXIiLCJzY29wZSI6WyJzZXJ2ZXIiXSwiZXhwIjoxNjA0MjEwMTY3LCJ1c2VyRGV0YWlscyI6eyJwYXNzd29yZCI6bnVsbCwidXNlcm5hbWUiOiJyZWFkZXIiLCJhdXRob3JpdGlcyI6W3siYXV0aG9yaXR5IjoiUkVBRCJ9XSwiYWNjb3VudE5vbkV4cGlyZWQiOnRydWUsImFjY291bnROb25Mb2NrZWQiOnRydWUsImNyZWRlbnRpYWxzTm9uRXhwaXJlZCI6dHJ1ZSwiZW5hYmxlZCI6dHJ1ZX0sImF1dGhvcml0aWVzIjpbIlJFQUQiXSwianRpIjoiNzE0MGRkOTItMDkxMy00MjBhLTk1ZTItZmFlOGM1YjA4ZjZhIiwiY2xpZW50X2lkIjoiY2xpZW50LTEifQ.MgLinSY_fxBW5D6LU6I9XlYcnvo4KIEuyPdAuCmb5krKtLAei3xZ17U-2zEx0YQiyj0uGGNBreZVJysT0tn_dUmPpMIAdFYaEVd6lsa3gF4-s1hBPHH4VF54txb4-CD28eXnXi8x6jN7GS1tF5q8XR6Xx5TcEzQeBfyuNdvomRluWVXVU5lwJ55DTDRzFGK6n5BSqTSlmLlIn7N8ZwewNSUbEBJnXOG6OvZWgQz-ubo3J3CIDYvBe1KuQWvFje8Y5GcAAPWIR9ltJeoi5YNeBhRD2EjDnvmXWLDCe55E_osaLE1qo2kd0j-Pbz00vsnRjg34LP02rQIKhlcWtZc2wQ",
    "token_type": "bearer",
    "refresh_token": "eyJhbGciOiJSUzI1NiIsInR5cCI6IkpXVCJ9.eyJhdWQiOlsidXNlci1yZXNvdXJjZS1pZCJdLCJ1c2VyX25hbWUiOiJyZWFkZXIiLCJzY29wZSI6WyJzZXJ2ZXIiXSwiYXRpIjoiNzE0MGRkOTItMDkxMy00MjBhLTk1ZTItZmFlOGM1YjA4ZjZhIiwiZXhwIjoxNjA2Nzk0OTY3LCJ1c2VyRGV0YWlscyI6eyJwYXNzd29yZCI6bnVsbCwidXNlcm5hbWUiOiJyZWFkZXIiLCJhdXRob3JpdGllcyI6W3siYXV0aG9yaXR5IjoiUkVBRCJ9XSwiYWNjb3VudE5vbkV4cGlyZWQiOnRydWUsImFjY291bnROb25Mb2NrZWQiOnRydWUsImNyZWRlbnRpYWxzTm9uRXhwaXJlZCI6dHJ1ZSwiZW5hYmxlZCI6dHJ1ZX0sImp0aSI6IjZWQxMjcxMzMtMWNjNS00YTEwLWExNjctZDY4OWY4ZGFiY2Q5IiwiY2xpZW50X2lkIjoiY2xpZW50LTEifQ.qxzrCpWTGuwgOF_N5Mp-KSMW3DlhIeFiskW_kxEg2nQJB4a0-eT9bC1XE2c-thmkHuFRxpuj5tYZzhIcN0rULtjJ5wuiAa73b9boBjvB_r4Ej2ZgfeKrvHVUkTNRt6iKqcfxwvUMPHHoIKqpd20vSlCLrlwB0ATASMk4czWV3z5H4x1MbBwT0SWF0s8CRX5Id0efUqoSTu8PvcORYujo3F9666mjGEbooF9qdneh-pddsz6AXEZ-bEo5ljt0dIM6Hf_G0Su3XzvJcHKblWtx0Ingzi6PXjYkK6A-JRQJWzSnebAnjWUQ7qQIp92HIlwpzwOQXQrBqVCsIuvbI3__zw",
    "expires_in": 7199,
    "scope": "server",
    "userDetails": {
        "password": null,
        "username": "reader",
        "authorities": [
            {
                "authority": "READ"
            }
        ],
        "accountNonExpired": true,
        "accountNonLocked": true,
        "credentialsNonExpired": true,
        "enabled": true
    },
    "jti": "7140dd92-0913-420a-95e2-fae8c5b08f6a"
}
```

由于登录用户 reader 只有 reader 权限，因此获取到的数据也只有读的权限。

步骤 07 验证权限控制。在浏览器中输入请求地址 "http://localhost:8081/user/name"，访问失败，

/user 需要身份认证。我们通过之前任何一种方式（密码模式/授权码模式/客户端模式）拿到访问令牌，用具有读权限的访问令牌访问资源服务器的 http://localhost:8081/user/ 地址即可。

> **注　意**
>
> 请求头加入 Authorization: Bearer XX-XX-XX，其中 XX-XX-XX 代表访问令牌。

步骤 08 之前都是通过 curl 或者 Postman 来模拟客户端获取 Token，接下来搭建 OAuth 客户端程序自动实现这个过程。快速创建 Spring Boot 模块，模块名称为 oauth2-client。在模块的 pom.xml 文件中添加 spring-cloud-starter-oauth2、spring-boot-starter-thymeleaf 以及 spring-boot-starter-web 等依赖。实现 MVC 的配置，具体代码如下：

```java
/**
 *
 * @author ay
 * @since 2020-10-26
 */
@Configuration
@EnableWebMvc
public class WebMvcConfig implements WebMvcConfigurer {

    /**
     * 配置 RequestContextListener 用于启用 session scope 的 Bean
     *
     * @return
     */
    @Bean
    public RequestContextListener requestContextListener() {
        return new RequestContextListener();
    }

    /**
     * 配置 index 路径的首页 Controller
     * @param registry
     */
    @Override
    public void addViewControllers(ViewControllerRegistry registry) {
        registry.addViewController("/")
                .setViewName("forward:/index");
        registry.addViewController("/index");
    }
}
```

WebMvcConfigurer 配置类其实是 Spring 内部的一种配置方式，采用 JavaBean 的形式来代替传统的 XML 配置文件形式进行针对框架个性化定制，可以自定义一些 Handler、Interceptor、ViewResolver、MessageConverter。基于 java-based 方式的 Spring MVC 配置需要创建一个配置类并实现 WebMvcConfigurer 接口。

实现安全方面的配置，具体代码如下：

```java
/**
 * 安全配置
 * @author ay
 * @since 2020-11-01
 */
@Configuration
@Order(200)
public class WebSecurityConfig extends WebSecurityConfigurerAdapter {
    /**
     * /路径和/login路径允许访问，其他路径需要身份认证后才能访问
     *
     * @param http
     * @throws Exception
     */
    @Override
    protected void configure(HttpSecurity http) throws Exception {
        http
                .authorizeRequests()
                .antMatchers("/", "/login**")
                .permitAll()
                .anyRequest()
                .authenticated();
    }
}
```

创建一个控制器，具体代码如下：

```java
/**
 * 描述：TestController
 * @author ay
 * @since 2020-10-25
 */
@RestController
public class TestController {

    @Resource
    OAuth2RestTemplate restTemplate;

    @GetMapping("/projectPage")
    public ModelAndView securedPage(OAuth2Authentication authentication) {
        return new ModelAndView("projectPage")
                .addObject("authentication", authentication);
    }

    @GetMapping("/getUserName")
    public String getUserName() {
        ResponseEntity<String> responseEntity =
restTemplate.getForEntity("http://localhost:8081/user/name", String.class);
        return responseEntity.getBody();
    }
}
```

protectPage 页面实现的功能是把用户信息作为模型传入视图，这样打开页面后就能显示用户名和权限。

remoteCall 接口实现的功能是通过引入 OAuth2RestTemplate 在登录后使用凭据直接从受保护资源服务器获取资源，不需要烦琐地实现获得访问令牌、在请求头里加入访问令牌的过程。

配置 OAuth2RestTemplate Bean，并启用 OAuth2Sso 功能：

```
@Configuration
//这个注解包含了@EnableOAuth2Client
@EnableOAuth2Sso
public class OAuthClientConfig {
    /**
     * 定义了 OAuth2RestTemplate
     *
     * @param oAuth2ClientContext
     * @param details
     * @return
     */
    @Bean
    public OAuth2RestTemplate oauth2RestTemplate(OAuth2ClientContext oAuth2ClientContext,
                                                  OAuth2ProtectedResourceDetails details) {
        return new OAuth2RestTemplate(details, oAuth2ClientContext);
    }
}
```

最后定义前端首页 index.html 和 protectPage.html。其中，index.html 的具体代码如下：

```
<!DOCTYPE html>
<html lang="en">
<head>
    <meta http-equiv="Content-Type" content="text/html; charset=utf-8"/>
    <title>Spring Security SSO Client</title>
    <link rel="stylesheet" href="https://maxcdn.bootstrapcdn.com/bootstrap/3.3.2/css/bootstrap.min.css"/>
</head>

<body>
<div class="container">
    <div class="col-sm-12">
        <h1>Spring Security SSO Client</h1>
        <a class="btn btn-primary" href="protectPage">Login</a>
    </div>
</div>
</body>
</html>
```

protectPage.html 的具体代码如下：

```
<!DOCTYPE html>
```

```html
<html lang="en">
<head>
    <meta http-equiv="Content-Type" content="text/html; charset=utf-8"/>
    <title>Spring Security SSO Client</title>
    <link rel="stylesheet" href="https://maxcdn.bootstrapcdn.com/bootstrap/3.3.2/css/bootstrap.min.css"/>
</head>
<body>
<div class="container">
    <div class="col-sm-12">
        <h1>Secured Page</h1>
        Welcome, <span th:text="${authentication.name}">Name</span>
        <br/>
        Your authorities are <span th:text="${authentication.authorities}">authorities</span>
    </div>
</div>
</body>
</html>
```

所有代码开发完毕后，在浏览器中输入访问地址"http://localhost:8082/ui/protectPage"，可以看到页面自动转到了授权服务器（8080端口）的登录页面，登录后显示当前用户名和权限。

再启动一个客户端网站，端口改为8083，然后访问 http://localhost:8083/ui/protectPage，直接显示登录状态，单点"登录"测试成功。

接下来演示客户端请求资源服务器资源。在浏览器中访问 http://localhost:8082/ui/getUserName，会先转到授权服务器登录界面，输入用户名和密码（reader/reader），登录后自动跳转回来，在浏览器中打印用户名。

第 13 章

Spring Cloud 组件容器化

本章主要介绍 Spring Boot 项目容器化、Spring Cloud Alibaba 组件容器化，其中包括 Nacos、Sentinel 以及 Seata 等组件。

13.1 Spring Boot 项目容器化

13.1.1 制作镜像

本小节将学习如何制作 Java 运行环境的镜像，并在此镜像上启动 Java 容器，具体步骤如下：

步骤 01 下载 JDK 安装包，下载地址为 https://www.oracle.com/technetwork/java/javase/downloads/jdk11-downloads-5066655.html。这里使用的 JDK 版本为 jdk-11.0.2_linux-x64_bin.tar.gz。

步骤 02 拉取 centos 镜像，并启动 centos 容器：

```
### 拉取 centos 镜像
$ docker pull centos
### 启动 centos 容器
$ docker run -i -t -v /c/Users/Ay:/mnt/ centos /bin/bash
### 查看路径是否挂载成功
[root@8326ca477b44 /]# ll /mnt/
total 950688
-rw-r--r-- 1 root root 179640645 Feb  1 06:00 jdk-11.0.2_linux-x64_bin.tar.gz

//省略大量代码
```

- -v 选项：-v 在 Docker 中称为数据卷（Data Volume），用于将宿主机上的磁盘挂载到容器中，格式为"宿主机路径:容器路径"。需要注意的是宿主机路径可以是相对路径，但是容器的路径必须是绝对路径。可以多次使用 -v 选项，同时挂载多个宿主机路径到容器中。

- /c/Users/Ay:/mnt/：/c/Users/Ay 为宿主机 JDK 安装包的存放路径,/mnt/为 centos 容器的目录。

步骤 03 在 centos 容器的/mnt/目录下解压安装包并安装 JDK：

```
### 将压缩包解压到 /opt 目录下
[root@8326ca477b44 /]# tar -zxf /mnt/jdk-11.0.2_linux-x64_bin.tar.gz -C /opt
```

配置环境变量：

```
### 设置环境变量，进入 profile 文件
[root@8326ca477b44 bin]# vi /etc/profile
```

在 profile 文件末尾添加如下配置：

```
### JAVA_HOME 是 java 的安装路径
JAVA_HOME=/opt/jdk-11.0.2
PATH=$JAVA_HOME/bin:$PATH:.
CLASSPATH=$JAVA_HOME/lib/tools.jar:$JAVA_HOME/lib/dt.jar:.
export JAVA_HOME
export PATH
export CLASSPATH
```

执行 source 命令，让配置生效：

```
[root@8326ca477b44 bin]# source /etc/profile
```

最后验证 JDK 是否安装成功：

```
### 查看 JDK 是否安装成功，从输出的信息可知 JDK 安装成功
[root@8326ca477b44 /]# java -version
java version "11.0.2" 2019-01-15 LTS
Java(TM) SE Runtime Environment 18.9 (build 11.0.2+9-LTS)
Java HotSpot(TM) 64-Bit Server VM 18.9 (build 11.0.2+9-LTS, mixed mode)
```

步骤 04 打开一个命令行终端，通过 docker commit 命令提交当前容器为新的镜像：

```
### 查看当前运行的容器
$ docker ps
CONTAINER ID   IMAGE    COMMAND     CREATED       STATUS       PORTS    NAMES
8326ca477b44   centos   "/bin/bash" 2 hours ago   Up 2 hours            infallible_kare
### 提交当前容器为新的镜像
$ docker commit 8326ca477b44 hwy/centos
sha256:324e55254ad9baa74477c08333bed2978e72051b7d3f77c957b773d25cfbe7c7
### 查看当前镜像，可知镜像已经成功生成
$ docker images
REPOSITORY      TAG        IMAGE ID        CREATED         SIZE
hwy/centos      latest     324e55254ad9    5 seconds ago   504MB
```

步骤 05 验证生成的镜像是否可用：

```
$ docker run --rm hwy/centos /opt/jdk-11.0.2/bin/java -version
java version "11.0.2" 2019-01-15 LTS
```

```
            Java(TM) SE Runtime Environment 18.9 (build 11.0.2+9-LTS)
            Java HotSpot(TM) 64-Bit Server VM 18.9 (build 11.0.2+9-LTS, mixed mode)
```

在上述命令中，我们在 hwy/centos 镜像上启动一个容器，并在容器目录/opt/jdk-11.0.2/bin/下执行 java -version 命令，可以看到命令行终端输出了 JDK 版本号相关信息。需要注意的是，上面的命令添加了一个 --rm 选项，该选项表示容器退出时可自动删除容器。

至此，手工制作 Docker 镜像已完成。

13.1.2　使用 Dockerfile 构建镜像

从上一小节的 Docker 命令学习中可以了解到，镜像的定制实际上就是定制每一层所添加的配置和文件。如果可以把每一层修改、安装、构建、操作的命令都写入一个脚本，那么用这个脚本来构建、定制镜像，实现整个过程的自动化，既可以提高效率，又能够减少错误。这个脚本就是 Dockerfile。

Dockerfile 是一个文本文件，其内包含了一条条的指令（Instruction），每一条指令构建一层，因此每一条指令的内容就是描述该层应当如何构建。下面我们创建一个空白文件（文件名为 Dockerfile），并学习 Dockerfile 指令。

1. FROM 命令

所谓定制镜像，一定是以一个镜像为基础，在其上进行定制。就像我们之前运行了一个 centos 镜像的容器，再进行修改一样，基础镜像是必须指定的，FROM 命令用于指定基础镜像。因此，一个 Dockerfile 中的 FROM 是必备的指令，并且必须是第一条指令。例如：

```
FROM centos:latest
```

FROM 命令的值有固定的格式，即"仓库名称：标签名"，若使用基础镜像的最新版本，则 latest 标签名可以省略，否则需指定基础镜像的具体版本。

2. MAINTAINER 命令

MAINTAINER 用于设置镜像的作者，具体格式为 MAINTAINER <author name>。例如：

```
### 建议使用"姓名+邮箱"的形式
MAINTAINER "hwy"<huangwenyi10@163.com>
```

3. ADD 命令

ADD 是复制文件命令，它有两个参数：<source>和<destination>。其中，source 参数为宿主机的来源路径；destination 参数为容器内路径，必须为绝对路径。ADD 命令的语法为 ADD <src> <destination>，例如：

```
### 添加jdk安装包到容器的 /opt 目录下
ADD  /c/Users/Ay/jdk-11.0.2_linux-x64_bin.tar.gz /opt
```

ADD 命令将自动解压来源中的压缩包，将解压后的文件复制到目标目录（/opt）中。

4. RUN 命令

RUN 命令用来执行一系列构建镜像所需要的命令。如果需要执行多条命令，可以使用多条 RUN 命令，例如：

```
### 执行 shell 命令
RUN echo 'hello ay...'
RUN ls -l
...
```

Dockerfile 中的每一个命令都会建立一层，RUN 命令也不例外。每一个 RUN 的行为就和我们手工建立镜像的过程一样，即新建立一层，在其上执行这些命令，构成新的镜像。上面的这种写法创建了多层镜像，但这是完全没有意义的，其结果就是产生非常臃肿、非常多层的镜像，不仅增加了构建部署的时间，也很容易出错。这是很多 Docker 初学者常犯的一个错误。

5. CMD 命令

CMD 命令提供了容器默认的执行命令。Dockerfile 只允许使用一次 CMD 命令，使用多个 CMD 命令会抵消之前所有的命令，只有最后一个命令生效。CMD 命令有以下 3 种形式：

```
CMD ["executable","param1","param2"]
CMD ["param1","param2"]
CMD command param1 param2
```

例如，使用如下命令在容器启动时输出 Java 版本：

```
### 容器启动时执行的命令
CMD /opt/jdk-11.0.2/bin/java -version
```

熟悉完 Dockerfile 文件命令后，下面我们使用 Dockerfile 构建一个 Java 镜像。

步骤 01 在用户目录下（C:\Users\Ay）创建 dockerfile 文件夹，在 dockerfile 文件夹中创建并编辑 Dockerfile 文件，同时把之前下载的 jdk-11.0.2_linux-x64_bin.tar.gz 文件放入 dockerfile 文件夹中。

步骤 02 在 Dockerfile 文件中添加如下命令，这些命令都是之前学习 Dockerfile 命令用到的。

```
FROM centos:latest
MAINTAINER "hwy"<huangwenyi10@163.com>
ADD jdk-11.0.2_linux-x64_bin.tar.gz /opt
RUN echo 'hello ay...'
CMD /opt/jdk-11.0.2/bin/java -version
```

> **注 意**
>
> 将 Dockerfile 文件与需要添加到容器的文件放在同一个目录下。

步骤 03 使用 docker bulid 命令读取 Dockerfile 文件，并构建镜像：

```
C:\Users\Ay\dockerfile>docker build -t hwy/java -f Dockerfile .
Sending build context to Docker daemon  179.6MB
Step 1/5 : FROM centos:latest
 ---> 9f38484d220f
```

```
Step 2/5 : MAINTAINER "hwy"<huangwenyi10@163.com>
 ---> Running in ca15590e3730
Removing intermediate container ca15590e3730
 ---> 74ae85fd1d25
Step 3/5 : ADD jdk-11.0.2_linux-x64_bin.tar.gz /opt
 ---> 4ab75fb4cd0f
Step 4/5 : RUN echo 'hello ay...'
 ---> Running in 61e64cc0d845
hello ay...
Removing intermediate container 61e64cc0d845
 ---> 3ea28a5e6b05
Step 5/5 : CMD /opt/jdk-11.0.2/bin/java -version
 ---> Running in 35d1544278ed
Removing intermediate container 35d1544278ed
 ---> 51e1a6888f9f
Successfully built 51e1a6888f9f
Successfully tagged hwy/java:latest
SECURITY WARNING: You are building a Docker image from Windows against a non-Windows
Docker host. All files and directories added to build context will have '-rwxr-xr-x'
permissions. It is recommended to double check and reset permissions for sensitive
files and directories.
```

- -t 选项：用于指定镜像的名称，并读取当前目录（.目录）中的 Dockerfile 文件。
- -f 选项：用于指定 Dockerfile 文件名称。

从输出信息可知，执行 docker build 命令后，首先构建上下文发送到 Docker 引擎中，随后通过 5 个步骤来完成镜像的构建工作，在每个步骤中都会输出对应的 Dockerfile 命令，而且每个步骤都会生成一个"中间容器"与"中间镜像"。例如步骤 5：

```
Step 5/5 : CMD /opt/jdk-11.0.2/bin/java -version
### 生成中间容器
 ---> Running in 35d1544278ed
### 删除中间容器
Removing intermediate container 35d1544278ed
### 创建一个中间镜像
 ---> 51e1a6888f9f
```

当执行完命令 CMD /opt/jdk-11.0.2/bin/java -version 后，将生成一个中间容器，容器 ID 为 9d10dbefcf6c。接着从该容器中创建一个中间镜像，镜像 ID 为 35d1544278ed。最后将中间容器删除。

> **注 意**
>
> 并不是每个步骤都会生成中间容器，但是每个步骤一定会产生中间镜像。这些中间镜像将加入到缓存中，当某一个构建步骤失败时，将停止整个构建过程，但是中间镜像仍然会存放在缓存中，下次再次构建时直接从缓存中获取中间镜像，而不会重复执行之前已经构建成功的步骤。

步骤04 查看生成的 Docker 镜像。

```
### 查看生成的 Docker 镜像
C:\Users\Ay\dockerfile>docker images
REPOSITORY        TAG           IMAGE ID            CREATED             SIZE
hwy/java          latest        51e1a6888f9f        3 minutes ago       504MB
hwy/centos        latest        39969e9d9569        27 hours ago        504MB
tomcat            latest        238e6d7313e3        4 days ago          506MB

centos            latest        9f38484d220f        4 months ago        202MB
hello-world       latest        fce289e99eb9        6 months ago        1.84kB
```

至此，完成了通过 Dockerfile 构建镜像的所有操作。

13.1.3 Spring Boot 集成 Docker

本小节我们开始学习如何在 Spring Boot 中集成 Docker，并在构建 Spring Boot 应用程序时生成 Docker 镜像。

步骤 01 创建一个 Spring Boot 项目，项目名为 spring-boot-docker。打开 pom.xml 文件，修改 artifactId 和 version，具体代码如下：

```xml
<?xml version="1.0" encoding="UTF-8"?>
<project xmlns="http://maven.apache.org/POM/4.0.0"
xmlns:xsi="http://www.w3.org/2001/XMLSchema-instance"
     xsi:schemaLocation="http://maven.apache.org/POM/4.0.0
http://maven.apache.org/xsd/maven-4.0.0.xsd">
    <modelVersion>4.0.0</modelVersion>
    <parent>
        <groupId>org.springframework.boot</groupId>
        <artifactId>spring-boot-starter-parent</artifactId>
        <version>2.1.6.RELEASE</version>
        <relativePath/> <!-- lookup parent from repository -->
    </parent>
    <groupId>com.example</groupId>
    <!-- 修改 artifactId -->
    <artifactId>spring-boot-docker</artifactId>
    <!-- 修改 version -->
    <version>0.0.1</version>
    <name>demo</name>
    <description>Demo project for Spring Boot</description>

    <properties>
        <java.version>1.8</java.version>
    </properties>

    <dependencies>
        <dependency>
            <groupId>org.springframework.boot</groupId>
            <artifactId>spring-boot-starter-web</artifactId>
        </dependency>
```

```xml
        <dependency>
            <groupId>org.springframework.boot</groupId>
            <artifactId>spring-boot-starter-test</artifactId>
            <scope>test</scope>
        </dependency>
    </dependencies>

    <build>
        <!-- spring boot maven 插件 -->
        <plugins>
            <plugin>
                <groupId>org.springframework.boot</groupId>
                <artifactId>spring-boot-maven-plugin</artifactId>
            </plugin>
        </plugins>
    </build>
</project>
```

Spring Boot 使用 spring-boot-maven-plugin 插件构建项目，通过使用 mvn package 命令打包后将生成一个可直接运行的 jar 包，默认文件名格式为 ${project.build.finalName}。这是一个 Maven 属性，相当于 ${project.artifacId}-${project.version}.jar，生成的 jar 包在/target 目录下。根据 pom 文件的配置，执行 mvn package 命令后，生成的 jar 包名为 spring-boot-docker-0.01.jar。

步骤 02 在 pom.xml 文件中加入 docker-maven-plugin 插件依赖，具体代码如下：

```xml
<properties>
        <java.version>1.8</java.version>
        <docker.image.prefix>springboot</docker.image.prefix>
    </properties>
<build>
        <plugins>
            <plugin>
                <groupId>org.springframework.boot</groupId>
                <artifactId>spring-boot-maven-plugin</artifactId>
            </plugin>
            <!-- Docker maven plugin -->
            <plugin>
                <groupId>com.spotify</groupId>
                <artifactId>docker-maven-plugin</artifactId>
                <version>1.0.0</version>
                <configuration>
                    <!-- 指定 Docker 镜像完整名称 -->
                    <imageName>${docker.image.prefix}/${project.artifactId}</imageName>
                    <!-- 指定 dockerfile 文件所在目录 -->
                    <dockerDirectory>src/main/docker</dockerDirectory>
                    <resources>
                        <resource>
```

```xml
                    <targetPath>/</targetPath>
                    <directory>${project.build.directory}</directory>
                    <include>${project.build.finalName}.jar</include>
                </resource>
            </resources>
        </configuration>
    </plugin>
    <!-- Docker maven plugin -->
</plugins>
</build>
```

- `<imageName>`：用于指定 Docker 镜像的完整名称。其中，`${docker.image.prefix}` 为仓库名称，`${project.artifactId}` 为镜像名。
- `<dockerDirectory>`：用于指定 Dockerfile 文件所在目录。
- `<directory>`：用于指定需要复制的根目录，其中 `${project.build.directory}` 表示 target 目录。
- `<include>`：用于指定需要复制的文件，即 Maven 打包后生成的 jar 文件。

步骤 03 在 src/main/docker/ 目录下创建 Dockerfile 文件，内容如下：

```
### 使用 Docker 提供的 Java 镜像
FROM java
### 作者信息：用户名 + 邮箱
MAINTAINER "hwy"huangwenyi10@163.com
### 复制文件并重命名为 app.jar
ADD spring-boot-docker-0.0.1.jar app.jar
### 将 8080 端口设置为可暴露的接口
EXPOSE 8080
### 使用 java -jar 启动项目
CMD java -jar app.jar
使用 mvn docker:build 命令构建项目
### 在项目 spring-boot-docker 目录下执行命令
→ spring-boot-docker >> mvn docker:build
```

命令执行后，可在控制台上看到相关的输出信息。执行 docker images 命令查看镜像是否成功生成。

```
D:\MyJob\spring-boot-docker>docker images
REPOSITORY                    TAG          IMAGE ID         CREATED            SIZE
springboot/spring-boot-docker latest       b705802f2dd4     48 seconds ago     660MB
hwy/java                      latest       51e1a6888f9f     23 hours ago       504MB
hwy/centos                    latest       39969e9d9569     2 days ago         504MB
tomcat                        latest       238e6d7313e3     5 days ago         506MB
centos                        latest       9f38484d220f     4 months ago       202MB
hello-world                   latest       fce289e99eb9     6 months ago       1.84kB
java                          latest       d23bdf5b1b1b     2 years ago        643MB
```

步骤 04 执行 docker run 命令启动容器。

容器在启动的时候会执行 Dockerfile 文件里的 CMD 命令：java -jar app.jar。该命令用来启动

Spring Boot 项目，之后就可以在控制台上看到 Spring Boot 的启动信息，具体内容如下：

```
  .   ____          _            __ _ _
 /\\ / ___'_ __ _ _(_)_ __  __ _ \ \ \ \
( ( )\___ | '_ | '_| | '_ \/ _` | \ \ \ \
 \\/  ___)| |_)| | | | | || (_| |  ) ) ) )
  '  |____| .__|_| |_|_| |_\__, | / / / /
 =========|_|==============|___/=/_/_/_/
 :: Spring Boot ::        (v2.1.6.RELEASE)

2019-07-23 14:57:37.624  INFO 5 --- [           main] com.example.demo.DemoApplication         : Starting DemoApplication v0.0.1 on e7db0dcedf96 with PID 5 (/app.jar started by root in /)
2019-07-23 14:57:37.658  INFO 5 --- [           main] com.example.demo.DemoApplication         : No active profile set, falling back to default profiles: default
2019-07-23 14:57:44.868  INFO 5 --- [           main] o.s.b.w.embedded.tomcat.TomcatWebServer  : Tomcat initialized with port(s): 8080 (http)
2019-07-23 14:57:45.059  INFO 5 --- [           main] o.apache.catalina.core.StandardService   : Starting service [Tomcat]
2019-07-23 14:57:45.060  INFO 5 --- [           main] org.apache.catalina.core.StandardEngine  : Starting Servlet engine: [Apache Tomcat/9.0.21]
2019-07-23 14:57:45.773  INFO 5 --- [           main] o.a.c.c.C.[Tomcat].[localhost].[/]       : Initializing Spring embedded WebApplicationContext
2019-07-23 14:57:45.773  INFO 5 --- [           main] o.s.web.context.ContextLoader            : Root WebApplicationContext: initialization completed in 7693 ms
2019-07-23 14:57:46.943  INFO 5 --- [           main] o.s.s.concurrent.ThreadPoolTaskExecutor  : Initializing ExecutorService 'applicationTaskExecutor'
2019-07-23 14:57:47.956  INFO 5 --- [           main] o.s.b.w.embedded.tomcat.TomcatWebServer  : Tomcat started on port(s): 8080 (http) with context path ''
2019-07-23 14:57:47.970  INFO 5 --- [           main] com.example.demo.DemoApplication         : Started DemoApplication in 12.77 seconds (JVM running for 15.003)
```

至此，Spring Boot 成功集成了 Docker 容器，并把 Spring Boot 项目打包成 Docker 镜像。

13.2 Spring Cloud Alibaba 组件容器化

13.2.1 Nacos Docker

Spring Cloud Alibaba 已经制作好 Nacos 镜像，我们只需要下载镜像启动即可，具体步骤如下：

步骤 01 下载 Nacos-docker 代码：

```
### 下载源码
> git clone https://github.com/nacos-group/nacos-docker.git
### 进入 nacos-docker 目录
> cd nacos-docker
```

步骤 02 启动 Nacos：

```
### 启动 Nacos
> docker-compose -f example/standalone-derby.yaml up
```

启动完成后，访问地址 http://127.0.0.1:8848/Nacos/，可以看到 Nacos 的首页，输入初始密码 nacos/nacos，进入 Nacos 控制台。

备注
需要自行安装 docker-compost。

使用 MySQL 5.7 的启动命令如下：

```
> docker-compose -f example/standalone-mysql-5.7.yaml up
```

使用 MySQL 8 的启动命令如下：

```
> docker-compose -f example/standalone-mysql-8.yaml up
```

集群模式的启动命令如下：

```
> docker-compose -f example/cluster-hostname.yaml up
```

步骤 03 验证 Nacos 是否可正常使用：

```
###服务注册
> curl -X POST 'http://127.0.0.1:8848/nacos/v1/ns/instance?serviceName=nacos.naming.serviceName&ip=20.18.7.10&port=8080'
### 服务发现
> curl -X GET 'http://127.0.0.1:8848/nacos/v1/ns/instance/list?serviceName=nacos.naming.serviceName'
### 发布配置
```

```
> curl -X POST
"http://127.0.0.1:8848/nacos/v1/cs/configs?dataId=nacos.cfg.dataId&group=test&co
ntent=helloWorld"
### 获取配置
> curl -X GET
"http://127.0.0.1:8848/nacos/v1/cs/configs?dataId=nacos.cfg.dataId&group=test"
```

详细内容参考官方文档：https://Nacos.io/zh-cn/docs/quick-start-docker.html。

13.2.2 Sentinel Docker

使用 Docker 容器启动 Sentinel，具体步骤如下所示：

步骤01 下载 Sentinel 镜像：

```
> docker pull bladex/sentinel-dashboard
```

步骤02 启动容器：

```
> docker run --name sentinel -d -p 8858:8858 bladex/sentinel-dashboard
```

步骤03 登录 Sentinel，账号/密码为 sentinel/sentinel：

```
> http://localhost:8858/#/login
```

13.2.3 Seata Docker

使用 Docker 容器启动 Seata，具体步骤如下：

步骤01 下载 Seata 镜像：

```
> docker pull seataio/seata-server
```

步骤02 启动 seata-server 实例：

```
> docker run --name seata-server -p 8091:8091 seataio/seata-server:latest
```

指定 seata-server IP 启动：

```
> docker run --name seata-server \
    -p 8091:8091 \
    -e SEATA_IP=192.168.1.1 \
    seataio/seata-server
```

自定义配置文件需要通过挂载文件的方式实现，将宿主机上的 registry.conf 和 file.conf 挂载到容器中相应的目录。

（1）指定 registry.conf：使用自定义配置文件时必须指定环境变量 SEATA_CONFIG_NAME，并且值需要以 file:开始，如 file:/root/seata-config/registry。

```
> docker run --name seata-server \
    -p 8091:8091 \
```

```
    -e SEATA_CONFIG_NAME=file:/root/seata-config/registry \
    -v /PATH/TO/CONFIG_FILE:/root/seata-config \
    seataio/seata-server
```

其中，-e 用于配置环境变量，-v 用于挂载宿主机的目录。

（2）指定 file.conf：如果需要同时指定 file.conf 配置文件，则需要在 registry.conf 文件中将 config 配置改为以下内容，name 的值为容器中对应的路径：

```
config {
  type = "file"

  file {
    name = "file:/root/seata-config/file.conf"
  }
}
```

步骤 03 可以使用 Docker compose 方式启动 Seata，docker-compose.yaml 示例如下：

```
version: "3.1"

services:

  seata-server:
    image: seataio/seata-server:latest
    hostname: seata-server
    ports:
      - 8091:8091
    environment:
      - SEATA_PORT=8091
    expose:
      - 8091
```

更多内容参考 Seata 官网：http://seata.io/zh-cn/docs/ops/deploy-by-docker.html。

第 14 章

使用 Spring Cloud 构建微服务综合案例

本章主要介绍如何使用 Spring Cloud、Spring Cloud Alibaba 以及开源技术框架一步一步搭建分布式微服务架构和服务治理平台，并提供了具体的架构图和原理图，帮助读者理解分布式架构的具体细节。

14.1 案例介绍

第 1 章～第 13 章主要介绍了 Spring Boot、Spring Cloud、Spring Cloud Alibaba 以及组件容器化等技术，并且在同一种场景中有不同的开源组件和解决方案，例如：

- 微服务注册中心：Eureka、阿里的 Nacos 组件等。
- 微服务配置中心：Spring Cloud Config、阿里的 Nacos 组件等。
- 熔断/降级：Hystrix 组件、阿里的 Sentinel。
- 网关：Zuul 网关、Spring Cloud Gateway 网关。

在众多技术当中，如何挑选适合自己企业的分布式微服务架构和服务治理平台并不是一件容易的事情。本章结合笔者的工作经验以及在公司中真实的分布式架构解决方案尽量为读者描绘出技术架构蓝图。

14.2 技术选型

14.2.1 Spring Boot 构建微服务

在 Spring Boot 还未火起来之前，传统的 Java Web 应用大多采用 SSM。SSM 即 Spring、Spring MVC 以及 MyBatis 三者结合起来：

- Spring 是一个轻量级的控制反转（IoC）和面向切面（AOP）的容器框架。
- Spring MVC 分离了控制器、模型对象、分派器以及处理程序对象的角色，这种分离让它们更容易进行定制。
- MyBatis 是一个支持普通 SQL 查询、存储过程和高级映射的优秀持久层框架。

SSM 在一定程度上限制你只能开发 Java Web 应用程序，MVC 框架必须是 Spring MVC，持久层必须使用 MyBatis。当然，也可以在此基础上加入其他的框架或者类库。

Spring Boot 的可选方案较多。Spring Boot 没有和任何 MVC 框架、持久层框架和其他业务领域的框架绑定。开发 Web 应用，只要引入 spring-boot-starter-web 依赖包，Spring Boot 就可以配置好 Spring MVC。不想用 Spring MVC，换成 Spring WebFLux（用 spring-boot-starter-webflux 依赖）写响应式 Web 应用也可以，而且是 Spring 5 主推的新 Web 框架。

数据持久层可以用 Spring Data 项目下的任何子项目（JPA/JDBC/MongoDB/Redis/LDAP/Cassandra/Couchbase/Noe4J/Hadoop/Elasticsearch 等），当然用非 Spring 官方支持的 MyBatis 也可以，只要引入对应技术或框架的 spring-boot-starter-xxx 依赖包即可。

Spring Boot 提供了很多 starters，这些 starter 依赖对应的框架或技术，但不包含对应的技术或框架本身。所以，项目中可以使用这些 starters 快速构建 Java Web 项目，省掉了很多配置的烦恼。对于传统的 SSM 架构项目，可以维持原样或者改造为以 Spring Boot 为基础框架；对于新的项目，推荐使用 Spring Boot 快速构建微服务。

业务应用基本都会使用到 MySQL 数据库来存储业务数据，使用 Redis 来作为缓存，只要引入 MySQL 和 Redis 相对应的 starter 即可，具体如图 14-1 所示。如果是前后端分离项目，可能会使用目前主流的 Vue 框架来构建前端页面；如果是前后端未分离项目，则会将静态页面存放在业务应用资源目录下，每个公司的情况不一样。

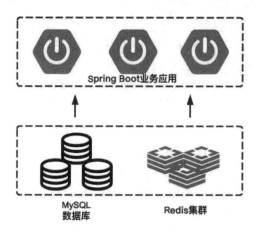

图 14-1　选用 Spring Boot 构建微服务

可以参考前面章节的内容快速构建 Spring Boot 微服务项目。

14.2.2 Nacos 注册/配置中心

有了 Spring Boot，就可以快速构建微服务了。接下来，我们选择合适的服务注册中心和配置中心：开源的服务注册中心技术框架非常多，比如 Zookeeper、Eureka、Consul、Nacos 等，开源的配置中心技术框架也很多，比如 Spring Cloud Config、Consul、Nacos 等。

这里推荐使用 Nacos 作为注册中心以及配置中心，原因如下：

- Nacos 既可以作为注册中心，也可以作为配置中心，一举两得。Eureka 只包含注册中心功能，Spring Cloud Config 只包含配置中心功能。
- Nacos 社区非常活跃，使用起来非常简单。
- 支持 Dubbo 、Spring Cloud、Kubernetes 集成。

当然，并不是说不能使用其他的技术作为注册中心或者配置中心，只是说如果要从头开始构建一套全新的分布式微服务框架和服务治理平台，那么 Nacos 是一个不错的选择。

有了 Nacos 这个服务注册中心和配置中心利器，我们就可以快速构建 Nacos 集群了。初学者可能会问为什么要用 Nacos 集群，单机的 Nacos 不行吗？当然可以，但是要看使用场景，比如纯粹个人学习 Nacos、企业的测试环境/开发环境、公司机器资源紧张等情况。如果是正式环境，Nacos 就不能单点部署了，必须考虑高可用，也就是 Nacos 集群部署。所以，现在架构变成如图 14-2 所示的形式。

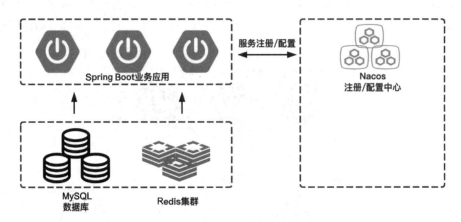

图 14-2　引入 Nacos 服务注册/配置中心

可以参考第 3 章的内容快速构建搭建 Nacos 集群，将 Spring Boot 微服务应用注入 Nacos。项目中经常是要分环境的，比如 dev、test、pre、prod 环境等。在不同的环境中，数据库配置或者 Redis 缓存配置等都不一样，可以使用 Nacos 的命名空间进行环境的区分，如图 14-3 所示。

第 14 章 使用 Spring Cloud 构建微服务综合案例

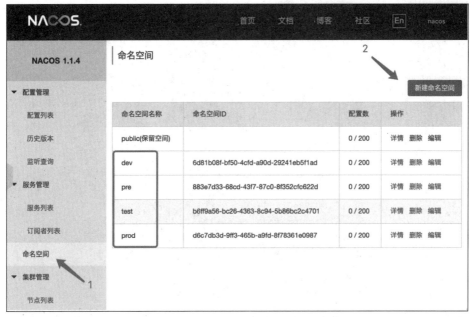

图 14-3 利用 Nacos 命名空间区分环境

通过 Nacos 的后台管理创建 dev、pre、test、prod 4 个环境，将微服务的配置抽到 Nacos 配置管理中。考虑到安全问题，不建议将数据库/redis 等密码明文直接配置到 prod 环境（线上环境），因此需要对这些密码进行加密。这里可以考虑使用 jasypt 加密包，jasypt 开发人员提供一种简单的方式来为项目增加加密功能。在 Spring Boot 中集成 jasypt 非常简单，具体步骤如下：

步骤 01 引入 jasypt-spring-boot-starter 依赖：

```xml
<dependency>
    <groupId>com.github.ulisesbocchio</groupId>
    <artifactId>jasypt-spring-boot-starter</artifactId>
    <version>3.0.3</version>
</dependency>
```

步骤 02 在 bootstrap.yml 配置文件中添加如下配置：

```yaml
jasypt:
  encryptor:
    ### 加密密码
    password: JH8AS90jasH
```

> **注 意**
>
> 线上环境可以将加密密码配置到服务器的全局变量，配置在应用的启动脚本命令上会更安全。

步骤 03 为配置文件中的明文密码加密。

加密前的数据：

```yaml
spring:
  cloud:
    config:
      server:
        git:
          uri: https://example.domain.com/helloworld.git
          username: yourname
          password: 123456
```

加密后的数据:

```yaml
spring:
  cloud:
    config:
      server:
        git:
          uri: https://example.domain.com/helloworld.git
          username: username
          password: ENC(DoyyHAMYaEyJBJHW496HiTT4VIazUYZo)
```

假设明文密码为 123456，创建一个测试类，加密后得到的加密密码为 ENC(DoyyHAMYaEyJBJHW496HiTT4VIazUYZo)。其中，ENC 是 jasypt 约定的关键字，代表字符串进行了加密。

```java
@RunWith(SpringRunner.class)
@SpringBootTest
public class BlogApplicationTests {

    @Autowired
    StringEncryptor stringEncryptor;

    @Test
    public void test() {
        System.out.println(stringEncryptor.encrypt("123456"));
    }
}
```

14.2.3　Spring Cloud Gateway 网关

本小节将选择一款合适的微服务网关。在第 4 章中，对比了 Spring Cloud Gateway 网关和 Zuul 网关，我们知道 Spring Cloud Gateway 的 RPS 是 Zuul 1.0 的 1.55 倍，平均延迟是 Zuul 1.0 的一半。从目前来看，Gateway 替代 Zuul 是未来趋势。因此，我们选择 Spring Cloud Gateway 作为内网微服务网关，如图 14-4 所示。

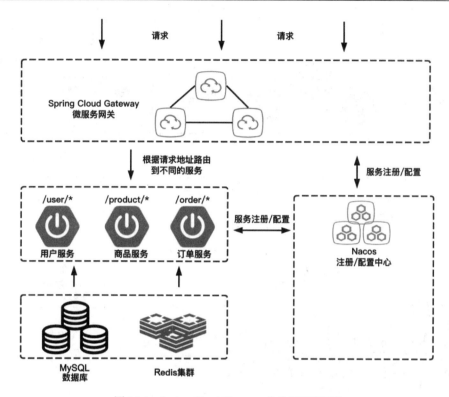

图 14-4　Spring Cloud Gateway 作为微服务网关

首先，需要将微服务应用的路由规则配置到配置文件或者 MySQL 数据库中，建议配置到 MySQL 数据库。Spring Cloud Gateway 网关启动时，可从数据库获取这些微服务路由规则加载到内存中，Spring Cloud Gateway 也需要注册到 Nacos 注册中心。这样，Spring Cloud Gateway 会代理注册中心的所有微服务应用。

例如，请求 http://you.domain.com/user/*，Gateway 网关会路由到用户服务；请求 http://you.domain.com/product/*，Gateway 网关会路由到商品服务；请求 http://you.domain.com/order/*，Gateway 网关会路由到订单服务。

Spring Cloud Gateway 除了路由转发功能外，还实现请求日志、请求过滤、权限校验、限流以及监控等。

读者可以参考第 4 章的内容设置 Spring Cloud Gateway 集群。之所以要做成集群方式，是考虑到高可用性，毕竟 Spring Cloud Gateway 作为内网流量入口，请求流量会非常大，单点网关无法保证服务高可用。

细心的读者可以发现使用的是 Spring Cloud Gateway 内网网关，因为公网的流量入口一般不会直接打到 Gateway 网关，而是先从 Nginx 进入，具体如图 14-5 所示。

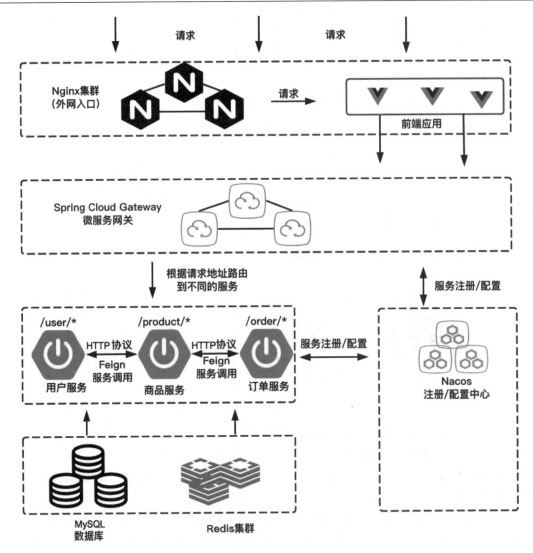

图 14-5 流量入口 Nginx+Vue 前端应用

请求从 Nginx 进入会代理到前端应用，前端应用会配置 Gateway 网关的域名地址，将请求达到 Gateway 内网网关，内网网关 Gateway 根据请求地址路由到不同的微服务应用，微服务可以直接通过 OpenFeign 进行调用。

> **注 意**
>
> 一般都会把前端应用部署到 Nginx 服务器，Nginx 也会采用集群的方式部署。但是，Nginx 作为外部的唯一访问入口是无法直接以集群的形式对外提供服务的，因为一个 Nginx 只能绑定一个公网 IP，中小型企业往往没有那么多的公网 IP 资源可用。在内网环境下，可以用 Nginx 集群，当然要有一个对外入口，所以需要在 Nginx 集群之上再加一层负载均衡器，作为系统的唯一入口。这个负载均衡器叫 LVS。

LVS 的内容已经超出本书的范围，这里不再深入讲解。

14.2.4　OpenFeign 服务调用

因为我们的分布式微服务框架和服务治理平台是基于 Spring Cloud 和 Spring Cloud Alibaba 的，所以服务之间的调用可以使用 OpenFeign 来实现远程过程调用（RPC）。

Feign 是 Spring Cloud 全家桶中推荐使用的 RPC 框架，同时使用 HTTP 作为传输层协议。总的来说，RPC 和 HTTP 并不是一个层面的内容。在 RPC 框架中，可以选择使用 HTTP 作为其传输层协议；在微服务体系中，无论使用 Feign 还是 RestTemplate，传输层都是基于 HTTP 协议进行传输的，具体内容参考图 14-5。

14.2.5　Ribbon 负载均衡

负载均衡主要考虑以下两点：微服务之间相互调用（比如商品服务调用用户服务的某个查询接口，用户服务有 5 个节点，商品服务如何选择用户服务进行调用）和网关将请求路由到应用服务（比如用户服务有 5 个节点，网关如何选择用户服务进行调用）。

1. 微服务之间相互调用

微服务之间的调用采用的是 OpenFeign 组件。OpenFeign 组件使用 Ribbon 进行负载均衡，所以 OpenFeign 直接内置了 Ribbon，即在导入 OpenFeign 依赖后无须再专门导入 Ribbon 依赖。

当商品服务通过 Feign 接口调用用户服务时，Ribbon 自动从 Nacos 中获取服务提供者地址列表，并基于负载均衡算法请求其中一个用户服务接口。

2. 网关将请求路由到应用服务

网关将请求路由到应用服务的负载均衡与第一种情况类似，因为 Spring Cloud Gateway 也是可以使用 Ribbon 作为负载均衡组件的，只是需要修改微服务路由配置规则，例如：

```
spring:
  cloud:
    gateway:
      routes:
      - id: myRoute
        ### 重点看这里，需要添加 lb://，代表需要负载均衡
        uri: lb://user-service
        predicates:
        - Path=/user/**
```

上述只是配置文件的写法，当把配置抽到 MySQL 数据库时，URL 地址带上 lb://即可，具体如图 14-6 所示。

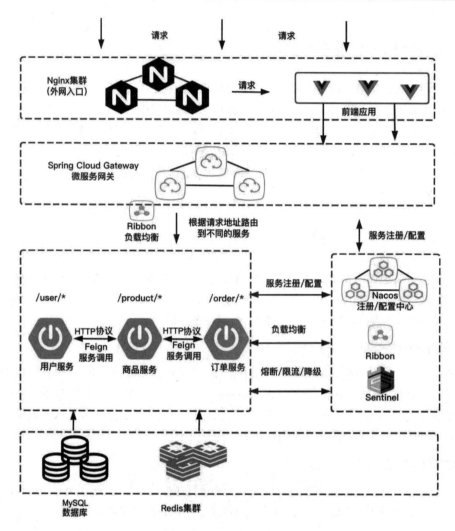

图 14-6 Ribbon 负载均衡组件

14.2.6 Sentinel 熔断/降级/限流

从图 14-7（来自网络图片）可以看出 Sentinel 和 Hystrix 的区别。Sentinel 不仅支持熔断和降级功能，还支持限流功能；而 Hystrix 不支持限流，且没有可视化的控制台。因此，在企业中可以使用 Sentinel 作为熔断/降级/限流的利器。

读者可以参考第 7 章的内容搭建 Sentinel 服务器。

	Sentinel	Hystrix
隔离策略	信号量隔离	线程池隔离/信号量隔离
熔断降级策略	基于响应时间或失败比率	基于失败比率
实时指标实现	滑动窗口	滑动窗口（基于Rxjava）
规则配置	支持多种数据源	支持多种数据源
扩展性	多个扩展点	插件的形式
基于注解的支持	支持	支持
限流	基于QPS，支持基于调用关系的限流	不支持
流量整形	支持慢启动，匀速器模式	不支持
系统负载保护	支持	不支持
控制台	开箱即用，可配置规则、查看秒级监控、机器发现等	不完善
常用框架适配	Servlet、Spring Cloud、Dubbo、gRPC等	Servlet、Spring Cloud Netflix

图 14-7　Sentinel 与 Hystrix 的对比

14.2.7　ELK+FileBeat 日志系统

随着系统服务数量越来越多，对于一次业务调用，可能需要经过多个服务模块处理。不同的服务可能由不同的团队开发，并且分布在不同的网络节点，甚至可能在多个地域的不同机房内。在异常发生时，排查代码错误会变得非常困难。

系统服务产生日志可以粗略分为非结构化日志和结构化日志两种。

1. 非结构化日志

非结构化日志主要面向人，也就是写给人看的，例如：

```
### 非结构化日志
logger.info("class:TraditionLogController and method:say and the param1 is :" +
param1 + " the param2 is :" + param2);
```

日志打印没有任何标准格式，基本都是字符拼接而成的字符串。

2. 结构化日志

结构化日志主要面向机器，是写给机器看的，例如：

```
{
    "log":{
        "http_cdn":"-",
        "remote_addr":"10.10.34.117",
        "request":"POST /strategy/collect/get HTTP/1.0",
        "first_byte_commit_time":"1",
        "body_bytes_sent":"6141",
        "server_addr":"10.238.38.7",
        "sent_http_content_length":"-",
        "@timestamp":"2019-02-03T12:28:16+0800",
        "request_time":"0.001",
        "host":"cpg.meitubase.com",
        "http_x_forwarded_for":"223.104.23.179, 10.10.34.45",
```

```
        "http_x_real_ip":"10.10.34.45",
        "category":"access",
        "content_length":"116",
        "status":"200"
    },
    "stream":"stdout",
    "@version":"1",
    "topic":"k8s_user-web_stdout",
    "time":"2019-02-03T04:28:16.181977586Z"
}
```

上述日志是标准的轻量级的 JSON 结构化数据，统一的日志格式便于日志系统的统一处理。

传统的应用日志由于日志量小，更多的是面向人，而分布式微服务架构背景面对大数据日志处理，已超过人工所能处理的范围，更多的是面向机器。

日志系统的架构很多，目前流行的开源解决方案（自研除外）不外乎以下几种：

- ELK：Elasitcsearch + Logstash + Kibana。
- EFK：Elasitcsearch + Fluentd + Kibana。
- EBK：Elasitcsearch + Filebeat + Kibana。

日志收集的软件很多，比如 Logstash、Fluentd、Filebeat。本书不对比这几款软件之间的优缺点，只要求大家记住一句话：Filebeat 更轻量，占用资源更少，所占系统的 CPU 和内存几乎可以忽略不计。因此，可以采用 ELK+Filebeat 作为我们的日志系统，如图 14-8 所示。

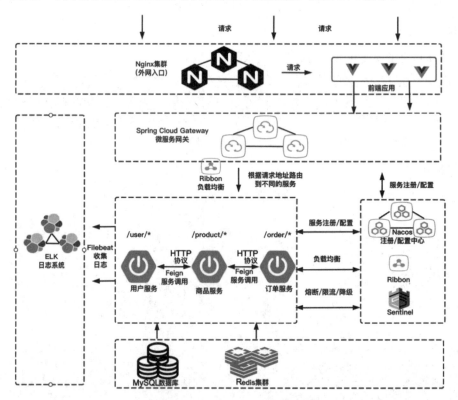

图 14-8　ELK+Filebeat 日志系统

Filebeat 代理部署在机器上，用来收集服务日志并将日志推送到 Logstash 组件，Logstash 将收集的日志进行解析和过滤，并保存到 Elasitcsearch，最后通过 Kibana 查询日志和相关报表统计，具体架构如图 14-9 所示。

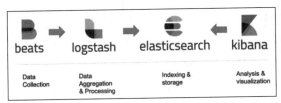

图 14-9　ELK+Filebeat 简单原理

14.2.8　Promethous+Grafana+InfluxDB 监控系统

在分布式的微服务架构中，当系统从单个节点扩张到很多节点的时候，如果系统的某个节点出现问题，对于运维和开发人员来说定位可能会变成一个挑战。当新的业务进来以后，我们可以通过监控手段对系统进行衡量或者做一个数据支撑。另外，还要理解分布式系统是什么样的拓扑结构、如何部署、系统之间如何通信、系统目前的性能状况如何以及出了问题怎么去发现。这些都是分布式系统需要面对的问题。出现这些问题后，监控就是一个比较常用、有效的手段。总的来说，监控主要解决的是感知系统的状况。

监控系统的组成主要涉及 4 个环节：数据收集、数据传输、数据处理和数据展示。不同的监控系统的实现方案在这 4 个环节所使用的技术方案不同，适合的业务场景也不一样。

目前业界主流的监控系统主要使用 Promethous+Grafana+InfluxDB 来打造企业监控系统，当然还要结合企业自身的业务情况选择合适的监控平台。

- Prometheus（普罗米修斯）：是一套开源的监控&报警&时间序列数据库的组合。Prometheus 由两部分组成：一部分是监控报警系统，另一部分是自带的时序数据库（TSDB）。官方架构如图 14-10 所示。

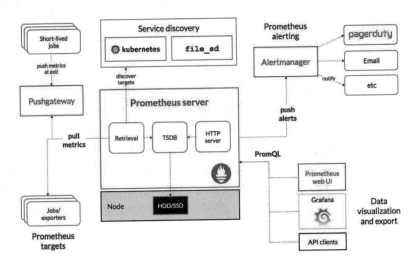

图 14-10　Prometheus 架构图

- Prometheus Server：负责定时在目标上抓取Metrics数据，每个抓取目标都需要暴露一个HTTP服务接口用于Prometheus定时抓取。这种调用被监控对象获取监控数据的方式被称为Pull方式。

某些现有系统是通过push方式实现的，为了接入这个系统，Prometheus提供对PushGateway的支持，这些系统主动推送metrics到PushGateway，而Prometheus只是定时去Gateway上抓取数据。Prometheus甚至可以从其他的Prometheus获取数据，组建联邦集群。

- AlertManager：独立于Prometheus的一个组件，在触发了预先设置在Prometheus中的高级规则后，Prometheus便会推送告警信息到AlertManager。
- Prometheus WebUI：数据展现除了Prometheus自带的WebUI，还可以通过Grafana等组件查询Prometheus监控数据。

Prometheus内部主要分为3大块（见图14-11）：Retrieval负责定时到暴露的目标页面上去抓取采样指标数据，Storage负责将采样数据写入磁盘，PromQL是Prometheus提供的查询语言模块。

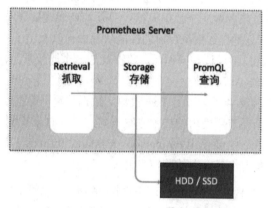

图 14-11 Prometheus Server 简单架构

Prometheus 通过 HTTP 接口的方式从各种客户端获取数据，这些客户端必须符合 Prometheus 监控数据格式，通常有两种方式：一种是侵入式埋点监控，通过在客户端集成（如 Kubernetes API 直接通过引入 Prometheus go client），提供/metrics 接口查询 kubernetes API 的各种指标；另一种是通过 exporter 方式在外部将原来各种中间件的监控支持转化为 Prometheus 的监控数据格式，如 Redis exporter 将 Redis 指标转化为 Prometheus 能够识别的 HTTP 请求。

Prometheus 为了支持各种中间件以及第三方的监控提供了 exporter（见图 14-12），大家可以把它理解成监控适配器，将不同指标类型和格式的数据统一转化为 Prometheus 能够识别的指标类型。譬如 Node exporter 主要通过读取 Linux 的/proc 以及/sys 目录下的系统文件获取操作系统运行状态；Redis exporter 通过 Redis 命令行获取指标；MySQL exporter 通过读取数据库监控表获取 MySQL 的性能数据。将这些异构的数据转化为标准的 Prometheus 格式，并提供 HTTP 查询接口。

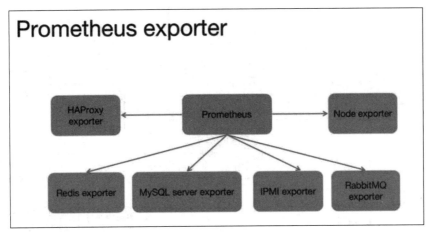

图 14-12　Prometheus exporter

Prometheus 支持以下两种存储方式。

1. 本地存储

通过 Prometheus 自带的时序数据库 TSDB 将数据保存到本地磁盘。为了性能考虑，建议使用 SSD。本地存储的容量毕竟有限，建议不要保存超过一个月的数据。Prometheus 本地存储经过多年改进，自 Prometheus 2.0 后提供的 V3 版本 TSDB 性能已经非常高，可以支持单机每秒 1000 万个指标的收集。Prometheus 本地数据存储能力一直为大家诟病，但 Prometheus 本地存储设计的初衷就是为了监控数据的查询，Facebook 发现 85% 的查询是针对 26 小时内的数据，所以 Prometheus 本地时序数据库的设计考虑更多的是高性能而非分布式大容量。

2. 远程存储

Prometheus 适用于存储大量监控数据，目前支持 OpenTSDB、InfluxDB、Elasticsearch 等后端存储，通过适配器实现 Prometheus 存储的 remote write 和 remote read 接口，便可以接入 Prometheus 作为远端存储使用。

InfluxDB 是一个开源的时序型数据，由 Go 语言编写，用于高性能地查询与存储时序型数据。InfluxDB 被广泛应用于存储系统的监控数据。因此，可以使用 InfluxDB 存储监控数据。

Grafana 是一款由 Go 语言编写的开源应用，主要用于大规模指标数据的可视化展现，是网络架构和应用分析中最流行的时序数据展示工具，目前已经支持绝大部分常用的时序数据库。

监控系统的总体流程为：根据要监控到的对象资源选择不同的exporter。例如，Redis使用Redis exporter，MySQL使用MySQL exporter。exporter将获取到的监控指标转化为Prometheus的监控数据格式，并转化为Prometheus能够识别的HTTP请求。Prometheus通过Pull方式定时拉取监控指标并存入InfluxDB时序数据库，最后通过Grafana进行可视化展示，具体如图14-13所示。

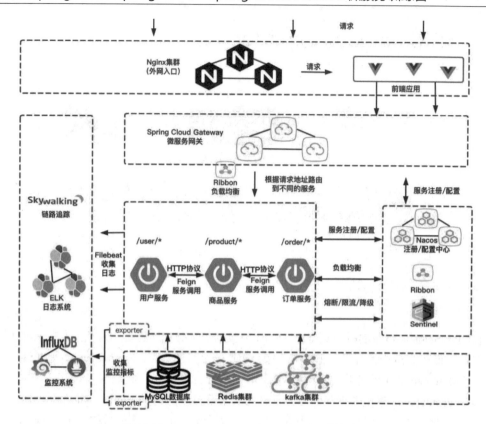

图 14-13　Promethous+Grafana+InfluxDB 监控系统

对于 Spring Boot 微服务应用，可添加如下依赖包：

```xml
<dependencies>
    <dependency>
        <groupId>io.micrometer</groupId>
        <artifactId>micrometer-core</artifactId>
        <version>1.6.1</version>
    </dependency>
    <dependency>
        <groupId>io.micrometer</groupId>
        <artifactId>micrometer-registry-prometheus</artifactId>
        <version>1.6.1</version>
    </dependency>
    <dependency>
        <groupId>org.springframework.boot</groupId>
        <artifactId>spring-boot-actuator-autoconfigure</artifactId>
    </dependency>
</dependencies>
```

micrometer-registry-prometheus 包对 Prometheus 进行封装。

同时，在应用中添加配置类 MonitorAutoConfig，具体代码如下：

```java
@Configuration
public class MonitorAutoConfig {
```

```
@Value("${spring.application.name}")
private String applicationName;

@Value("${spring.profiles.active}")
private String applicationProfile;

@Bean
public MeterRegistryCustomizer<MeterRegistry> metricsCommonTags() {
    return registry -> registry.config().commonTags("application",
applicationName + "-" + applicationProfile);
    }
}
```

最后，在 bootstrap.yml 中添加如下配置：

```
management:
  metrics:
    tags:
      application: ${spring.application.name}
```

14.2.9 SkyWalking 链路追踪系统

业界比较有名的服务追踪系统实现有美团 CAT、阿里的鹰眼、Twitter 开源的 OpenZipkin，还有 Naver 开源的 Pinpoint，它们都是受 Google 发布的 Dapper 论文启发而实现的。

服务追踪系统的实现主要包括 3 个部分：

- 在服务端进行埋点，收集埋点数据。
- 收集到的链路信息进行实时数据处理，按照 TraceId 和 spanId 进行串联和存储。
- 为处理后的服务调用数据，按照调用链的形式展示出来。

OpenZipkin 在第 10 章已有详细介绍，这里不过多描述。Pinpoint 是开源的支持 Java 语言的服务追踪系统，简单架构如图 14-14 所示。

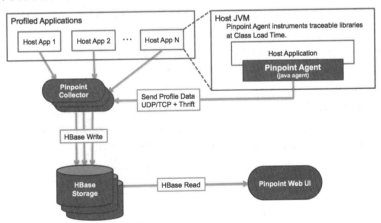

图 14-14 Pinpoint 架构图

Pinpoint 的主要组成部分如下：

- **Pinpoint Agent**：通过 Java 字节码注入的方式来收集 JVM 中的调用数据，通过 UDP 协议传递给 Collector，数据采用 Thrift 协议进行编码。
- **Pinpoint Collector**：收集 Agent 传过来的数据，然后写到 HBase Storage。
- **Pinpoint Web UI**：通过 Web UI 展示服务调用的详细链路信息。

我们该如何选择链路追踪系统呢？

- 从支持语言来说，OpenZipkin 提供了不同语言的库，例如：C#、Go、Java、JavaScript、Ruby、Scala、PHP 等，而 Pinpoint 目前只支持 Java 语言。
- 系统集成难易程度上看，Pinpoint 要比 OpenZipkin 简单。因为 Pinpoint 是通过字节码注入的方式来实现拦截服务调用，从而收集 trace 信息的，所以不需要代码做任何改动。
- 从调用链路数据的精确度上看，Pinpoint 要比 OpenZipkin 精确得多。OpenZipkin 收集到的数据只到接口级别，Pinpoint 不仅能够查看接口级别的链路调用信息，还能深入到调用所关联的数据库信息。

总而言之，如果你的业务采用的是 Java 语言，那么可以采用 Pinpoint 作为链路追踪系统；如果你的业务不是用 Java 语言实现的或者采用了多种语言，那么应该选择 OpenZipkin。

除此之外，Pinpoint 不仅能看到服务与服务之间的链路调用，还能看到服务内部与资源层的链路调用，功能更为强大。

除了 OpenZipkin 和 Pinpoint 外，国内的一款开源服务追踪系统 SkyWalking 的功能也很强大。SkyWalking 是一款应用性能监控（APM）工具，对微服务、云原生和容器化应用提供自动化、高性能的监控方案。

SkyWalking 提供了分布式追踪、服务网格（Service Mesh）遥感数据分析、指标聚合和可视化等多种能力，项目覆盖范围从一个单纯的分布式追踪系统扩展为一个可观测性分析平台（observability analysis platform）和应用性能监控管理系统。它包括以下主要功能：

- 基于分布式追踪的 APM 系统，满足 100%分布式追踪和数据采集，同时对被监控系统造成极小的压力。
- 云原生友好，支持通过以 Istio 和 Envoy 为核心的 Service Mesh 来观测和监控分布式系统。
- 多语言自动探针，包括 Java、.NET 和 NodeJS。
- 运维简单，不需要使用大数据平台即可监控大型分布式系统。
- 包含展示 Trace、指标和拓扑图在内的可视化界面。

网络上也有人对 OpenZipkin、Pinpoint 以及 SkyWalking 这几个工具做过测试比对，得到的结论是每个产品对性能的影响都在 10% 以下，其中 SkyWalking 对性能的影响最小。

因此，企业使用 SkyWalking 作为链路追踪系统是一个不错的选择。

14.3 总　结

通过前面几节内容，我们构建出一套分布式微服务架构和服务治理平台，这套框架对于中型或者大型企业都具有一定的参考价值。这里再次强调，任何脱离业务的架构都是不合适的。例如：文件存储系统既可以选择开源的 FastDFS 文件存储系统，也可以选择购阿里云的 OSS 存储；消息中间件既可以使用目前主流的 Kafka，也可以使用 RabbitMQ，不同的消息中间件使用的场景也不一样；邮件/短信服务既可以购买阿里云的产品，也可以购买腾讯云的产品，需要结合企业自身来决定使用哪个云厂商。

图 14-15 所示的最终架构图仅仅是从宏观上面做一次总结，在大公司里面每一块内容可能都有单独专门的研发团队，比如 Kafka 团队、Redis 团队、日志系统团队、监控系统团队等。很多细节没有办法一一列出，而且并不是所有的企业都适用于这种分布式和服务治理架构，需要根据企业的特点以及具体业务情况而定。

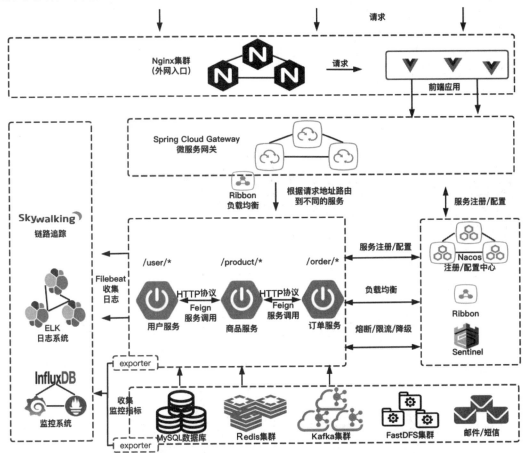

图 14-15　最终分布式微服务架构和服务治理平台

参 考 文 献

[1] 翟永超. Spring Cloud 微服务实战[M]. 北京：电子工业出版社，2017.
[2] 陈韶健. Spring Cloud 微服务架构实战[M]. 北京：电子工业出版社，2020.
[3] 倪炜. 分布式消息中间件实践[M]. 北京：电子工业出版社，2011.
[4] Leader-us. 架构解密：从分布式到微服务[M]. 北京：电子工业出版社, 2017.
[5] 杨保华，戴王剑，曹亚仑. Docker 技术入门与实战[M]. 北京：机械工业出版社，2018.
[6] 朱林. Elasticsearch 技术解析与实战[M]. 北京：机械工业出版社，2017.